Praise for *Quality Function Deployment and Six Sigma*

"Joe Ficalora has done an excellent job of updating Cohen's classic text on Quality Function Deployment to show how Six Sigma tools can be effectively used in applying the QFD methodology and how valuable QFD can be in Design for Six Sigma projects. I highly recommended it to practitioners of both disciplines. They will be well rewarded by using this book to guide them in delighting their customers."

—John L. Schoonover
Director of Quality
Global Tungsten and Powders

IMPROVING A CLASSIC

"For good reasons, the saying 'Don't mess with success' applies to many things in life; thus Lou Cohen and Joe Ficalora have accomplished a feat often fraught with risk. The original book on QFD, by Lou Cohen, was excellent, with ideas and detailed examples for applying QFD to many types of problems. The new edition builds on the original by integrating material on Design for Six Sigma, making it highly likely that it will remain a key reference. Whether you are working on developing a new product or a transactional business process, this updated edition will help you focus your efforts on what customers really want and need—so important to delivering something customers will pay for."

—John P. King
Jewett & King Associates

Quality Function Deployment and Six Sigma

SECOND EDITION

Quality Function Deployment and Six Sigma

A QFD HANDBOOK

SECOND EDITION

Joseph P. Ficalora

Louis Cohen

PRENTICE HALL

Upper Saddle River, NJ · Boston · Indianapolis · San Francisco
New York · Toronto · Montreal · London · Munich · Paris · Madrid
Capetown · Sydney · Tokyo · Singapore · Mexico City

The publisher offers excellent discounts on this book when ordered in quantity for bulk purchases or special sales, which may include electronic versions and/or custom covers and content particular to your business, training goals, marketing focus, and branding interests. For more information, please contact:

U.S. Corporate and Government Sales
(800) 382-3419
corpsales@pearsontechgroup.com

For sales outside the United States, please contact:

International Sales
international@pearson.com

Visit us on the Web: informit.com/ph

Library of Congress Cataloging-in-Publication Data is on file.

ISBN-13: 978-0-13-336443-9
ISBN-10:   0-13-336443-7
This product is printed digitally on demand. This book is the paperback version of an original hardcover book.
First printing, July 2009

THIS BOOK IS DEDICATED TO MY WIFE LYNN AND OUR CHILDREN ANNE, KATHERINE, AND MARK FOR THEIR UNDERSTANDING AND PATIENCE WHILE I WAS DOING LATE-NIGHT EDITS AND EARLY MORNING RE-WRITES OVER COFFEE WHEN I SHOULD HAVE BEEN SPENDING TIME WITH ALL OF YOU. WITHOUT YOU, MY LIFE WOULD HAVE LITTLE MEANING, SO THANK YOU FOR ALL YOUR LOVE AND SUPPORT.

—JPF

Contents

Foreword

One of the byproducts of the old Total Quality Management ideology was the term Voice of the Customer (VOC). While this term became ubiquitous in the marketing and product-development worlds, it was a relatively empty term because there was no clear methodology defining the Voice of the Customer, nor any roadmap to obtaining valid information about customer wants and needs. That is, until Lou Cohen's first edition of this book.

Before then, VOC was overstated and not well understood. The mantra was, "We've got to get the Voice of the Customer." Everyone in the room would simultaneously nod their heads and readily agree. Then someone would wander up to marketing to get the VOC.

What resulted was the VOC that marketing thought represented the new market. It usually didn't. This phenomenon is readily apparent in the U.S. automotive industry. The major car companies are now in trouble largely because they completely missed the real VOC, which, by the way, the Japanese car companies nailed.

Lou Cohen's first edition of this book put Quality Function Deployment (QFD) on the map as an important method and, more importantly, as a useful tool by which VOC could be clearly defined as information upon which action could be readily taken.

So, why does there need to be a second edition of a fine original? Well, there's more to defining VOC than just QFD. QFD became fairly commonly used among design teams. While at AlliedSignal in the early 1990s, I worked with a team that was developing an innovative new application. The developers showed me their completed QFD with pride and announced, "We have the Voice of the Customer." When I asked how much of the QFD input was derived from actually talking to the prospective customers (mostly from Japan), I was met with blank stares.

The final result: The developer completely missed a new emerging market, because when they actually talked with customers, they found that the customer requirements they *thought* were real in fact weren't. There are a lot of things that need to happen before and after the QFD to create effective results.

Joe Ficalora has been able to integrate QFD with product design. He did this by first mastering Design for Six Sigma (DFSS), which includes QFD. The wonder of combining QFD and DFSS emerges when QFD tells designers what they need to accomplish and DFSS provides the tools by which new designs are executed to create a product that can actually be manufactured.

Joe has done an incredible job of (1) making QFD even more accessible, and (2) linking QFD to Six Sigma and DFSS. I was first exposed to QFD while at Motorola in the late 1980s. I found in my own use of QFD that I had to start with simple 2×2 matrices and graduate to the more complex House of Quality. If this second edition had been available in the late 1980s, I believe my evolution in QFD would have been light-years faster. Use this book as written and you will see immediate improvement in your product development design targets. And, of course, that will lead to growth, which is a good thing!

—Dr. Steve Zinkgraf
CEO, Sigma Breakthrough Technologies, Inc.

Preface

Over the last 40 years, companies in the United States have moved toward new styles of doing business in the face of overseas competitive pressures, the needs of global economics, and the ever-accelerating pace of new technologies. U.S. companies have taken many steps to become more competitive. Early on came the adoption of the Total Quality Management (TQM) approach or one of its many aliases, all of which have stressed customer-driven planning, continuous improvement, and employee empowerment.

A key component of TQM was the adoption of "tools" to assist in creative thinking and problem solving. These were not physical tools, such as computers or micrometers; rather, they were *methods* that relate ideas to ideas, ideas to data, and data to data; that encourage team members to communicate more effectively with each other; and that helped teams to effectively formulate business problems and their solutions.

Quality Function Deployment (QFD) was an adaptation of some of the TQM tools. In Japan in the late 1960s, QFD was invented to support the product-design process (for designing large ships, in fact). As QFD itself evolved, it became clear to QFD practitioners that it could be used to support service development as well.

Today, its application goes beyond product and service design, although those activities comprise most applications of QFD. QFD has been extended to apply to any planning process where a team has decided to systematically prioritize possible responses to a given set of objectives. The objectives are called the Whats,[1] and the responses are

---

1. Credit is generally given to Harold Ross (General Motors) and Bill Eureka (American Supplier Institute) for the What/How terminology, which is now widely used by QFD practitioners in the U.S.

called the Hows. QFD provides a method for evaluating How a team can best accomplish the Whats.

The basic problems of product design are universal: Customers have needs that relate to using products; the needs must be addressed by designers who have to make hundreds or thousands of technical decisions; and there are never enough people, time, and dollars to put everything that could be imagined into a product or service.

These problems confront the developers of automobiles, cameras, hot-line service centers, school curricula, and even software. QFD can be used to help development teams decide how best to meet customer needs with available resources, regardless of the technology underlying the product or service.

Customers have their own language for expressing their needs. Each development team has *its* own language for expressing its technology and its decisions. The development team must make a translation between the customer's language and its technical language. QFD is a tool that helps teams systematically map out the relationships between the two languages.

The QFD experience has been generally glossed over, as if the process can run itself. Nothing could be further from the truth! Many QFD horror stories have at their root the uninformed decisions of an inexperienced QFD facilitator. The possibilities for wasting time and leading a team into a cul-de-sac are endless.

On the other hand, many QFD success stories have at their root the creative decisions of a capable QFD facilitator. The opportunities for helping teams save time and arrive at breakthroughs are abundant.

Who Should Read This Book

This second edition is meant to bridge the gap between traditional Quality Function Deployment (QFD) and the rise of Lean Six Sigma over the 15 years since the first edition was written. This book is written as a QFD reference for the many Black Belts and Green Belts in Six Sigma and Design for Six Sigma (DFSS).[2] It also fulfills a need for the many engineers, managers, and engineering supervisors who wish to improve their companies' new-product development processes by including QFD, by itself or with deployment of Lean Six Sigma or DFSS. The original great work by Lou Cohen has been retained for the most part, and I have updated it to include some of the key tools, approaches, linkages and methodologies of Six Sigma, including DFSS.

---

2. The author's blog on DFSS can be useful if you have not read about this before: See www.dfss4u.com. Also recommended is the SBTI Web site's page on DFSS: www.sbtionline.com/OurServices/GrowthExcellence/DFSS.php.

For the novice, it might be worthwhile to read the last chapter first, as it contains a case study involving both QFD and elements of DFSS. That example is one I am intimately familiar with, for it represents my first venture into energy alternatives, namely solar power. For those versed in the QFD side of the methodologies, this second edition presents some of the links to Six Sigma and DFSS. For Six Sigma and DFSS practitioners, this second edition will help show how QFD should be put to use in your projects and tool sequences.

My Experience with QFD, Product Development, and DFSS

I first came across QFD as a methodology in the late 1980s, during my design and development efforts while at AlliedSignal, now Honeywell Corporation. At that time, I had over 10 years' engineering experience and was an optical designer leading a design team for a new family of products. QFD was introduced and taught to us. The matrix approach seemed like a great way to keep track of the multitude of requirements and relationships that drive design decisions during the course of product development. However, when it was introduced, we were taught that Marketing could fill out the customer needs and weight them as to importance. It became obvious to me that such a course of action could possibly get us off-track from the outset of the design and development efforts. In those days, most engineering personnel were not allowed to travel with the Marketing folks to the customers and find out first-hand what customers wanted or needed. I was fortunate to work in an organization that was not so limited in its thinking. At AlliedSignal, Dr. David Zomick, the late William J. Mitchell, and Ernest Lademann created an atmosphere of cooperation between Engineering and Marketing wherein engineers could travel with marketing personnel to customer locations. I learned a lot from these three men about how to link customer needs to design decisions. QFD then helped me personally to understand how to strengthen and document the linkages between customers needs and design decisions; we ultimately created a new product family that flourished and grew in size, employing many people.

When I left corporate employment for management consulting in 1998, I found many of the tools from Six Sigma being deployed into engineering and development activities. At some point, people began calling this Design for Six Sigma. When Sigma Breakthrough Technologies, Inc. (SBTI) created its DFSS program in 1997, QFD was a key part of the comprehensive approach. It has remained so ever since. I have been the product-line owner for our extensive DFSS offerings and customizations for the last few years. In each customized deployment of DFSS, QFD has been retained due to its inherent value in product and service development. I have enjoyed seeing how clients apply the QFD aspect of DFSS to their design and development efforts, and witnessing the many valuable outcomes.

The chance to add to the fine first-edition QFD book that Lou Cohen had written was one I did not wish to pass up. Lou's book highlighted many of the practical how-tos that were missing for me when I was first introduced to QFD as a product-design engineer. The QFD Handbook portion of this book focuses on the *doing* of QFD. This book also addresses other aspects of QFD as well: Parts of this text explain what QFD is, and how QFD can fit in with other organizational activities.

The QFD Handbook section is fairly unique in that it provides detailed information on how QFD can successfully be implemented, along with a wide range of choices for customizing QFD, with their advantages and disadvantages. My original goal for this work was to make the QFD implementation path easier and more accessible to practitioners. If this text continues to serve that purpose, then QFD will be used more widely in the United States, and we can hope for better products, better services, and greater market share as a result.

How to Read This Book

This book is divided into five parts. Each part looks at QFD from a different perspective.

Part I, About QFD and Six Sigma, provides motivation for QFD and puts it into perspective within the framework of the global business environment. It gives the briefest of overviews of what QFD is. It's equivalent to the type of fifteen-minute management summary often given to describe QFD. Read this part if you are new to QFD, or if you would like to find out where it came from and how it has become so well known among product developers.

Part II, QFD at Ground Level, explains QFD in an expository fashion. It's a kind of textbook within a textbook. It explains each portion of the House of Quality (HOQ) in considerable detail. I hope it will be a useful reference for QFD implementers who are looking for detailed information about the HOQ. Read this part if you would like to be fully informed about the House of Quality.

Part III, QFD from 10,000 Feet, assumes the reader has some familiarity with QFD. It provides an organizational perspective on the way product and service development occurs. It shows how QFD can help organizations become more competitive by developing better products and services. Chapter 13 paints a picture of the way the world ought to be, and Chapter 14 discusses the world the way you may have experienced it. If you want to introduce QFD into an organization, you'll find the ideas in Part III to be helpful for developing your strategy for organizational change.

Part IV, QFD Handbook, is where the rubber meets the road. It assumes you have decided to implement QFD, and it shows you how to start, what to anticipate, and how to finish successfully. Part IV assembles the accumulated lessons from the authors' mistakes and from those of many of our colleagues. Read this section before you try to implement QFD. We think it will be a good investment of your time.

Part V, Beyond the House of Quality, points the way to extensions of the HOQ that can take the development team all the way to the completion of its project. Although it is used predominantly for product planning, QFD has the potential to help the development team deploy the VOC into every phase of development. Part V describes some of the possible paths teams can take after completing the HOQ. It also includes some topics that may be of interest to only some readers. These include specialized adaptations of QFD for software and service development, as well as for organizational planning. Read this part if you are comfortable with QFD concepts and are ready to make QFD a key part of your development process. This section ends with a solar energy case study showcasing one of the many ways to integrate the VOC, QFD, and DFSS.

Acknowledgments

This book would not have been possible without the help and guidance of a great many people. First thanks go to Lou Cohen for writing a great first edition and allowing me to modify his excellent work. I thank Dr. Stephen Zinkgraf, whose generosity in all things allowed for my participation on this project. Mr. Dan Kutz was instrumental in keeping things running smoothly at SBTI while I was preoccupied with this book. The same may be said of the entire executive team at SBTI—Ian Wedgwood, Joe Costello, Debby Sollenberger, Dick Scott, our departed friend Mike Brennan, Jesse Ferguson, and Roger Hinckley—who provided invaluable support while I was writing.

To the contributors to and reviewers of this second edition, a great debt is owed for invaluable additions, comments, critiques, and many great discussions. I consider Joe Kasabula as "Mr. VOC" for his lifelong devotion to the art and processes involved therein. I owe Randy Perry a great deal of thanks for his kind words, edits, suggestions, and contributions. For contributions and case examples, I again thank Bill Rodebaugh, Randy Perry, and Roger Hinckley.

I also need to thank God for the life I have led, which in many ways seems so miraculous, looking back. Some would choose to call it luck, but I never have believed that life is just a series of fortunate accidents.

—JPF

About the Authors

Joe Ficalora is currently the President of Global Services at Sigma Breakthrough Technologies, Inc. (SBTI), a consulting firm for manufacturing, quality, and engineering services, with specialties in Six Sigma and Lean Enterprise applications. In this role, Mr. Ficalora is responsible for the worldwide deployment of SBTI methodologies and for managing the international partners of SBTI.

Mr. Ficalora has more than 20 years of industrial experience in project management, engineering, manufacturing, and quality control. He first came across QFD while at AlliedSignal, now Honeywell, in 1991. His involvement with Quality tools and techniques has been continuous since that time. He designed and developed the highly acclaimed SBTI Six Sigma Master Black Belt Program. He has led Six Sigma deployments in Operations and Engineering at several clients, including Executive and Champion rollout sessions. Mr. Ficalora holds an MEE from Stevens Institute of Technology, and a BS in physics from Rensselaer Polytechnic Institute.

Mr. Ficalora has worked and is certified as a Black Belt and Master Black Belt. He has consulted for clients in the industries of aerospace, medical devices, beverages, health care, and in manufacturing of food packaging, electronics, metal, glass, and plastics. His current interests include DFSS in renewable energy and energy conservation, as well as future energy options, including nuclear fusion. His current interests also include investment performance predictions, economics, crime statistics, and crime prevention within the USA. He has mentored, and designed and taught workshops to Executives, Champions, Master Black Belts, Black Belts, and Green Belts in Design for Six Sigma, Six Sigma in Manufacturing, and Transactional Business projects worldwide. Mr. Ficalora is a top-rated and sought-after instructor and speaker in these areas.

Mr. Ficalora also holds several patents in lasers and optical devices and a patent in process improvement. He is an active member in IEEE and ISSSP.

Lou Cohen was a product developer, computer and software development manager, quality manager, and consultant during 41 years of professional life.

He lived and worked in Japan in 1984, where he extensively studied quality and productivity methods, including a detailed study of the theories of Dr. W. Edwards Deming. Upon his return to the U.S., he became a nationally known expert in the use of Quality Function Deployment. He has helped his clients use QFD in many diverse industries, including electric-power utilities, financial services, medical instruments, software, communications and telecommunications, laundry detergents, aerospace components, and office furniture.

Now in retirement, he lives in Cambridge, Massachusetts.

PART I

About Quality Function Deployment and Six Sigma

What Are QFD and Six Sigma?

This chapter provides an orientation to Quality Function Deployment (QFD) and Six Sigma. It begins with a brief overview of each of these topics. It then provides a chronology of events that led to the current state of QFD and its integration with Six Sigma in the United States. Finally, the chapter discusses some of the ways QFD is being used, independently and together with Six Sigma methodologies today, to provide a sense of the applicability and flexibility of the methods. Current best practices have evolved in the Six Sigma community to include Lean Manufacturing Methods and tools, and therefore is sometimes called Lean Six Sigma or Lean Sigma. For the sake of avoiding certain definition issues, the term Six Sigma will be used for a broad description of the Six Sigma methodologies described herein.

1.1 BRIEF CAPSULE DESCRIPTION

1.1.1 QUALITY FUNCTION DEPLOYMENT DEFINED

Just what is **Quality**? In English, as an adjective it means "excellent." The original Japanese term for QFD has meanings of features or attributes too. Quality is determined by customer expectations, so you cannot have a quality product or a quality service without identifying your customers and discovering their expectations. The second word, **Function**, means how you will meet customer expectations, or how your products or services will function to meet them. The third word, **Deployment**, defines how you will manage the flow of development efforts to make certain that customer expectations drive the development of your new products and services.

This book provides both formal and informal examples of QFD. In many ways, successful businesses large and small are applying QFD in formal and informal ways. No business can have continued success in the 21st Century without listening and responding to customers. Enabling the reader to learn how to do this better is the fundamental purpose of this book. In the words of W. Edwards Deming: "Learning is not compulsory, but neither is survival."

QFD is a method for structured product or service planning and development that enables a development team to specify clearly the customer's wants and needs, and then evaluate each proposed product or service capability systematically in terms of its impact on meeting those needs. QFD is fundamentally a quality *planning and management* process to drive to the *best possible product and service solutions*. A key benefit of QFD is that it helps product-introduction teams communicate to management what they intend to do, and to show their strategy in the planned steps forward. Management can then review these plans and allocate budget and other resources. QFD helps enable management to evaluate whether the product plans are worth the investment. Working through the QFD process together provides the important benefit of alignment—within the project team, and to management desires.

The QFD process tends to vary from practitioner to practitioner. In all cases, however, successful QFD work requires accurate assessment of customer needs. This begins with gathering the Voices of Customers (VOC), and ends with validating their needs. Some variant of the VOC process is included in most definitions of QFD, Six Sigma, Design for Six Sigma (DFSS), Marketing for Six Sigma (MFSS), Six Sigma Process Design (SSPD), and Technology for Six Sigma (TFSS). An illustration of this front-end work is provided in Figure 1-1.

Clearly, several steps must be followed to determine customer needs before beginning the matrix work that is often associated with the QFD process. QFD is a flawed exercise if it does not acquire well-defined and validated customer voices.

As mentioned above, the QFD process includes constructing one or more matrices (sometimes called **Quality Tables**). The first of these matrices is called the **House of Quality** (HOQ), shown in Figure 1-2. It displays the customer's wants and needs (the VOC) along the left, and the development team's technical response for meeting those wants and needs along the top. The matrix consists of several sections or sub-matrices joined together in various ways, each containing information that is related to the others (see Figure 1-3). As I have often said when teaching various Six Sigma tools, "Some tools flag the gaps, and others fill them." QFD is a method that flags gaps in knowledge, capability, and understanding as the design team works through the various QFD elements. It also keeps track of how key product and process design decisions relate to customer needs.

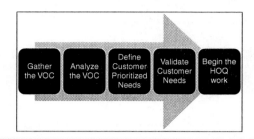

Figure 1-1 Front End of the QFD Process

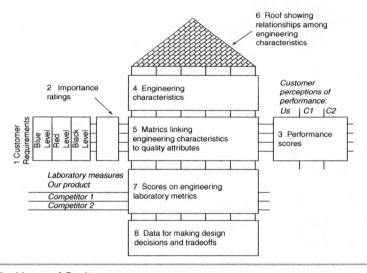

Figure 1-2 The House of Quality

Each of the labeled sections (1 through 8) is a structured, systematic expression of a product- or process-development team's perspective on an aspect of the overall planning process for a new product, service, or process. The numbering suggests one logical sequence for filling in the matrix.

Section 1 contains a structured list of customer wants and needs. The structure is usually determined through qualitative market research. The data is presented in the form of a tree diagram (defined and explained in Chapter 3).

Section 2 contains relative-importance ratings as determined by sampling customers for their priorities.

Section 3 contains two main types of information:

- Quantitative market data, including the customer's satisfaction levels with the organization's and its competition's current offerings
- Strategic goals for the new product or service

Section 4 contains, in the organization's technical language, a high-level description of the product or service it plans to develop. Normally, this technical description is generated from the customer's wants and needs in Section 1.

Section 5 contains the development team's judgments of the strength of the relationship between each element of its technical response and each customer want or need.

Section 6, Technical Correlations, is half of a square matrix, split along its diagonal and rotated 45°. Its resemblance to the roof of a house led to the term House of Quality becoming the standard designation for the entire matrix structure. Section 6 contains the development team's assessments of the implementation interrelationships between elements of the technical characteristics.

Section 7 contains a comparison of product or service metrics between your current offerings and that of competitors.

Section 8 contains relative weighting of customer needs versus how well you are performing on the relevant product and service metrics.

Section 8 contains three types of information:

- The computed rank ordering of the technical responses, based on the rank ordering of customer wants and needs from Section 2 and the relationships in Section 5
- Comparative information on the competition's technical performance
- Technical-performance targets

QFD authorities differ in the terminology they associate with the various parts of the House of Quality. I have tried to be consistent throughout this book, but in day-to-day usage, most people use various terms for sections of the HOQ. I have tried to indicate common alternate terminology whenever I have introduced a new term. There is no reason to be concerned about the lack of standardization; it rarely causes confusion.

Beyond the House of Quality, QFD optionally involves constructing additional matrices to further plan and manage the detailed decisions that must be made throughout the product or service development process. In practice, many development teams don't use these additional matrices. They are missing a lot. The benefits that the House of Quality provides can be just as significant to the development process after the initial planning phase. I urge you to use Part V of this book to become familiar with the later stages of

QFD, and then, with that knowledge available to you, to make an informed decision about how much of QFD your development project needs.

Figure 1-3 illustrates one possible configuration of a collection of interrelated matrices. It also illustrates a standard QFD technique for carrying information from one matrix into another. In Figure 1-3 we start with the HOQ, in this instance labeled 1: Product Planning. We place the Whats on the left of the matrix. Whats is a term often used to denote benefits or objectives we want to achieve. Most commonly, the Whats are the customer needs, the VOC data, but the development team's own objectives could also be represented as Whats. As part of the QFD process, the development team prioritizes the Whats by making a series of judgments based in part on market-research data. Many different techniques for determining these priorities are described later in this book.

Next, the development team generates the Hows and places them along the top of Matrix 1. The Hows are any set of potential responses aimed at achieving the Whats. Most commonly, the Hows are technical measures of performance of the proposed product or service. The relating of the Whats to the Hows is critical, because assumptions can sneak in that are unwarranted. This is why the VOC work preceding the definition of Whats is so crucial!

Based on the weights assigned to the Whats and the amount of impact each How has on achieving each What, the Hows are given priorities or weights, written at the bottom of the HOQ diagram. These weights are a principal result of the HOQ process.

In this simplified view, the second matrix is labeled 2: Product Design. This second matrix could as easily be labeled Service Design; or, if you are developing a complex Product System, it could comprise a sequence of matrices for System Design, Subsystem Design, and finally Component Design. To link the HOQ to Matrix 2, the development team places all, or the most important, of the HOQ Hows on the left of Matrix 2, and the

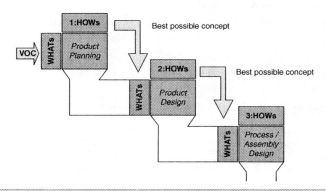

Figure 1-3   Interrelated Matrices

priorities of those Hows on the right. These HOQ Hows now become the Whats of Matrix 2, their relative importance to the development team having been determined in the HOQ process.

To achieve the Matrix 2 Whats, the development team needs a new, more technical or more detailed set of Hows, which are put at the top of Matrix 2. As before, the team uses the weights of the Matrix 2 Whats, and estimates the degree of relationship between the Matrix 2 Hows and the Matrix 2 Whats, to arrive at weights or priorities for the Matrix 2 Hows.

To link Matrix 2 to Matrix 3, the Matrix 2 Hows are transferred to the left of Matrix 3, becoming the Matrix 3 Whats. The weights of the Matrix 2 Hows are transferred to the right side of Matrix 3, and new Matrix 3 Hows are generated.

Each matrix in the chain represents a more-specific or more-technical expression of the product or service. In the classical model for QFD,[1] which was designed for the development of hardware products, the relationship of Whats to Hows in each matrix is as shown in Figure 1-3.

While this model mirrors the process of designing and manufacturing a physical product, similar models exist for developing services, for designing processes, and for developing software products. They are all covered in other parts of this book.

Other multiple-matrix QFD schemes are considerably more elaborate than the three-matrix scheme illustrated in Figure 1-3.[2] Some QFD schemes involve as many as thirty matrices that use the VOC's priorities to plan multiple levels of Design Detail, Quality Improvement Plans, Process Planning, Manufacturing Equipment Planning, and various Value Engineering plans. Some QFD experts believe that the use of additional matrices is not optional, and that the process should not be called QFD unless a series of matrices is constructed. In this book, however, we take a more liberal view; each team must make decisions on how far to go, considering the costs of function deployment versus the value delivered. We take the position that QFD delivers many possible benefits at many levels, and should be implemented at a level of detail appropriate to the task at hand.

Depending on the benefits a development team needs or is willing to work for, it will construct just the initial House of Quality, a large collection of interrelated matrices, or something in between. Teams will further customize their matrices to solve the problems they need to solve. From our point of view, it's all QFD. Adapting the old adage "science is what scientists do," we believe that "QFD is what QFD practitioners do."

---

1. Don Clausing, *Total Quality Development* (New York: ASME Press, 1994).

2. Bob King, *Better Designs in Half the Time: Implementing QFD Quality Function Deployment in America* (Methuen, Massachusetts: GOAL/QPC, 1987).

Matrix	What	How
House of Quality	Voice of the Customer	Technical Performance Measures
Subsystem Design Matrix	Technical Performance Measures	Piece-Part Characteristics
Piece-Part Design Matrix	Piece-Part Characteristics	Process Parameters
Process Design Matrix	Process Parameters	Production Operations

Figure 1-4   Classical Model for QFD Matrices

As we see in Figure 1-4, QFD is a tool that enables us to develop project priorities at various levels in the development process, given a set of priorities at the highest level (customer needs). QFD is not only a prioritization tool; it is also a **deployment tool.** What we mean by "deployment" is that QFD helps us to start with the highest level of Whats, generally the Voice of the Customer, and to deploy, or translate, that voice into a new language that opens the way for appropriate action. Every developer is familiar with this translation process and probably performs the translation informally all the time. QFD helps make the translation process explicit and systematic.

Finally, QFD provides a repository for product planning information. The repository is based on the structure of the QFD matrices. The matrices allow for entering the Voice of the Customer (and, in subsequent matrices, other deployed What and How information) and all related quantitative information, the Voice of the Developer and all related quantitative information, and the relationships between these voices. This information represents a succinct summary of the key product planning data. To be sure, very detailed elaboration of this information cannot fit in the QFD matrices and must be stored elsewhere. Documenting the matrices provides a golden thread throughout the design process, tracing the design decisions and tradeoffs and potential alternatives for enhanced offerings. The matrices can be viewed as a top-level view or directory to all the rest of the information. In fact, QFD has sometimes been described as a "visible memory of the corporation."[3]

There are many other views of QFD, and many will be explored elsewhere in this book. For now, it will be useful to keep in mind these main ideas: QFD provides a formal linkage between objectives (What) and response (How); it assists developers in developing or deploying the Hows based on the Whats; it provides a systematic method of setting priorities; and it provides a convenient repository of the information.

---

3. Quote attributed to Max Jurosek, Ford Motor Company, in *Enhanced Quality Function Deployment* (Cambridge, Mass.: MIT Center for Advanced Engineering Study, n.d.). Series of five instructional videotapes.

1.2 WHAT IS SIX SIGMA?

If you are familiar with this term and methodology, you may want to skip this section. Readers new to this business strategy—and it is a business process—will want to read this section carefully, keeping in mind that it is only an introduction.

Six Sigma can be read about and still greatly misunderstood. In 2006, a shareholder of Honeywell proposed an item to be voted upon at the annual shareholder meeting that Honeywell drop a related use of the term as misleading stockholders and the general public. There are many detailed and descriptive texts on the subject so we will attempt to be brief here, mostly for the benefit of any readers who are new to these topics. I will attempt to describe Six Sigma in terms of its three most widely applied and published interpretations, namely as a metric, a project methodology, and finally as a company initiative.

1.2.1 SIX SIGMA: THE METRIC

We will describe Six Sigma as a metric in a very brief manner here, as other sources contain more-detailed descriptions.[4,5] When Six Sigma was developed at Motorola in the 1980s, products were measured in terms of quality by counting their defects. For brevity's sake, we will define a defect for our purposes as anything not meeting customer expectations or requirements. Suppose for example that a customer requires on-time deliveries, specified as "no sooner than two days before promise date nor later than 1 day after promise date." Any delivery made to that customer outside of that range is unacceptable, considered to have a **delivery defect.** Other examples include **packing defects**, such as missing or incorrect items; **shipping defects**, such as arrival with damage or missing paperwork; and, of course, **product defects** in performance or appearance. All defects should be counted if they create customer dissatisfaction or add cost to the business for inspection and correction. However, not all defects are equal: Some have more impact on customer satisfaction and cost than do others. Sound business prioritization must take precedence over simple defect counting as if all were equal.

The goal of Six Sigma can be stated as simple as: "to define, measure, analyze, improve, and control the sources of variations that create defects in the eyes of customers and the business." The count of defects is a key metric. In fact, Motorola production operations focused for a time almost exclusively on reducing defects by an order of

---

4. Stephen A. Zinkgraf, *Six Sigma: The First 90 Days* (Upper Saddle River, N.J.: Prentice-Hall, 2006).

5. Ian Wedgwood, *Lean Sigma: A Practitioner's Guide* (Upper Saddle River, N.J.: Prentice-Hall, 2006).

magnitude over a specified period of time. If we declare that any defective item has at least one defect, then we can (loosely) state that the yield of any ongoing process is equal to 1 minus the sum of the defects created in that process:

$$\text{Yield} = 1 - \Sigma(\text{defects})$$

We can also count opportunities to create defects in a process. This might be as simple as counting the number of major steps in the process or as complex as looking in-depth at the number of opportunities for errors, mistakes, and non-conformities in the entire process. Whatever method is used, it is prudent to standardize and not change it. For assembled products, one of the best and simplest methods I have seen is to count the number of components for a product and multiply the total by three. The logic of this method is that each component can (1) be purchased correctly or incorrectly, (2) be assembled correctly or incorrectly, and then (3) perform correctly or incorrectly. Each purchased subassembly counts as only one component.

To normalize across different products and processes, the sigma metric involves counting defects and opportunities for defects, and then calculating a Sigma Value. An interim step is to calculate the **Defects per Million Opportunities** (DPMO) as:

$$\text{DPMO} = \Sigma(\text{defects}) \div \Sigma(\text{opportunities}) \times 10^6$$

We use DPMO to arrive at the Sigma Value. The higher the Sigma Value, the higher the quality. The table in Figure 1-5 equates Yield, equivalent defects per million opportunities or DPMO, and the Sigma Value.

We can see that when yields are in the range of 60 percent to 95 percent, we need few other measures to view improvement. However, as we approach 95 percent, our yield measure becomes fairly insensitive and it makes sense to find a more sensitive metric — hence Motorola's use of defect counting. If we further need to compare across vastly different product and process complexities, it makes sense to normalize by opportunity counts. Ultimately, as there are fewer defects to count, it is more important to move away from counting flaws to continuous-scaled measures of performance that tie well to customer satisfaction and product value.

So how do different products and services compare on Sigma Values? Figure 1-6 illustrates DPMO levels in parts-per-million (PPM) versus the Sigma Value.

How is the Sigma Value calculated? In short, it is the number of standard deviations from the mean of an equivalent normal distribution from a specified limit having the same defect rate as the one reported as DPMO. If we assume a defect occurs when any output is beyond the specification limit, then the odds of a defect at a Sigma Value of 6 are approximately 1 in 1 billion. *However,* defect measurements made in the short term are not entirely correlated with results over the long term. Over time, if left unchecked,

99.99966% Yield → 3.4 DPMO → 6 Sigma
99.9767% Yield → 233 DPMO → 5 Sigma
99.379% Yield → 6210 DPMO → 4 Sigma
93.32% Yield → 66807 DPMO → 3 Sigma
69.1% Yield → 308,537 DPMO → 2 Sigma

Figure 1-5   Yield, DPMO, and Sigma Values

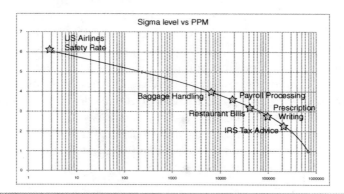

Figure 1-6   Sigma Value versus DPMO, Scaled in PPM

process quality degrades with shifts and drifts in the average output value, inflation of the Sigma Value, or both. A rule of thumb is needed to account for output shifts or degradation over time. Though this is somewhat controversial, George Box has commented that he is glad some accounting for long-term shift is being done.

The rule that is usually applied is that over time, the average change is 1.5 standard deviations. This shift arises originally from work done on tolerance stack-ups and shifts.[6,7,8] The switch to universal acceptance of this amount of shift is often based on empirical experience and not a theoretical construct.[9] I have personally seen long-term shifts of the mean over time between 0 and 4 Sigma, and my best recommendation is to

---

6. A. Bender, "Benderizing Tolerances—A Simple Practical Probability Method of Handling Tolerances for Limit Stack-Ups," *Graphic Science* (Dec. 1962).

7. H. Evans, "Statistical Tolerancing: The State of Art, Part III: Shifts and Drifts," *Journal of Quality Technology* 7:2 (April 1975): 72.

8. M.J. Harry and R. Stewart, *Six Sigma Mechanical Design Tolerancing* (Scottsdale, Ariz.: Motorola Government Electronics Group, 1988).

9. Davis Bothe, "Statistical Reason for the 1.5 Sigma Shift," *Quality Engineering* 14(3) (2002), 479–487.

aim for 1.5 Sigma. If we reduce our Six Sigma distance by this amount, we get a 4.5 Sigma distance to the upper specification, and the resulting odds of being out of specification rise to 3.4 defects per million items. This is the defect rate number most often touted in Six Sigma quality descriptions. See Figure 1-7 for an illustration of the shift between short-term and long-term performance.

So now we have a Six Sigma metric description, albeit in an oversimplified way for those who are experienced in the craft. This metric can be used in many ways, from counting simple defects to performing simple measurements of product or service output. Knowing these measurements is a key step in delivering quality as measured in the eyes of customers.

1.2.2 SIX SIGMA: THE PROJECT METHODOLOGY

Six Sigma is more than just a metric; it is a project methodology as well. Often QFD is utilized as one of the tools within a Six Sigma project methodology. I view the difference between a tool and a methodology as follows:

- a methodology comprises several steps to achieve an aim or purpose, using multiple tools
- a tool comprises a single function, or multiple functions that may be applied in several ways

When Six Sigma is viewed as a process-improvement methodology, it typically follows a five-phase approach. This approach is sequenced as Define, Measure, Analyze, Improve, and Control (DMAIC). Every Six Sigma process-improvement project follows this sequence, although it may involve taking several loops back, as it is a discovery process in search of causes and controls.

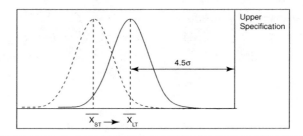

Figure 1-7   The Shift from Short-Term to Long-Term Performance

Six Sigma earns its clout by completing business projects aimed at significant process problems. These problems are identified by an organization in two ways: either top down or bottom up. The bottom-up approach is to listen to customer complaints and/or list internal chronic problems. The top-down approach is to set aggressive business cost-reduction goals and then refine them to a manageable level. Projects are defined around these issues, and all follow the general phases DMAIC. For more on this "project-linked-to-business-goals" approach, see for example *Six Sigma: The First 90 Days* (2006) by Dr. Stephen Zinkgraf.[10]

Within each of the DMAIC phases, key tools are applied to solve the process problems encountered in the project. Project leaders are called either Black Belts or Green Belts in general. Black Belts usually lead the higher-value, higher-risk, higher-visibility, more-complex projects in an organization and receive extensive training in process-improvement tools and statistical analysis, along with some team-building and presentation training depending upon organizational needs. A Black Belt may be deployed in many different areas of the company to solve a process problem. Green Belts typically lead smaller projects, usually within their local areas of expertise, and receive less training. Industry body-of-knowledge standards exist for Black Belt practitioners; and these leaders typically receive four weeks of training spread out over four or five months. Green Belts typically receive nine or ten days of training in two or three sessions spread out over two or three months. Both types of training are excellent for developing your people into more valuable, more versatile company resources.

Much has been written on the tools topic, and an excellent practical text written by my colleague Dr. Ian Wedgwood is called *Lean Sigma: A Practioner's Guide.*[11] This book is very straightforward regarding the tools involved in the methodology.

While consulting at Sylvania in 1998, I was challenged by some excellent engineers to explain how all the tools fit together. I started with a flipchart, and eventually put the major tools into a PowerPoint presentation that many have commented is helpful, so I will repeat it here as a series of figures.

The first step is to begin with the end in mind. We seek potential causes of our primary process issue (Y) and the relationship of the causes to Y. The issues may be multiple but we look at them separately and consider each issue or defect as a process result or Y. This can be stated mathematically as:

---

10. Zinkgraf, *Six Sigma.*

11. Wedgwood, *Lean Sigma.*

$$Y = f(x_1, x_2, x_3, x_4, \ldots, x_n)$$

Y can be a delivery time, a processing time, a product defect, a service error, an invoicing error or anything that we can explicitly state which can be measured. This $Y = f(x)$ transfer function is the ultimate objective in all good product development work. It is necessary to move forward in QFD as we can associate the Y with Whats and the Xs with Hows. These transfer functions are needed to fill in each QFD matrix. Regarding the equation, some of the tools in the roadmap work on the Y, some help find Xs, and some help establish the relationships to the Y. In the Define and Measure phases we are applying tools at some point that work on Y. These tools are: Measurement System Analysis (MSA), Process Control Charts, and Capability Analysis. MSA helps us determine how much error exists in measuring Y. Control Charts tell us if the Y is stable or predictable, and Capability Analysis helps determine the extent of variation in Y relative to a requirement established during the Define phase. Together these two are called an Initial Capability Assessment. So now our picture looks like Figure 1-8.

The next steps in the Measure and Analyze phases require identifying all the process Xs that are associated with this Y and then funneling them down to the potential few independent variables that truly influence Y. A Process Map is a tool to collect and

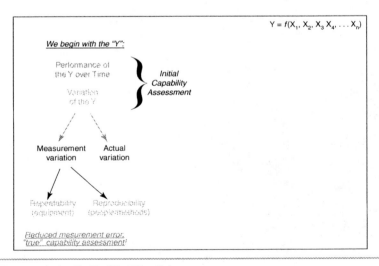

Figure 1-8   Begin with the End in Mind, Y = f(x)

identify these Xs, and a variant of a QFD matrix called the Cause and Effect matrix (C&E Matrix) combines the teams knowledge to "funnel" the larger list to known or observed influencers. A Failure Mode and Effects Analysis, or FMEA, helps establish how each independent X variable can be wrong, how severe is its impact, how often it occurs, and whether it can be detected. This helps make some "quick hits" to improving independent potential actors or Xs, and begins stabilizing the process. We now have a smaller number of variables, and we study these Xs passively (Multi-vari) to see if any have statistical correlation with actual variation observed in the Y variable, as shown in Figure 1-9.

In the Improve phase, we take the output of our funnel and begin Design Of Experiments (DOE) to establish true cause and effect relationships of Y = f(x) on our reduced list of suspect independent variables. From there we identify the true actors causing the variability in Y and establish proper controls with a control plan (Figure 1-10).

Now we have the major tools of the Process Improvement Methodology in sequence and relationships to establishing our goal of Y = f(x)!

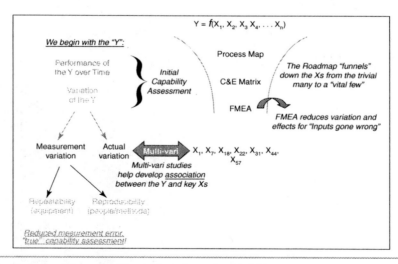

Figure 1-9   Finding and Funneling Down the Independent Variables

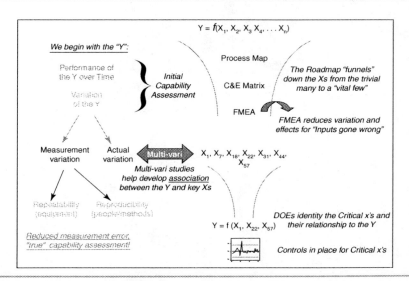

Figure 1-10   Establishing Y = f(x), Finding the Key Xs, and Setting Proper Controls

Six Sigma applied to process problems has as its fundamental aim or goal to define the equation that describes where your desired output gets its dependency: i.e., Y = f(x). So when applying the tools our goal is to develop this key equation. In addition to the Six Sigma Tools, Lean Manufacturing tools were integrated as part of the combining of two separate initiatives while I was at AlliedSignal, now Honeywell, thanks to Larry Bossidy's wisdom on initiatives. There are certainly many ways to combine these initiatives. As a set of tools, it makes sense to combine them with the phased approach of DMAIC. Early on at SBTI, I wrote the DMAIC tools for Six Sigma in the form of the M-A-I-C loop shown in Figure 1-11. We then integrated the Lean tools into these four phases, as seen in Figure 1-12. Following this "Lean Sigma Roadmap" of tools within the DMAIC phases, one can work a project towards a result with less variation in process output, process times, and process flow. Please recall the earlier comments in the methodology introduction that each individual tool may be applied in many ways. This is just one approach to showing how Lean tools and Six Sigma tools may be integrated into a unified methodology of MAIC.

1.2.3 SIX SIGMA: DEPLOYING AS AN INITIATIVE

To begin a Six Sigma initiative, top management or executive leaders must create the vision of what the initiative is to do for the company, customers, employees, and owners.

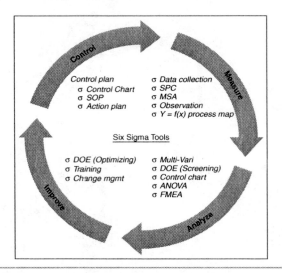

Figure 1-11   Six Sigma Tools in the M-A-I-C Sequence

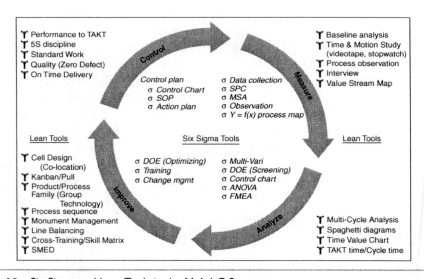

Figure 1-12   Six Sigma and Lean Tools in the M-A-I-C Sequence

They need to foster acceptance with communication, involvement, and leading from the front of the deployment. They must set goals, timing, and financial expectations for the program if it is to be effective. These are often done in off-site meetings or executive workshops. The previously cited work by Zinkgraf[12] provides more depth on these subjects. From these workshops, a deployment plan is developed for the Six Sigma initiative, and it is refined further by champions and sponsors of this effort. The refinement is around people, projects, and training timetables.

Refinement around people and projects means getting the right people on the bus, and then dropping them into the right project areas, armed with the right tools and training. To have great project work on significant business problems, a company needs sponsorship by the business ownership, management, or executive force, or all three depending upon the size of the organization. This requires initiative alignment with strategy, people, processes, business metrics, and customer needs. Alignment with customer needs, strategy, internal stakeholder needs, and business metrics is a great application of QFD and will be discussed further in Chapter 2.

The Black Belts or Green Belts need to know they have support and access to whatever is needed to conduct their project work. It is generally management's job to set expectations and then inspect for progress on tasks, measured against agreed-upon goals. Management must also support and coach the project leaders through the first few projects, and to do so must understand the basics of Six Sigma. People who fill this role in a Six Sigma initiative context are usually described and trained as either Champions or Sponsors. Champions or Sponsors identify the business areas and rough out the projects that are best linked to the business goals and needs. They also identify individuals for training as Green Belts and Black Belts. Green Belts and Black Belts take the projects to training and follow a Plan-Train-Apply-Review cycle through two, three, or four training sessions, depending upon the focus, level, and deployment plan within a company.

Due to the scope of a Six Sigma Initiative and the nature of this book, we will delve no further into the initiative topic. The previously cited work by Zinkgraf[13] will provide more depth in this area.

1.3 HISTORY OF QFD AND SIX SIGMA

1.3.1 ORIGINS OF QFD

The people identified in this chronology have made special efforts in the best interests of U.S. industry. They could have kept QFD and Six Sigma as proprietary secrets, not to be

---

12. Zinkgraf, *Six Sigma.*

13. Zinkgraf, *Six Sigma.*

shared with the competition. Instead, they shared their experiences with others, including their competitors, and everyone has gained.

QFD became widely known in the United States through the efforts of Don Clausing, of Xerox and later MIT, and Bob King of GOAL/QPC. These two worked independently, and likely first came into contact in October 1985, when Clausing presented QFD at a GOAL/QPC conference in Massachusetts. By that time, both men had already made significant contributions toward promoting QFD.

The Japanese characters for QFD are:

- 品質 (hinshitsu), meaning "quality," "features," "attributes," or "qualities"[14]
- 機能 (kino), meaning "function" or "mechanization"[15]
- 配置 (tenkai), meaning "deployment," "diffusion," "development," or "evolution"

Any of the English words could have been chosen by early translators of Japanese articles. It's little more than a matter of chance that QFD is not called Feature Mechanization Diffusion today. In the early days, when Lou Cohen explained QFD to audiences, he attempted to rename it Structured Planning, or Quality Feature Deployment, in the hope that people would be able to tell from its name what QFD was all about. For better or for worse, "Quality Function Deployment" has stuck in the United States, and no alternative name is likely to survive. None of the thirty-two possible combinations of English equivalents really denotes what QFD actually is. We must be content with a name for the process which is not that self-explanatory.

1.3.2 EARLY HISTORY OF QFD IN JAPAN

Yoji Akao[16] cites the rapid growth of the Japanese automobile industry in the 1960s as a driving force behind the development of QFD. With all the new product-development drives in the Japanese auto industry, people there recognized the need for design quality and that existing QC process charts confirmed quality only after manufacturing had begun. Mr. Akao's work with Kiyotaka Oshiumi of Bridgestone led to "Hinshitsu Tenkai" or "Quality Deployment," which was taken to various companies with little public attention. The approach was later modified in 1972 at the Kobe Shipyards of Mitsubishi

---

14. www.dictionary.reference.com/browse/quality.

15. www.dictionary.reference.com/browse/function.

16. Yoji Akao, "QFD: Past, Present, and Future," paper presented at the International Symposium on QFD (1997).

Heavy Industry to systematically relate customer needs to functions and the quality or substitute quality characteristics. The first book on the topic, *Quality Function Deployment* by Akao and Mizuno, was published by JUSE Press in 1978.[17]

1.3.3 HISTORY OF QFD IN THE USA

In 1983, the first article on QFD by Akao appeared in *Quality Progress* by ASQC,[18] and from there things spread quickly. Don Clausing first learned about QFD in March 1984, during a two-week trip to Fuji-Xerox Corporation, a Xerox partner in Japan. Clausing, a Xerox employee at that time, had already become interested in the Robust Design methods of Dr. Genichi Taguchi, who was a consultant to Fuji-Xerox. While in Japan, Clausing met another consultant to Fuji-Xerox, a Dr. Makabe of the Tokyo Institute of Technology.

At an evening meeting, Dr. Makabe briefly showed Clausing a number of his papers on product reliability. "After fifteen minutes," relates Clausing, "Dr. Makabe brushed the papers aside and said, 'Now let me show you something *really* important!' Makabe then explained QFD to me. I saw it as a fundamental tool that could provide cohesion and communication across functions during product development, and I became very excited about it."

In the summer of that same year, Larry Sullivan of Ford Motor Company organized an internal company seminar. Clausing was invited to present QFD. Sullivan quickly grasped the importance of the QFD concept and began promoting it at Ford.

Clausing continued to promote QFD, Taguchi's methods, and Stuart Pugh's concept-selection process at conferences and seminars. When Clausing joined the faculty at MIT, he developed a semester-length graduate course that unified these methods along with other concepts into a system for product development that eventually became called "Total Quality Development." Many of his students, already senior managers and engineers at large U.S. companies, returned to their jobs and spread his concepts to their coworkers.

In June 1987, Bernie Avishai, associate editor of *Harvard Business Review*, asked Don Clausing to write an article on QFD. Clausing felt that the paper should be given a marketing perspective, and he invited John Hauser to coauthor it. Hauser had become intrigued with QFD after learning about it from a visit to Ford. The article, published in

---

17. Shigeru Mizuno and Yoji Akao, eds., *Quality Function Deployment: A Company-wide Quality Approach* (Tokyo: JUSE Press, 1978).

18. Masao Kogure and Yoji Akao, "Quality Function Deployment and CWQC in Japan," *Quality Progress* 16, no. 10 (1983): 25.

the May–June 1988 issue of the *Harvard Business Review,* has become one of the publication's most frequently requested reprints. That article probably increased QFD's popularity in the United States more than any other single publication or event.

Larry Sullivan founded the Ford Supplier Institute. This was a Ford Motor Company organization aimed at helping Ford's suppliers improve the quality of the components they developed for Ford. Sullivan and others at Ford gained a detailed understanding of QFD by working with Dr. Shigeru Mizuno and Mr. Akashi Fukahara from Japan. Eventually Ford came to require its suppliers to use QFD as part of their development process, and the Ford Supplier Institute provided training in QFD (along with other topics) to these suppliers.

The Ford Supplier Institute eventually became an independent nonprofit organization, the American Supplier Institute (ASI). ASI has become a major training and consulting organization for QFD. It has trained thousands of people in the subject.

Bob King, founder and executive director of GOAL/QPC, first learned of QFD from Henry Klein of Black and Decker. Klein had attended a presentation on QFD given by Yoji Akao and others in Chicago in November 1983. The following month, Klein attended a GOAL/QPC course on another TQM topic, where he told King about this presentation and about QFD. King began offering courses on QFD starting in March 1984. In the summer of that year, King learned more details about QFD from a copy of a 1978 book by Akao and Mizuno, *Facilitating and Training in Quality Function Deployment.* In the fall of 1984, King began offering a three-day course on QFD, based on the understanding of the tool he had gained from the Akao and Mizuno book. In November 1985, King traveled to Japan and met with Akao to "ask him all the questions he couldn't answer." Akao provided King with his course notes on QFD, and he gave GOAL/QPC permission to translate the notes and use them in his GOAL/QPC courses.

Based on these notes, GOAL/QPC offered its first five-day course on QFD in February 1986. Lou Cohen attended that course and learned about QFD there for the first time.

At the invitation of Bob King, Akao came to Massachusetts and conducted a workshop on QFD in Japanese with simultaneous translation into English. Akao conducted a second workshop in June 1986, also under the auspices of GOAL/QPC. For this second workshop, GOAL/QPC translated a series of papers on QFD, including several case studies. This translation was later published in book form.[19] Eventually this collection of QFD papers became what remains the standard advanced book on QFD.

In 1987, GOAL/QPC published the first full-length book on QFD in the United States: *Better Designs in Half the Time,* by Bob King.[20] In this book, King described QFD as a

---

19. Yoji Akao, *Quality Function Deployment: Integrating Customer Requirements into Product Design,* trans. by Glenn H. Mazur and Japan Business Consultants, Ltd. (Cambridge, Mass.: Productivity Press, 1990).

20. King, *Better Designs.*

"matrix of matrices" (see Chapter 18). King relates that in June 1990, Cha Nakui, a student of Akao's and later an employee in Akao's consulting company, "comes to work for GOAL/QPC and corrects flow of QFD charts." Among Nakui's contributions to our understanding of QFD is his explanation of the Voice of the Customer Table (see Chapter 5).

I first saw QFD while at AlliedSignal, now Honeywell, circa 1990. Early examples at that time included Toyota Motors' QFD processes. At the same time, Value Engineering was being taught at AlliedSignal, along with Robust Design methods from the American Supplier Institute. John Fox's seminal work, *Quality Through Design: Key to Successful Product Delivery,* was published in 1993, and included aspects of Design Process Flow, QFD, Design for Manufacture, and Critical Parameter Management, among many other methods. This work is one of the earliest to set the stage for Design for Six Sigma (DFSS) and QFD.

Other important early publications in the United States include

- "Quality Function Deployment and CWQC in Japan," by Professors Masao Kogure and Yoji Akao, Tamagawa University, published in *Quality Progress* magazine, October 1983
- "Quality Function Deployment," by Larry Sullivan, published in *Quality Progress* magazine, June 1986
- Articles on QFD by Bob King and Lou Cohen in the spring and summer 1988 editions of the *National Productivity Review*
- A series of articles on QFD in the June 1988 issue of *Quality Progress* magazine
- A course manual on QFD to supplement ASI's three-day QFD course
- *Annual Proceedings* of QFD symposia held in Novi, Michigan, starting in 1989

QFD software packages first became available in the United States around 1989. The most widely known package was developed by International TechneGroup Incorporated. Some organizations heavily committed to QFD, such as Ford Motor Company, have developed their own QFD software packages.

Early adapters of QFD in the United States included Ford Motor Company, Digital Equipment Corporation, Procter and Gamble, and 3M Corporation. Many other companies have used QFD, and the tool continues to grow in popularity. More than fifty papers were presented at the Sixth Symposium on Quality Function Deployment in 1994. Only a few of these papers were not case studies. The majority of companies using QFD are reluctant to present their case studies publicly, since they don't want to reveal their strategic product planning. Therefore, it is likely that the fifty papers presented at the QFD Symposium represent just the tip of the iceberg in terms of QFD implementation.

Figure 1-13 identifies many key events in the development of QFD, both in Japan and the United States. In some cases, where exact dates are not known, approximate months or years have been provided.

Date	Source	Event
1966	*Facilitating and Training in Quality Function Deployment,* Marsh, Moran, Nakui, Hoffherr	Japanese industry begins to formalize QFD concepts developed by Yoji Akao
1966	*Facilitating and Training in Quality Function Deployment*	Bridgestone's Kurume factory introduces the listing of processing assurance items: "Quality Characteristics"
1969	*Facilitating and Training in Quality Function Deployment*	Katsuyoshi Ishihara introduces QFD at Matsushita
1972	*Facilitating and Training in Quality Function Deployment*	Yoji Akao introduces QFD quality tables at Kobe Shipyards
1978	*Facilitating and Training in Quality Function Deployment*	Dr. Shigeru Mizuno and Dr. Yoji Akao publish Deployment of the Quality Function (Japanese book on QFD)
1980	*Facilitating and Training in Quality Function Deployment*	Kayaba wins Deming prize with special recognition for using Furukawa's QFD approach for bottleneck engineering
1983	*Facilitating and Training in Quality Function Deployment*	Cambridge Corporation of Tokyo, under Masaaki Imai, introduces QFD in Chicago along with Akao, Furukawa, and Kogure
10/83	*Quality Progress* magazine	"Quality Function Deployment and CWQC in Japan," by Professors Masao Kogure and Yoji Akao, Tamagawa University
11/83	Bob King	Akao and others introduce QFD at a U.S. workshop in Chicago, Illinois
3/84	Don Clausing	Professor Makabe of Tokyo Institute of Technology explains QFD to Don Clausing
3/84	Bob King	Bob King begins offering a one-day course on QFD
7/17/84	Don Clausing	Don Clausing presents QFD to a Ford internal seminar organized by Larry Sullivan
1985	*Facilitating and Training in Quality Function Deployment*	Larry Sullivan and John McHugh set up a QFD project involving Ford Body and Assembly and its suppliers
10/30/85	Don Clausing	Don Clausing presents QFD at GOAL/QPC's annual conference
11/85	Bob King	King meets with Akao in Japan. Akao gives GOAL/QPC permission to translate his classroom notes and use them in GOAL/QPC's training
1/27/86	Don Clausing	Don Clausing presents QFD to Ford's Quality Strategy Committee #2, chaired by Bill Scollard of Ford
2/86	*Facilitating and Training in Quality Function Deployment*	GOAL/QPC introduces Akao's materials in its five-day QFD course (the author attended this course in February 1986)

Figure 1-13   History of QFD in the United States[21]

---

21. Stephen A. Zinkgraf, personal communication, Autumn 2008.

6/86	Don Clausing	Larry Sullivan sponsors Dr. Mizuno who gives a three-day seminar on QFD
6/86	*Quality Progress* magazine	"Quality Function Deployment" by Larry Sullivan
10/86	Don Clausing	Larry Sullivan launches QFD at Ford
6/87	Don Clausing	Bernie Avishai, associate editor of *Harvard Business Review,* asks Don Clausing to write an article on QFD. Don invites John Hauser to co-author it. It is published in May–June 1988
1989 - present	ASI, GOAL/QPC	Sponsorship of QFD Symposia at Novi, Michigan
2/6/91	Don Clausing	Don Clausing and Stuart Pugh present "Enhanced Quality Function Deployment" at the Design and Productivity International Conference, Honolulu, Hawaii
1993	John Fox	Publication of *Quality Through Design*, a book linking QFD to the pre- and post-design aspects around QFD, including Kano's Model, Design for Manufacture, Value Engineering, Reliability Growth, Critical Parameter Management, Failure Mode and Effects Analysis, Taguchi Methods, and Statistical Process Control
1995	Steve Zinkgraf	Steve Zinkgraf utilizes a QFD matrix in setting up the Cause & Effect Prioritization Matrix later used by many in the Six Sigma community when comparing multiple process inputs to multiple process outputs

Figure 1-13   Continued

1.3.4 HISTORY OF QFD WITH SIX SIGMA

Dr. Stephen Zinkgraf [22] observes that at Motorola in the late 1980s, QFD's use was frequently at the front end of designing new products. The focus was initially on two-way matrices, until Fernando Reyes used QFD in a very innovative fashion—to develop the strategic plan for manufacturing automotive electronic applications based on customer requirements.

In 1995, Zinkgraf was leading Six Sigma Deployment at the AlliedSignal Engineered Materials Sector, and was putting together the Six Sigma operations roadmap. He concluded that Six Sigma should be based on understanding the interaction between process inputs (Xs) and process outputs (Ys). Since the Ys were to reflect satisfying customer needs, it seemed that a two-way matrix mapping the process inputs generated by the process map focused on inputs and outputs to process requirements. It fit perfectly into the final roadmap. The process map generated the inputs to the Cause & Effect matrix. The C&E Matrix focused the FMEA on only the important inputs, thereby shortening and focusing the FMEA process. The roadmap—including the process map, the C&E

---

22. Zinkgraf, personal communication.

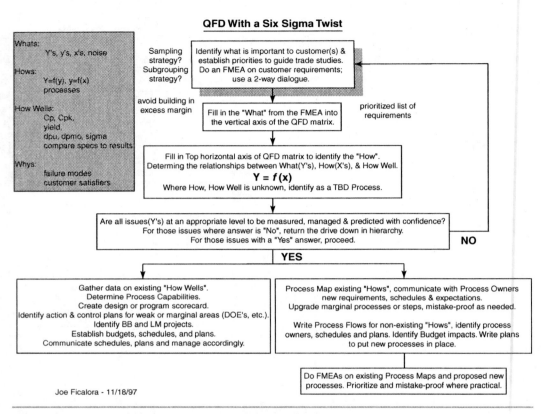

Figure 1-14 A Historical Flow Map of QFD and Six Sigma Tools from 1997

Matrix, and the FMEA—opened the door to the Analysis phase of MAIC. The C&E Matrix, when done properly, yielded a process focus that really hadn't existed before, outside of the archaic fishbone diagram. The C&E Matrix is essentially the result of quantitative generation of multiple fishbone diagrams quickly. The limitation of the fishbone diagram had been that it was not focused on process inputs and allowed analysis of only one output at a time. With multiple outputs, it was not possible to aggregate the results into a single plan of action.

In 1998, I gave a talk entitled "QFD with a Six Sigma Twist" at an Annual Black Belt conference. In this talk, I commented that the QFD applications I had seen previously needed more of a VOC emphasis on the front end with FMEA, as well as some sampling strategies to better detail what they really needed. I also commented that more-detailed data and analysis on the linkages between matrices could be obtained with process maps and other Six Sigma tools, to enable users to truly understand the

$Y = f(x)$ relationships implied when connecting Hows to Whats. Additionally, I commented on the "How wells"—that true process capabilities needed to be determined in scoring what was possible. I received positive feedback on turning what was then some large variation of QFD application into a flow map that could be followed. That flow map is illustrated in Figure 1-14.

1.4 WHAT IS QFD BEING USED FOR TODAY?

As with any versatile tool, the applications of QFD are limited only by one's imagination. The original intent of QFD was to provide product developers with a systematic method for "deploying" the Voice of the Customer into product design. The requirement to evaluate potential responses against needs is universal, however, and in the United States a wide range of applications sprang up quite rapidly.

Following are some typical QFD applications that do not fit the model of product development.

- Course design: Whats are needs of students for acquiring skills or knowledge in a certain area; Hows are course modules, course teaching style elements. Curriculum design is a natural extension of this application and has also been done using QFD[23]

- Internal corporate service group strategy: Whats are business needs of individual members of service group's client groups; Hows are elements of the service group's initiatives

- Business group five-year product strategy: Whats are generic needs of the business group's customers; Hows are product offerings planned for the next five years

- Development of an improved telephone response service for an electric utility company:[24] Whats are the needs of the customer; Hows are critical measures of performance of the telephone answering center

---

23. Mahesh Krishnan and Ali A. Houshmand, "QFD In Academia: Addressing Customer Requirements in the Design of Engineering Curricula," presented at the Fifth Symposium on Quality Function Deployment (Novi, Mich., June 1993).

24. Amy Tessler, Norm Wada, and Bob Klein, "QFD at PG&E: Applying Quality Function Deployment to the Residential Services of Pacific Gas & Electric Company," presented at the Fifth Symposium on Quality Function Deployment (Novi, Mich., June 1993).

In my experience, those who become enthusiastic about QFD are generally very creative in conceiving new applications. Those who dislike the "matrix" approach to planning are generally very creative in producing reasons why QFD doesn't work.

Of course, QFD itself does not "work." Just as we cannot say that hammers "don't work," so also we cannot say that problem-solving tools such as QFD "don't work." A more accurate statement would be: "I can't work it." This book is intended to raise your confidence that, when it comes to QFD, you can say, "I can make it work for me."

1.4.1 QFD USES WITHIN SIX SIGMA METHODS

Any place there are voices to be heard, analyzed, and driven into development activities, QFD can be utilized. In general, there are four broad categories of voices in the Six Sigma community that may be linked to development activities involving QFD:

1. Voice of the Customer (VOC): the voices of those who buy or receive the output of a process, clearly an important voice in product and process development
2. Voice of the Business (VOB): the voices of funding and sponsoring managers for marketing and development activities
3. Voice of the Employee (VOE): the voices of those employees who work in your company and/or develop new products, services, and technologies
4. Voice of the Market (VOM): the voices of trendsetting lead-user or early-adopter market segments, or market-defining volume purchasers

Figure 1-15 illustrates how four different Six Sigma methodologies interact with these four broad voice categories. Design for Six Sigma (DFSS) employs both VOC and VOB, so that products are designed to meet and exceed customer and business requirements.

Six Sigma Process Design (SSPD) requires VOC and employee voices (VOE), so that any new process meets both customer needs and the needs of the users of the new or re-designed process. Marketing for Six Sigma (MFSS) needs the business requirements and

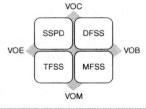

Figure 1-15   Six Sigma Methodologies and Four Voice Categories

the market requirements to succeed in planning new product and market-segment launches. Finally, Technology for Six Sigma (TFSS) requires the technologist's voices on technology capabilities coupled with market requirements in order to certify and release new technologies to the product designers.

1.5 DISCUSSION QUESTIONS

- Who is your customer? Try writing down your answer.
- Think about your current development process for products, services, courses, strategies, or anything else. Do you know your customers' needs? How do you know them? What form do you use to represent those needs?
- What does your first formulation of Hows look like? What form does it take? How do you generate your Hows?
- How do you determine the relationship of your Hows to your customers' needs?
- How many levels of Hows do you have in your development process? Write down a brief description of each of these levels and show these to your colleagues. Notice how long it takes to get agreement.
- Besides your development process, where else in your daily work might QFD apply? Define the Whats and Hows for those additional applications.

How QFD Fits in the Organization

This chapter sets a context for QFD in terms of the problems QFD helps us solve. While the technical aspects are covered in detail, one must keep in mind the additional value of QFD as a team alignment and communication methodology. This is a team sport, and the value it brings to the members of a cross-functional team while they work through the various aspects of QFD can be enormous. Additionally, the fundamental business challenges of maintaining competitiveness, increasing revenues, decreasing costs, adjusting to market shifts, and reducing the time to produce new products and services can all be enhanced by QFD. With the rapid changes in today's marketplaces, businesses big and small often change their business models, organizational structures and physical layouts. Designing and implementing these changes can be planned and managed through QFD as well. This chapter explains QFD's role with respect to each. One of the most important product development process concepts, specifically aimed at reducing product development time, is Concurrent Engineering (CE). This chapter will introduce CE and show how QFD is critically connected to it.

Finally, we'll explore the Kano model for customer satisfaction. The model expands our traditional views of how customers are satisfied by products and services. It provides some important food for thought as we explore the way in which QFD uses the Voice of the Customer.

2.1 THE CHALLENGE TO THE ORGANIZATION

The challenges to every company are varied, but almost all organizations must compete with others in terms of value creation, value delivery, and value management to be successful in the short- and long-term horizons. **Value creation** includes the development of all new products and services. **Value delivery** is defined by the processes that operate continuously in every organization that directly move goods and services to customers. **Value management** is defined as all the managerial and support processes that operate in a company to keep the organization functioning properly—for example, personnel management, invoicing, or billing processes. These processes may be complex, dynamic and diverse, depending on the size of the organization. QFD has traditionally been applied in the value-creation area, but also has more recently been applied to rapidly redesign how value delivery and value management are done.

The needs of three groups of people or stakeholders must be met or exceeded during value creation, delivery, and management (Figure 2-1). Customers, owners, and employees have varied and changing needs and expectations. Without customers, no organization can exist for very long. Most organizations have a wide range of customers with changing expectations. Owners or stockholders usually have expectations of increasing success in profits, company performance, and so on. Employees also have expectations for success: a good work environment, pay for performance, and job security, among others. Additionally, outside groups regulating finances, record-keeping, privacy, and environmental impacts have voices that must be taken into account. All these voices need to be organized and addressed. A tool like QFD becomes paramount when addressing multiple, broad-ranging needs while creating value or changing how value is managed or delivered.

All companies, be they public, private, or non-profit, must provide ever-increasing value to customers and markets they serve in order to have stability and growth. Managing, owning, and working in a growing company or organization is a lot more enjoyable than in a stagnant or shrinking organization. QFD enhances the ability to drive customer, owner, and employee needs into an organization's development and design activities. These activities are a key step to revenue growth.

Figure 2-1   Organizational Stakeholder Groups

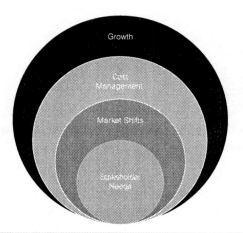

Figure 2-2 Organizational Challenges

As illustrated in Figure 2-2, the foremost challenge in addressing these complex and multiple needs is to **grow the revenue**, by providing equal or greater value than competing alternatives. The second but no less daunting challenge is to **keep costs in step with market pricing**, by designing and managing the value-delivery and value-management processes appropriately. The third major challenge organizations regularly face is **dealing with market shifts**. Ideally, these market shifts are anticipated and planned, but occasionally they are a surprise. In order for an organization to react appropriately and quickly to developments, whether anticipated or surprised, a plan (or playbook or score) must be written, and followed throughout the effort. QFD helps plan, align tasks, direct actions, and manage deployment of resources for success.

With the rapid changes in today's dynamic marketplaces, businesses big and small must evolve their business models, organizational structures, and physical layouts to keep up with the pace of external change. Designing and implementing these evolutions can be planned and managed through QFD as well. One way to view these challenges is to align and compare External Realities, Internal Activities, and Financial Targets, as described by Larry Bossidy.[1] When re-designing various aspects of a business—such as the business model, organizational structure, or physical layout—QFD may be applied to

---

1. Larry Bossidy and Ram Charan, *Confronting Reality: Doing What Matters to Get Things Right* (New York: Crown Business, 2004).

align, plan, and manage the efforts. The key parts here are to (1) obtain the voices of internal and external customers and stakeholders, and (2) align your functions and activities to deploy "quality" appropriately for success.

2.2 INCREASING REVENUES

Growing the top line of a company's income statement is the foremost challenge in addressing the multitude of needs of various organizational stakeholder groups. Increased revenues are normally achieved by selling more of a product or service, creating new products or services to sell, or charging more for existing products or services. None of these desirable results can be attained without products or services that have more **inherent value** to customers than competing alternatives.

Value is decided upon by customers as exceeding other options or not. There is only one purchase or selection by any individual customer and it is either/or, there is no middle ground.

Value Equation: Value = (Qualities or Features) - Price Paid

Value is measured differently by each customer, but in all cases includes the components of Qualities on the positive side and Price on the negative side. Qualities can have many aspects, including convenience, lead time, availability, features, colors, flexibility, and so on. Developing new products, or redeveloping old ones to increase desired qualities and remove costs, increases value overall. QFD plays an important role in planning, managing, and directing new value propositions from idea to product or service launch. The immediate related challenge to an organization is to be certain all new value propositions, products, and services will exceed those already in the marketplace as well as extensions of existing alternatives likely to be in the marketplace at the time of introduction. The right side of the House of Quality helps with this challenge.

Since alternatives change often in modern marketplaces, the challenges to the organization are dynamic; the company must continue to enhance the value propositions it brings to market as time passes, and must communicate those value propositions well. The only ways to accomplish all these things are to

- offer new products, services, and/or features in existing markets, or
- find new markets or customers, or
- design out costs from existing products or services

Virtually every value-based stock-picking method screens for new products or services when selecting stocks to buy. Clearly, new products or services are related to enhancing company value over time, which satisfies owners and stockholder needs. QFD contributes to increased revenues by helping organizations to concentrate their efforts on customer needs, and to accurately and effectively manage and plan those customer needs into the right product design or the right service characteristics.

In today's world of **growth through acquisition**, finding the best ways to deploy merger activities can be critical to a successful acquisition. QFD can be a guiding management tool to make certain that the merger teams doing the work are meeting the needs of the diverse voices of all stakeholders. In a growth-through-acquisition environment, those voices will likely include:

- Voice of the Business: Cost targets, timings, realized efficiencies, etc.
- Voice of Employees: Parties on both sides of the merger, communications, etc.
- Voice of Owners/Stockholders: What will change, when, financials, etc.

As plans are created to grow a business, organically or through acquisitions, QFD can play a role. Capturing important voices to add value to planned activities will define the quality of those activities. Deploying those voices into the design functions will make certain that they add value to the planning endeavor. QFD facilitates this process, helps manage it, and documents how and why it was done.

2.3 DECREASING COSTS

A difficult challenge for all organizations is to meet profit expectations (or in a non-profit setting, to have excess monies available to add more services or benefits to constituents). The preceding section showed how QFD can enhance revenues. QFD can also help in the long term and short term with managing and planning the reduction of the cost side of the profit-and-loss equation:

$$Profit \text{ (before taxes)} = Revenues - Costs$$

Some of the costs are non-recurring development costs, while others are contained in recurring value-delivery and value-management processes. Costs can be designed out of products, delivery processes, and support processes. In fact, these are focus areas of Six Sigma too. If part of the Six Sigma effort involves redesign, or reorganization, QFD can play the useful role of managing and planning cost reduction in a Six Sigma project.

As a management tool, QFD can focus development efforts in the right areas, thus avoiding wasted time and materials on non-customer-valued pursuits. This added reduction in development costs helps meet profit expectations by enhancing the development process, getting the "right things" to market, and avoiding the cost of redevelopment to weed out the "wrong things." QFD thus enhances **design-effort efficiencies,** leading to cost savings in the long term.

Decreased overall costs can be achieved by such actions as lowering the cost of purchased materials or services, reducing overhead costs for the office or plant, and reducing payroll. When competing in mature markets, designing out product and delivery costs to stay ahead of expected price erosion is one key competitive strategy.

All of these cost-cutting efforts may be Six Sigma projects. Where redesign is the approach, QFD should be included. QFD contributes to decreased costs in a redesign effort by:

- Increasing the likelihood that a product or process design will not have to be changed or redone. This dampening effect comes about because QFD allows developers to evaluate proposed mid-project changes against the same criteria used to evaluate all design decisions at the beginning of the project. Developers have simply to add the new proposed change to their QFD matrices and apply the same analysis that they applied to all the earlier decisions. This systematic analysis helps the team avoid panicky, rushed decisions that fail to take the entire product and all customer needs into account. Most midcourse corrections are easily rejected or postponed when QFD analysis is applied to them.

- Focusing product and process development on the work that matters the most to the customer, rather than on work that means little or nothing to the customer. This is another way of saying the work that gets done is what QFD analysis has shown to be most clearly related to meeting customer needs.

2.4 DEALING WITH MARKET SHIFTS AND CYCLE-TIME REDUCTION

Nearly all organizations have competition. Each price change, each product announcement, each reorganization—in fact, each change an organization makes—is like a move in a game between competitors. Each move affects an organization's competitiveness, for better or for worse. Often the results cannot be observed immediately, nor can an organization regain lost competitiveness rapidly. Markets can shift fairly quickly once a new innovation or new competitor enters the market. This is because the Internet now amplifies and accelerates communication of new developments. Suddenly, customer expectations are changed, and raised accordingly. The stakes of the game are high.

One of the most important "moves" in the "game" is to make new products or services available before the competition does. By introducing a new product or service before a competitor can offer an equivalent one, your organization can rob the competition of market share that may be impossible to regain. This is especially true if the new product or service is on target in meeting key customer needs. Many believe today that rapid development of new products and services is the single biggest factor in competitiveness—in other words, it's the winning "move."

Since most organizations begin designing new versions of products or services as soon as the previous ones have been released, the product development process is usually viewed as a "cycle" (as shown in Figure 2-3). In fact, the development process, like all other processes, is not quite as neat as a simple circle. Often, work on the *next* product begins well before the *current* product is ready to be sold. Many other simultaneous activities are usually happening as well. Nevertheless, the "cycle" model is a helpful way of viewing the activity.

QFD is an important key to cycle-time reduction. Throughout this book, we'll be seeing how QFD helps development teams make key decisions early in the development process, at a time when the cost of a decision is relatively low.

Some of the ways QFD contributes to reduced cycle time are the following.

- QFD helps reduce midcourse changes. Midcourse changes, such as shifts in priorities, key vendor replacements, or replacement of key technologies, wreak havoc on development schedules. When a project is well underway and a major change in

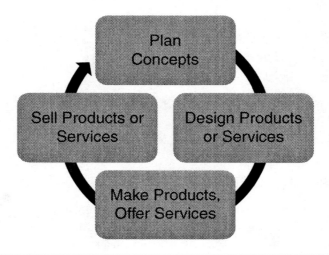

Figure 2-3 Product-Development Cycle

direction is proposed, most development teams are unable to evaluate the proposal properly. This is because there is enormous schedule pressure on teams, and that pressure normally increases as development proceeds. Consequently, midcourse changes are often made rapidly, without due consideration of their overall impacts.

- When a project has been planned using QFD, the House of Quality and other matrices provide a summary-at-a-glance of the project strategy. A well-done QFD planning process will clearly lay out all the needs of the customer, and will show how the strategic and design decisions of the project relate to those needs. Midcourse proposals can rapidly be added to the QFD matrices and evaluated in the context of all the previously made decisions. Very few midcourse changes actually survive this type of thorough analysis.

- QFD helps reduce errors in implementation. By working through the QFD process, the development team ensures a common vision of customer needs and responses to them. This common vision, along with detailed planning starting from the same place (the House of Quality), results in consistent follow-through across the entire development team.

2.5 CHALLENGES TO RAPID PRODUCT DEVELOPMENT

A key to competitiveness is the ability to respond to the competition by producing new products and services rapidly. Among the many challenges to rapid production of products and services are

- Properly understanding customer needs
- Strategically prioritizing efforts
- Properly assessing risk
- Establishing accurate supply-chain and service capabilities
- Communicating customer needs well
- Establishing reliance on testing to find defects

2.5.1 UNDERSTANDING CUSTOMER NEEDS

Inadequate research into customer needs leads to varied, uninformed opinions. This in turn leads to disagreement among the developers as to customer needs. Such disagreements create delays in decision-making, reversals of decisions, and individuals working at cross-purposes.

QFD provides a standardized method of representing customer needs. This is by no means a method of **learning** what the customer needs are (acquiring the Voice of the Customer is discussed in other parts of this book), but it does provide a way of systematically representing those needs. The standard representation can then be used as a basis for itemizing differences of opinion, which can be researched as needed. (The best approach, of course, is to perform credible market research up front to eliminate the chance for these differences of opinion. However, not all developers do so.)

QFD's standardized method of **mapping** these customized needs to product or service development decisions helps further reduce differences of opinion.

2.5.2 STRATEGICALLY PRIORITIZING EFFORTS

One key to cycle-time reduction is to invest in what's important and resist investing in what's not important.

Failure to take time to work on capabilities that are important to customers leads to noncompetitive products. Such failure, in turn, often leads to longer cycle times, because the overlooked capabilities must be added after the product or service has been designed. Generally, this type of midcourse correction is very error-prone, because it is done hastily, and because it is attempted as an overlay on an existing design.

Taking time to work on product capabilities that matter little to customers can only increase cycle time with no benefit to the customer or to revenues. Often these unnecessary efforts add cost to products or services and delay their introduction, which makes them less competitive.

QFD helps developers decide on the relative importance of their choices by deriving *their* priorities from their *customers'* priorities.

2.5.3 DEVELOPING PROPER RISK ASSESSMENT

All too often, product developers plunge into projects that have unnecessary risks built in. Typically these risks relate to unmanufacturable and unserviceable designs. Often these risks are assumed in desperation. Developers feel forced to commit to schedules they can't meet. Manufacturing managers and engineers feel forced to commit to volumes, quality levels, and cost limitations they can't meet. QFD helps make these risks more visible and easier for development teams to plan for up front. Once the risks are made visible, Failure Mode and Effects Analysis (FMEA) may be applied for risk reduction. Also, Pugh's Concept Selection Process (see Chapter 4, Section 4.6.7), an important adjunct to QFD, helps teams to identify or synthesize less-risky alternatives.

2.5.4 ACCURATE SUPPLY CHAIN AND SERVICE CAPABILITIES

Typically, a product or service is conceived by a core team. After developers have worked out the system-level details to a certain point, they will normally decompose the design into subsystems or subservices. They may decide that certain subsystems or subservices will be purchased from external or internal suppliers.

The normal method of requisitioning is based on conveying a specification for the desired subsystem or subservice to the supplier organization, which is held at arm's length. The supplier works on an assigned task, and when finished, or at designated intermediate points, presents its results to the core team.

Such a "throw it over the wall" method often leads to unsatisfactory, even disastrous results, because the supplier's subsystem or subservice, while meeting the specification, will not completely meet the ultimate customer's needs.

The underlying problem lies in the myth that a well-written specification is all that is needed to ensure the desired results. This leads us to our next challenge.

2.5.5 GOOD COMMUNICATION OF CUSTOMER NEEDS

A specification by itself, whether verbal or written, whether a page of text or a thousand pages, can never express all that is required. Deming taught us this many years ago.[2] We often operate under the myth that we should be able to write down everything that is needed in a subsystem, but this is rarely the case.

Developing a subsystem by simply meeting a specification is a lot like driving a car without looking at the road (Figure 2-4). Imagine simply turning left or right when told to by a navigator sitting next to the driver. Too many details about the road, the speed of the car, and the positions of other cars, people, and objects on the road must be taken into account when a driver makes a turn. These things can't be communicated accurately or in time for a safe journey.[3]

Don Clausing has pointed out that product and service developers make many decisions every day, perhaps thousands of decisions over the course of the project. To be sure, most of these detailed decisions are small in scope individually, but the sheer quantity of

---

2. W. Edwards Deming, *Quality, Productivity, and Competitive Position* (Cambridge, Mass.: MIT Center for Advanced Engineering Study, 1982).

3. Despite the ability of some movie characters to drive cars while being impaired (for example, Al Pacino's character Frank Slade in *Scent of a Woman*), readers of this book will be well advised not to try such a stunt.

Figure 2-4   Specifying How to Drive

them dramatically affects the final result. Even if the answer to every question were to be found in some voluminous specification, we cannot imagine that a developer would take the time to consult the specification for every one of thousands of decisions.

In order to ensure that developers in the core team and in supplier teams are making daily decisions that all come together in a unified whole that meets customer needs, some other approach to communicating requirements is necessary.

QFD helps in this area by providing a focus for the team's discussion and for translation of customer needs to each level of the developer and supplier organization, using terminology and language appropriate to that level. The discussion is as important as the resulting specification, because it carries with it all the subtle nuances of meaning that qualify and elaborate on the static language of the specification. In a sense, the discussions inherent in QFD make a specification come alive through dynamic interpretation and questioning around customer needs and how they are being addressed as design decisions are made. This extends all the way down to process designs, and can be further viewed as an opportunity to bring in representative participants from the entire supply chain. It may be one of the best ways to assure that suppliers get involved early enough to influence the design decisions, and to communicate customer needs throughout the supply chain.

2.5.6 RELIANCE ON TESTING TO FIND KEY DEFECTS

Testing programs can find plenty of defects in products and services. Developers trying to meet tight deadlines who must submit work to a tester know this from experience.

In complex hardware and software products there are usually many defects found during internal testing, and developers spend a great deal of time debating which to fix by process changes and which to contain through inspection. However, testing often comes only at the end of the design-and-development cycle. More than 70 percent of defects are design related, and by the time they are found, it can be too late to fix the design. Assembly or delivery goals must often be compromised in lieu of implementing a design change.

Despite all the internal testing, all too often we see customers complain about the defects they have just purchased. Given the embarrassment of defects[4] awaiting the unsuspecting customer, what's needed is a proactive strategy for testing and defect-fixing that concentrates on eliminating those defects most important to the customer—in other words, prioritizing the defects.

Notwithstanding Deming's remonstration to cease dependence on mass inspection, abundant defects exist in complex designs and complex services. Therefore, design-defect management remains an unhappy fact of life. Six Sigma methods create the means to measure, analyze, and improve the incidence of defects caused by faulty design and processes and to verify when they are eliminated. Defect elimination must be considered as an **input** to the next design iteration or any future related designs to cease dependence on inspection, thus lowering costs for each design iteration.

QFD provides a method for linking customer needs, and their relative importance, to all development activities, including testing and defect elimination. QFD can be applied to prioritize testing and repair activities so as to best meet customer needs for known design defects.

2.5.7 QFD: AN OBSTACLE TO RAPID DEVELOPMENT?

QFD can be time-consuming, as are all planning and management activities. Worse than that, it can be *explicitly* time-consuming, in the sense that QFD makes obvious and visible the need for several long meetings, attended by quite a few people. For groups that have never used QFD before, this appears as time added to their already-crowded schedules. What's not as explicit or visible is the time and cost that QFD saves by avoiding redesign, relaunch, and warranty problems. Organizations new to QFD should take care to observe that QFD is a cultural change in development communities, and usually must be deployed in phases. The reason for this is that it takes time to realize QFD's benefits and to learn to do it well. Many organizations start with just the first matrix or first three

---

4. This play on the common phrase "embarrassment of riches" is dedicated to my good friend and QFD pioneer Russ Doane.

matrices on one development project. The team can see the value when the product launches smoothly with good marketplace acceptance. So the organizational learning will take place in cycles.

The first step in any development activity is a **Requirements Phase**. In this phase, key decisions are made that determine the path of the rest of the development process. In most development environments, this process is experienced by the development team as unstructured, endlessly recycling, mysterious, and unsatisfactory if not tied explicitly to the New Product Review or Gate Processes. The QFD process systematically guides a development group through answering questions and making judgments that are exactly those they should be dealing with in order to determine the requirements. The difference between QFD and less-structured requirements-setting processes is that QFD forces development teams to consider *all* the issues, especially all the customer needs, in a systematic fashion.

It might seem that a comprehensive look at all issues could take longer than most development groups customarily spend. However, experience has shown that the QFD process generally covers more ground faster than do less-structured methods. Not only can QFD reduce the overall development process by providing more complete planning, it also takes less time than less-structured planning methods.

This is because QFD provides a *process,* a clear set of steps, for making the upfront decisions that constitute the requirements phase of development. QFD can be seen as a planning road map that informs the development team of what decisions must be made at each step, and what information is needed to make those decisions. This road map helps developers plan in such a way that they will have the information they need when it's time to make decisions, thus reducing the likelihood that they will need to revisit previous decisions. When tied to the New Product Development (NPD) Gate Processes as *required elements*, QFD matrices take on more added meaning to both the development team and management.

2.6 QFD's Role as Communication Tool

In the preceding section we reviewed many of the challenges in rapid development of products and services. Obviously, many factors could explain why these challenges occur at all. One factor, poor communication, emerges frequently as the enemy of all efficient group processes. It is so common a theme that it must be regarded as a major cause of delays in product development.

QFD provides a method for individuals involved at various steps of the development process to communicate with one another. It does this by effectively translating the language of one phase of development into the language of the next, and leaves a "golden thread" to follow for any questions that may arise. The fact that representatives from each phase perform these translations together increases the likelihood that the translation will be understood by everyone.

Once the development team has completed its translation and recorded it in a QFD matrix, the matrix can be presented and explained to others who were not present during the QFD process. These people can examine the translation process in minute detail if they need to, focusing on areas that are of special interest to them and skipping over other areas.

QFD's contribution to improving communication is then threefold:

- It provides a standard format for translating Whats to Hows
- It helps people focus on facts, rather than feelings
- It allows the decision process to be recorded in the matrices, where it can be reexamined, revisited, and even modified at any time

2.7 CONCURRENT ENGINEERING: A PARADIGM SHIFT

Concurrent Engineering (CE) is the ideal of planning and implementing all product development steps, from early product conceptualization to delivery and service, as early as possible. Team members are responsible for each step, working together throughout.[5] Consider the schematic of a typical product-development process that does *not* employ Concurrent Engineering (Figure 2-5).

In the traditional product-development process, each step is conceived of as a unit with clear inputs and outputs. Steps further downstream, such as Manufacturing Process Development, are not supposed to start until the results of previous steps, such as Component Design, are well defined. This "production-line" view of the development process assumes that time is wasted in downstream steps if upstream steps have not yet been completed, with plans for later steps solidified.

Although it is true that downstream work must take upstream decisions into account, the major problem not dealt with in the traditional model is that upstream steps may arrive at results that are unrealistic, impractical, or otherwise not optimal for downstream implementation.

For example, the engineer may unwittingly choose a design that is unnecessarily difficult to manufacture, or that is expensive to repair in the field. Manufacturing and

---

5. Don Clausing, *Total Quality Development* (New York: ASME Press, 1994).

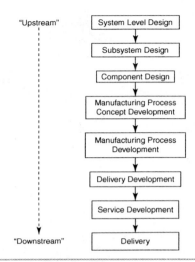

"Upstream"

System Level Design

Subsystem Design

Component Design

Manufacturing Process
Concept Development

Manufacturing Process
Development

Delivery Development

Service Development

"Downstream"

Delivery

Figure 2-5   Traditional Product-Development Schematic

field-support interaction with the designers could influence those decisions and thereby lower the company's downstream costs.

The best approach to making upstream *and* downstream decisions is an interactive one, in which representatives from all functions collaborate, sharing their decision-making processes throughout. In such a dynamic, give-and-take scenario, realistic decisions can be made that achieve the best, most workable results for all functions. In today's interconnected world, with online databases and shared server access, all functions, regardless of location or even company affiliation, may participate in interactive, concurrent product development.

Process and software development strongly resemble, or can be made to resemble, product-development processes, except that the manufacturing steps refer to the development of tangible objects that support, but are somewhat incidental to, the intangible software or service. These tangible objects might be computer memory storage devices or computer user manuals in the case of software. For services, such as a credit-card service or a home appliance repair service, the tangible objects might be signs, leaflets, or tools that help the service organization to deliver its service.

In some cases, development of these tangible objects may be straightforward, risk-free, or somehow separable from the main development work. In other cases, careful examination of customer needs may reveal that the tangible objects are intrinsically bound to the success of the service.

The Concurrent Engineering model aims at starting all development process steps as early as possible, even simultaneously (Figure 2-6). Its success comes from *each step*

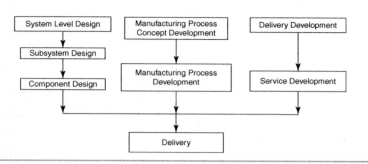

Figure 2-6   Concurrent Processes

influencing the others as the development process moves forward. With sufficient communication among people responsible for each step, practical and optimal results are more likely for all steps.

Communication can occur frequently enough and at a detailed-enough level to achieve these desirable results if the people responsible for each development step are treated as a single "multifunctional team," located together, and all with the same objective: success of their jointly developed product. More and more U.S. companies are adopting this style of product development. It is often referred to as "project," rather than "functional," organization. The objectives of such "project" organization are

- Earlier start of each development step, leading to earlier completion of all steps
- Optimization of decisions at every step through dialogue and collaboration among representatives from all disciplines, leading to optimal product design and lower production and delivery costs
- Overall lower development cost

All these product-development advantages are equally desirable in the development of software or services. The competitive pressures are no different. The need to produce software or services rapidly, at low cost, and optimized for customer satisfaction is just as urgent as for tangible products. Hence, the principals of Concurrent Engineering are equally applicable, even if somewhat less familiar to professionals in the software and service domains.

QFD is a central tool in support of Concurrent Engineering. It brings the multifunctional team together in the first place, to develop the top-level House of Quality matrix. At each step in the process, it helps keep the team focused on customer satisfaction, the primary ingredient of product success and revenue growth.

QFD further provides the team a method for prioritizing development actions, which in turn leads them to concentrate most on the actions that will have greatest impact on customer satisfaction. This optimization strategy is a key to reduced development costs. Without an effective optimization strategy, the team would be at risk of optimizing product characteristics that are unimportant to the customer, and thereby raising costs.

In product and manufacturing process design, a key optimization tool is Taguchi's Robust Design method, which is utilized in Design for Six Sigma (DFSS). QFD's prioritization capabilities help development teams decide where to apply key DFSS tools and methods. In service and software development, DFSS methods have not been widely used, although ingenious practitioners have had some successes. Other optimization techniques, such as Statistical Process Control and the Seven Management and Planning Tools, are available to developers of all products and services. QFD provides a "pull" for application of these powerful tools by identifying where more knowledge or capabilities are required to satisfy customer needs.

2.8 KANO'S MODEL

The Japanese TQM consultant Noriaki Kano,[6] has provided us with a very useful model of customer satisfaction as it relates to product characteristics. We'll be using the term *characteristics* in a fairly precise way later on, and we'll be drawing sharp distinctions between *customer needs* and *product characteristics*. For now, however, we'll use the term *characteristics* to refer to features or capabilities of a product or service

Kano's model divides product/service characteristics into three distinct categories, each of which affects customers in a different way. The three categories are

- Fitness to Standard, also known as "must-be," "basic," or "expected" characteristics
- Fitness to Use, also known as "competitive", "more the better", "one-dimensional" or "straight-line" characteristics
- Fitness to Latent Expectations, also known as "delighter", "attractive" or "exciting" characteristics

The horizontal axis in Figure 2-7 shows the actual performance or state of physical fulfillment in delivering each of these product characteristic categories to the customer, while the vertical axis indicates the customer's level of satisfaction.

---

6. Noriaki Kano, Nobuhiko Seraku, Fumio Takahashi, and Shinichi Tsuji, "Attractive Quality and Must-Be Quality," *Hinshitsu* 14:2 (February 1984), p. 147. Published by the Japan Society for Quality Control. The author's access to this Japanese-language article was through a translation by Glenn Mazur.

A competitive strategy for developing products and services must take into account these three categories of product characteristics. It must determine what the current levels of satisfaction are for each of these categories, and it must decide what proportion of project resources to allocate to product or service characteristics in each of the categories.

2.8.1 FITNESS TO STANDARD

A **Fitness to Standard** feature is a product or service characteristic that the customer takes for granted when it is present, but causes dissatisfaction when it is missing or incorrect. Fitness to Standard features are things that customers don't normally ask about, because they *expect* them to be taken care of in the course of normal product or service delivery. Fitness to Standard features are **assumed quality,** in the sense that customers expect products or services to be "right" in these areas; if they are not, the customers are dissatisfied. On the other hand, efforts to "exceed the need" in Fitness to Standard features do not create improved customer satisfaction beyond a certain level of assumed quality. Examples of "Fitness to Standard" problems or dissatisfiers are scratches or blemishes on the surface of a product, broken parts, missing instruction booklets, and missing features that are routinely supplied in similar products or services. (See Figure 2-8.) Customers don't tell us they want assumed quality because they take for granted that we will provide it.

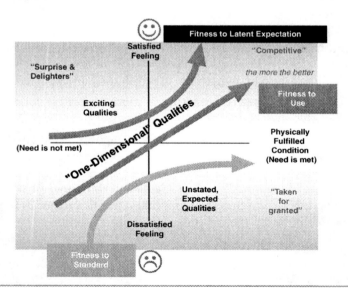

Figure 2-7   Kano's Diagram

Expected Quality	Dissatisfiers
Smooth surface	Scratches, blemishes
All parts work	Broken parts
Product comes with instructions	Missing instruction book
Products of this type normally perform function X	Function X not provided
Product is safe to use	Product is unsafe
Product conforms to local standards	Product is nonconformant

Figure 2-8 Fitness to Standard Qualitied and Related Dissatisfiers

If we deliver a product or service that has many dissatisfiers, customers will be extremely unhappy. Obviously, we will have happier customers if we can eliminate these dissatisfiers. However, eliminating dissatisfiers doesn't by itself achieve a very high level of customer satisfaction. Our customers will hardly notice all the work we've done to eliminate the dissatisfiers; they will just start noticing other aspects of the product or service.

In other words, our best possible performance in reducing dissatisfiers can raise customer satisfaction to a "not dissatisfied" state, but no further. Often the first focus of Six Sigma is to eliminate these "defects," which is necessary work but will rarely increase customer satisfaction.

As we have noted, although customers won't ask for expected quality, they will be dissatisfied if they don't get it, and they will tell us by complaining. Therefore, customer complaints are a primary source of information on dissatisfiers existing in our current products and services.

Many traditional product-quality programs focus upon complaint management. In fact, quality control is sometimes erroneously defined as a process to respond to, and perhaps eliminate, customer complaints. The Kano model shows us that this approach is not enough, although it is perhaps the first place to start continuous improvement efforts.

Complaint management deals only with expected quality, ignoring the other two categories of quality, **competitiveness** and **delighters**. Merely eliminating dissatisfiers cannot result in competitively high levels of customer satisfaction.

2.8.2 FITNESS TO USE OR COMPETITIVENESS

A **Fitness to Use** or competitiveness characteristic is something that customers want in their products and usually ask for. The more we provide such satisfiers, the happier customers will be. Satisfiers are sometimes called **desired quality** because they represent the aspects of a product that define it for the customer. Examples of satisfiers are increased capacity, lower cost, higher reliability, greater speed, and easier use (Figure 2-9). Satisfiers

Desired Quality (Customer Need)	Performance Measurement	Direction of Goodness
Capacity	Cubic feet of storage	Larger the better
Price	Dollars	Smaller the better
Reliability	Mean time between failure	Larger the better
Speed	Transactions per second	Larger the better

Figure 2-9 Example of Desired Quality

are the attributes that tend to be easy to measure, and therefore they become benchmarks used for competitive analysis. In other words, you can expect the satisfiers to be present in all competitive products or service alternatives, to a greater or lesser extent.

2.8.3 FITNESS TO LATENT EXPECTATIONS OR DELIGHTERS

Delighters are product or service attributes or features that are pleasant surprises to customers when they first encounter them. If Delighters are not present, customers will not necessarily be dissatisfied, since they may be unaware of what they are missing. Delighters are sometimes called **exciting quality** or **unexpected quality**. A typical customer reaction to a delighter is to say to a friend, "Hey! Take a look at this!"

As with dissatisfiers, customers don't tell us they want delighters, but for the opposite reason. Customers expect expected quality, so they don't bother to mention it, but customers can't expect "exciting quality" or "unexpected quality" *by definition*. Hence, we cannot learn about product or service delighters by directly asking our customers. We instead must observe their use of our or similar products and services to discover what we can do to address unstated needs. Chapter 17 covers this topic in detail.

Examples of delighters are not as instructive as examples of satisfiers and dissatisfiers. Each delighter is unique, and as a group there are no patterns. Some delighters are entire products that have created new markets (a typical consequence of building delighters into products). One very famous delighter is the Sony Walkman. Before its introduction, who could have told you that they wanted a play-only portable cassette player without speakers? After its introduction, everyone wanted one, and a new market was created. The same goes for Apple's iTunes service, which catapulted Apple's iPod to a leading market position and grew the company rapidly.

The 3M Post-It Note is another example of a delighter. It's a product that filled user needs that had not previously been filled satisfactorily.

Other delighters are more subtle. They may create not entire new markets, only temporary competitive advantages. Some examples are anti-lock brakes and intermittent windshield wipers in automobiles; random or shuffle play on CD players; instant-skip buttons on Digital Video Recorders; and DVD video recording and editing on an increasing number of desktop computers.

The needs that delighters fill are often called **latent** or **hidden** needs, either because they cannot be explicitly identified or because customers don't say that these needs are important to them. Some indirect methods for identifying latent or hidden needs and product delighters are discussed in Part IV, the QFD Handbook part of this book. These hidden needs are sometimes intimately linked to customers' perceptions of the limits of technology. As such, they are very difficult to separate from technical solutions.

Let's look at an example. In 1840 in the United States, coast-to-coast travel was an arduous and dangerous experience. In hindsight, we might say that the need for a rapid (five-hour) travel method, during which a passenger might read a book or watch a movie, existed even in 1840. However, no one was likely to express the need, because the technology to achieve it was unimaginable. Clearly, if a five-hour transportation method from New York to San Francisco had suddenly become available in 1840, it would have to have been classified as a "delighter," even without the movie!

Since no one in 1840 could imagine the possibility of jet aircraft, we must ask the question: Did the need for a five-hour transcontinental transport really exist then? Technically, one might say that if the need for a three-month transport method (covered wagon) existed, then it would be natural that travelers of the day would have articulated a need for a two-and-a-half-month transport method, or even a one-month transport method. However, a five-hour transport method would have been more than 400 times faster than the prevailing method—so much faster that it would have necessarily been considered fanciful. Did the need exist? Would anyone have articulated such a need?

There is no answer to this question that holds true for all customers and all products. Indeed, there is no clear method for discovering delighters that is guaranteed to work in all cases.

One of the disciplines that QFD helps us to maintain is to separate customer needs from technical solutions. It is consistent with the intent of QFD to search first for customer needs and only afterwards for technical responses to those needs, including delighters. Whoever is listening to customers must be knowledgeable and vigilant enough to identify customer needs when customers mention them, and to identify brilliant technical solutions when anyone happens to mention them. Clearly, the needs and solutions must eventually be sorted out, so that the technical solutions can truly be innovated and then evaluated in terms of how they relate to customer needs.

2.8.4 MARKET DYNAMICS OF DISSATISFIERS, SATISFIERS, AND DELIGHTERS

Delighters often create new markets or new market segments, thereby giving their creators a temporary competitive advantage. Once the novelty of a delighter wears off and competitors include the delighter or some equivalent solution in their own products or services, customers begin to *expect* the delighter, since it's available in many if not all competing products. When this happens, the delighter is no longer unexpected quality; instead, it has become expected quality. In other words, delighters become demoted to satisfiers. After awhile, customers assume these satisfiers will be included in the product. Once this has happened, the satisfier has become an element of Fitness to Standard.

This migration of quality attributes happens all the time, with all products and services. This is the excitement of competition! In order to remain competitive, the product or service developer must continually search for new delighters, provide more satisfiers than anyone else, and see to it that no dissatisfiers reach the customer. QFD is an excellent planning tool for sorting out and managing these different qualities.

2.9 THE LESSONS OF KANO'S MODEL

There are two major lessons that Kano's model teaches us.

First, not all customer satisfaction attributes are equal. Not only are some more important to the customer than others, but some are important to the customer *in different ways* than others. For example, Fitness to Standard attributes matter not at all when they are met, but seriously detract from overall satisfaction when they are not met. In contrast, Competitive qualities contribute to overall satisfaction overtly.

Second, the old product-quality strategy of responding to customer complaints can now be seen to be inadequate for competing in today's world. Customer complaints are for the most part linked to dissatisfiers. A quality strategy based solely on removing dissatisfiers can never result in satisfied customers.

Responding to customer complaints can be thought of as a passive quality strategy. A strategy that will lead to customer satisfaction and to a leadership product or service must be far more proactive. The strategy must be based on a deliberate policy of seeking out customers and potential customers to discover and characterize their needs, met and unmet. It must aim at breaking old thought patterns and finding creative ways of meeting those needs and exceeding customers' expectations. Finally, it must be based on a clear, reliable way of estimating the efficacy of each potential method for meeting customer needs, so that the best ways can be exploited.

2.10 SUMMARY

Any organization's most basic goals are to increase revenues, decrease costs, and produce new products and services rapidly. QFD can be used strategically and tactically to help in these endeavors.

With regard to increasing revenues, QFD provides a key tool for linking customers' needs to product, service, process and organizational design.

With regard to decreasing costs, a principal benefit of QFD is to help development teams plan projects well enough to reduce the likelihood of midcourse corrections, the most devastating source of costs in most development projects.

With regard to rapid product development, the most important way to stay competitive is to respond rapidly to changes in the market environment. Rapid response is best achieved by reducing the development time of new products and services. QFD helps reduce development-cycle time by reducing implementation errors, improving communication, and supporting Concurrent Engineering.

Kano's model of customer satisfaction teaches us that competitive products must not only be free of **dissatisfiers**, but must have a generous allotment of **competitive** and **delightful** qualities. A competitive product development strategy cannot simply respond to customer complaints, but must actively determine and address customers' needs. Further, while some needs of customers are fairly easy to discover, the needs that are keys to *delighters* are the unmet, latent, and sometimes unspoken needs that require special methods to uncover, and often require engineering brilliance to meet.

Now that we understand why QFD might be important for any organization that produces products or services, we're ready to judge for ourselves how QFD's contributions fit in to the broader enterprise, discussed in Chapter 3. Part II then provides a detailed description of the most fundamental part of QFD, the House of Quality.

2.11 DISCUSSION QUESTIONS

- What are your organization's key strategies for increasing revenues? Reducing costs? What role does reducing rework play?

- Have you observed mid-project shifts in direction? What have been the benefits of such shifts? What have been their disadvantages?

- How do project members communicate project goals to each other today? What works well, and what doesn't? How would communication be different if you were using QFD?

- For some existing product or service you have been associated with, which characteristics are **delighters**? Which are **satisfiers**? Which are Fitness to Standard? Try the same classification for your competition's product or service. How did you decide on the correctness of your classifications? What data do you have about your customers to confirm your classifications?

Tying QFD to Design, Marketing, and Technology

With Roger Hinckley, Joe Kasabula

This chapter explains in detail how Quality Function Deployment (QFD) ties into Six Sigma Methods. Each of the various Six Sigma methods will be described along with the unique roles that QFD plays within that particular methodology. Six Sigma often begins with a focus on bottom-line results, addressing those defects associated with customer complaints as discussed in relation to Kano's model in Chapter 2. As Kano's model teaches us, addressing customer complaints is not enough to keep a thriving business healthy. What must also happen is further development and understanding of customer needs, both stated and unstated. These understandings lead to products and services with characteristics that make them more competitive and delightful to customers in their markets.

As a quality management tool, QFD plays the roles of guardian-advocate and linkage manager for those needs during the creative and developmental processes for new products and services. As such QFD is a tool included in nearly all Design for Six Sigma (DFSS) roadmaps, which are varied.[1] QFD is also included in other customized versions of Six-Sigma focused upon the marketing and sales functions, R&D or technology-development functions, and in process or service development. In each of these situations, QFD's guardian-advocate and linkage-manager roles are present, but the voiced needs come from different groups. Since DFSS is the most widespread Six Sigma methodology utilizing QFD, it will be covered in a more-detailed way compared to other Six Sigma methods.

---

1. See Randy C. Perry and David W. Bacon, *Commercializing Great Products with Design for Six Sigma* (Upper Saddle River, N.J.: Prentice-Hall, 2006).

QFD has a limited but key role to play in Operations Six Sigma. While a whole section is not devoted to it in this chapter, some comments are worthwhile before proceeding with QFD and other Six Sigma methods. As Operations Six Sigma projects proceed through the various phases of DMAIC, oftentimes questions about why key process variables are controlled or uncontrolled remain unanswered. This is the time to apply QFD. If a key process variable is controlled and the organization is expending time and money to do so without understanding why, then its relationships to product parameters should be carefully documented within a QFD context. If none of the matrices exist, or if there are significant gaps, then there is no better time to begin working upwards from the process variables back to product variables and ultimately customer needs. It is in this area of Operations Six Sigma that QFD plays a key role.

3.1 QFD AND DESIGN FOR SIX SIGMA (DFSS)

3.1.1 WHAT IS DFSS?

A DFSS initiative has as a primary goal to grow the revenues of, and attract more customers to, the organization. It is not unusual to see large corporations increasing revenues by double-digit percentages or smaller organizations nearly doubling their revenues. Companies experiencing such large returns after DFSS deployments include Sylvania, Tyco, Huber, and Samsung. In each of these cases, from $100 million up to $1 billion in savings were achieved. A significant proportion of these results is directly due to growth obtained with DFSS. The DFSS methodology achieves this project-by-project, developing products and services that better meet customer needs than do competing alternatives. QFD plays a managing role in DFSS by assuring that customer needs are foremost in development activities. This focus on customer needs must become the *way of doing business* as DFSS gets deployed. The key challenges addressed by DFSS are improving how new products get developed, removing any weaknesses or gaps in the development process, and driving the customer's needs and hence value into the products so designed.

In traditional Operations Six Sigma, the focus is on adding a **process improvement** toolset. One percent to three percent of employees get trained as Black Belts, with upwards of ten percent trained as Green Belts, all executing projects along the way. In DFSS, we do things differently. That is because we are changing How the work gets done in engineering. The integration of the DFSS tools into the product-development process must be non-negotiable if you want it to stick. Everyone gets training in some form or another, and all must participate in a development project utilizing DFSS tools, as this is the new way of developing products in the company going forward. This enhancement of product development, including consistent training, serves to reduce variation in product development and moves the whole process to a higher performance level.

If you are familiar with these terms and with DFSS methodology, you may want to skip the remainder of this section. Readers new to this aspect of business—and it is a business program—may want to read this section entirely, keeping in mind that it is only an introduction.

3.1.2 How Is QFD Tied into DFSS?

Meeting customer needs is not the only criterion for successful new-product launch. Business needs must also be met, so there are business-leader voices to be captured, analyzed, and driven into new-product development. This is often referred to as incorporating the Voice of the Business (VOB). Both VOC (Voice of the Customer) and VOB are required for successful new-product development. For the remainder of this section, wherever you see the word "customer," please remember that *internal* customers or business leaders are included in the meaning of "customer."

Engineers love to create new products and services. It is normally very satisfying to design ways of fulfilling perceived needs of customers and see a product through to completion. However, most seasoned engineers can tell you some stories about how their "baby" was called "ugly" at some point during the development process, with a few horror stories about such complaints coming at the end of the development process, from customers. This clearly indicates failure at the front end of the process to

- Identify the right customers
- Capture the right voices
- Analyze these voices in context
- Prioritize customer needs
- Validate customer needs, and
- Manage the fulfillment of customer needs throughout the development process with QFD

We refer again to a figure from Chapter 1, Front End of the QFD Process, shown here as Figure 3-1. Some practitioners call this "the fuzzy front end," and some insist this is part of modern QFD, while DFSS practitioners often call it the VOC step of Concept Development. Regardless of what you choose to call it, the Front End QFD Process can be the most difficult and important part of product, service, process, and technology development. Whether your project or business is big or small determines the work complexity or simplicity, but these front-end steps should not be unconsciously skipped.

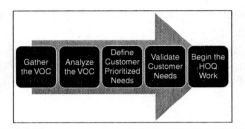

Figure 3-1   Front End of the QFD Process

Within DFSS, the QFD matrices are tied together as described in Chapter 1. However, to accelerate product development in existing core competencies, a modified approach is used. To speed the QFD process, the links that flow down from the House of Quality (HOQ), or first matrix, are those items that are *important, difficult to meet,* or *new.* These three categories broadly classify the items that represent the highest risk to overall success. By evaluating factors in this manner, we drive not every characteristic or feature into all the subsequent QFD matrices, but only those items that are critical. The intention is to *focus* on these key areas and find gaps in our knowledge or capability to meet them. This modification of QFD is shown schematically in Figure 3-2.

3.1.3 DESIGN FOR SIX SIGMA: THE INITIATIVE

A DFSS initiative may begin at any time in a company's Six Sigma journey, but usually it follows deployment of Six Sigma in the operations or manufacturing area. This happens for any number of reasons. Some of them may be

- Value delivery improvement: Defects attributable to the product designs become more obvious
- Logical time sequence: It is hard to design a successful product for an unstable supply chain
- Customer complaints: Fixing problems before attempting to market competitive or delightful products
- Skeptical leadership: Six Sigma in Operations yields cash to the bottom line relatively quickly, compared to applications in areas other than Operations
- Unpredictable project timelines: A build-test-fix process eventually becomes recognized as inefficient

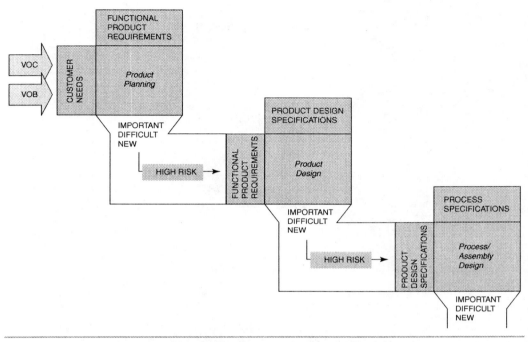

Figure 3-2   Flow of New, Important, and Difficult Factors within the QFD Matrices

Six Sigma in Operations creates new cash to the bottom line before DFSS, such that any worries like "oh, but we're different" are abated. DFSS may take more than a year, and up to 18 months, before any of the new products start getting to market. The long timeframe for achieving results requires patient leadership and the right metrics to guide the deployment.

There are many aspects, both well-known and subtle, to deploying a DFSS initiative. Covering them all is well beyond the scope of this short introduction.[2] In lieu of a full and comprehensive description, we will focus on one aspect, the type and usage of metrics, to provide a perspective on a DFSS initiative.

The ways DFSS initiative metrics translate, flow down,and move through the organization levels, from planning to results, are shown in Figures 3-3 and 3-4. The metrics throughout the initiative move from planning metrics flowing downward to results tallies flowing upward as DFSS projects are completed. In Figure 3-3, we see that the

---

2. See Perry and Bacon, *Commercializing Great Products.*

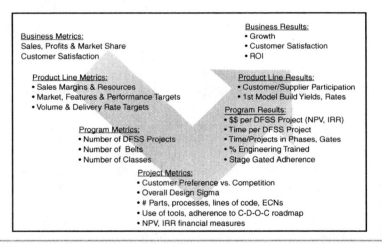

Figure 3-3   DFSS Metrics Flow through the Organization

Business Metrics:
Sales, Profits & Market Share
Customer Satisfaction

Business Results:
• Growth
• Customer Satisfaction
• ROI

Product Line Metrics:
• Sales Margins & Resources
• Market, Features & Performance Targets
• Volume & Delivery Rate Targets

Product Line Results:
• Customer/Supplier Participation
• 1st Model Build Yields, Rates

Program Metrics:
• Number of DFSS Projects
• Number of Belts
• Number of Classes

Program Results:
• $$ per DFSS Project (NPV, IRR)
• Time per DFSS Project
• Time/Projects in Phases, Gates
• % Engineering Trained
• Stage Gated Adherence

Project Metrics:
• Customer Preference vs. Competition
• Overall Design Sigma
• # Parts, processes, lines of code, ECNs
• Use of tools, adherence to C-D-O-C roadmap
• NPV, IRR financial measures

Figure 3-4   DFSS Initiative Metrics Timeline from Planning to Results

initiative begins at the organizational level, then flows down metrics-wise to the pro-
gram and project levels, before results flow up as time progresses.

In Figure 3-4 we illustrate sample metrics typically used throughout a DFSS Initiative
timeline. The metrics begin with business targets and expected results and end with
measures of how we did against these goals. As they move from the organizational level to
the program and project levels, these expectations get translated into measures of appro-
priate activities in the planning phase, and appropriate results in the execution phase.

In addition, traditional project-management metrics such as planned versus actual schedule and cost metrics should be considered for each project.

Due to the broad scope of a DFSS Initiative and the QFD focus of this book, we will delve no further into the initiative part of DFSS.

3.1.4 DESIGN FOR SIX SIGMA: THE METHOD

DFSS involves a sequence of tools, including QFD, carefully linked and planned, and laid into the new-product development process. The philosophy of DFSS is that critical decisions should be based on customer needs, data, and proper analysis. Since we are now listening to and communicating with our customer, the linkage with QFD begins to appear! DFSS methodologies vary wildly throughout industry, with each group and company determining what DFSS means to it. In using DFSS methods, the thoroughness of application and the understanding of the tools and sequences by the product-development teams are most important. If those conditions are present, then good results typically follow. We will describe here the DFSS method we have used in one form or another since 1997.

Every DFSS methodology relies on some sort of roadmap and product-development process. The best DFSS methods we have seen rely on assessing the New-Product Development (NPD) process and customizing the DFSS tool sequences to enhance and support the NPD process. This requires a customized DFSS approach for every organization. The benefits to this are many, in that the NPD process is part of organizational culture and enhancing it represents evolutionary investment rather than revolution. The DFSS methodology and results will be no better than the NPD process already embedded. If that's not working well, it needs to be fixed before attempting to add DFSS tools and methods.

For many years, we have used a four-phase process to illustrate and explain the tools sequence. The process phases are Concept, Design, Optimization, Capability (CDOC). These four phases illustrate the challenge of new-product development. In the Concept phase, we need to develop concepts that more accurately reflect customer needs, that are more centered on customer value in a product. In the Design phase, we emphasize flowing customer needs into the design processes, using statistical tools to help determine how close or far we are from targets. In the Optimization phase, we iterate the design to optimize reliability and product robustness. Finally, in the Capability phase we demonstrate capability to meet requirements and verification and validation of product performance in the short and long term. The tool sequence that supports this, at a very high level, is illustrated in Figure 3-5.

As can be seen in Figure 3-5, QFD plays an early role in the tool sequence. In fact, it plays a great role, in that many elements found in the tools have direct linkages to QFD

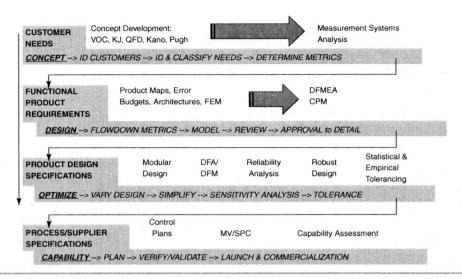

Figure 3-5   DFSS CDOC Methodology and High-Level Tool Sequence

for inputs and to complete some of the gaps identified by appropriate application of QFD. Since it is being applied to an early part of the process and is a key component in setting targets, priorities, and direction, QFD must be done well and thoroughly, or all work and tools that follow may be way off target.

A detailed description of all these tools is beyond the scope of this book, but may be found in Perry's book,[3] including definitions of the acronyms in Figure 3-5.

3.1.5 DESIGN FOR SIX SIGMA: THE METRIC

In DFSS, we want at some point to determine the "Sigma" value of the new-product design. Is it Three, Four, Five, or Six Sigma? How do you get to one number for an entire product design? Can we compare multiple products, and is it a fair comparison? A Design Scorecard is used to rate the new-product design in terms of its capabilities. It's a lengthy process, which will be only briefly described here.

The methodology used to score a design's Sigma is rooted in capturing the capability of the design's performance and summing up potential process and purchased-item defects. On the performance sigma, we identify the projected mean value, sigma value,

---

3. Perry and Bacon, *Commercializing Great Products.*

and expected requirements—upper, lower, or both if appropriate. We list each performance parameter against which the product must be measured to gauge acceptability to customers. Most performance parameters are measured on a continuous scale. The expected mean or average value is listed, along with expected sigma and appropriate specifications from the HOQ. We then Z-score each performance parameter according to the following equation, where USL stands for Upper Specification Limit and LSL stands for Lower Specification Limit:

$$Z_{USL} = (USL\text{-}Mean)/Sigma \quad \& \quad Z_{LSL} = (Mean\text{-}LSL)/Sigma$$

The resulting Z-scores tell us how many sigmas away the average value is from each specification limit. We now have the equivalent sigma levels as if the parameters were part of a normal distribution. With this information, we can predict the number of defects for each specification and for every parameter. This is but one way to make a predicted defect calculation for each performance parameter. These expected defects are summed up, normalized by the number of total performance parameters, and an expected Defects Per Million Opportunities (DPMO) can be calculated. We then reverse the process at the top level and for the entire design, calculate the equivalent Z-score, and arrive at a top-level sigma for product performance. A similar process is utilized for any purchased subassemblies, parts, and the processes expected to be used during manufacture. By doing this, we arrive at the sigma of the product design. This focuses the design team on improving expected performance, through early testing, similarity, simulation, or any combination of the three. The Design Scorecard also rewards the team for reducing complexity and for flagging less-than-good parts and processes. When it is tightly linked to the front end through appropriate measures and requirements from QFD, we have a good measure of how well we will be meeting customer expectations. The Design Scorecard has been utilized in DFSS efforts for more than 10 years, and it is *the* metric by which a DFSS product design team can be evaluated and rewarded for its efforts.

3.2 QFD AND SIX SIGMA PROCESS DESIGN

The challenge of designing a new process or service for customers is both simpler and more complex than designing a new product. Designing a new service process *may* involve less mathematical and scientific knowledge than designing a new aircraft engine. Therefore, some people might approach this task with the expectation of less effort required and less risk of failure. A jet engine can be simulated, tested, inspected, and measured along the assembly and final test paths. Rarely in my observations are similar investments made in most service process design efforts. Manufacturing processes are in general an exception, in that they are often well simulated and quite carefully designed.

Consider that when process failures occur in administrative, service, or manufacturing areas, they are subtle, can go undetected, and may have severe consequences due to the hidden nature of failure in these processes. An example in the news from September 16, 2005 was a power outage in Los Angeles.

> LOS ANGELES—*An inaccurate work order* led to the power outage that shut down elevators, traffic lights and ATMs across much of the city earlier this week, the Department of Water and Power said. "It was a case of miscommunication," said Henry Martinez, assistant general manager of the DWP. The outage, which affected about 2 million people, happened when *a utility crew cut several control lines* in the San Fernando Valley. Martinez said *DWP engineers had directed that the lines be left intact, but inaccurate work drawings called for the lines to be cut.* The DWP was trying to determine who drafted the order. *[emphasis added]*[4]

Not all process-design efforts should be as simple as our first impression might suggest. This is especially true for customer-facing processes like invoicing, complaint resolution, service or warranty support, and the ever-increasing Web-interfaced services like ordering, downloads, and shipment tracking. Web-based services allow for ease of use in theory, and when they don't work properly in practice the customer is already in a position to post complaints for many to see on the Web if there isn't a help button on the Web site. In fact, Forbes.com[5] has actually rated corporate complaint sites.

The role of QFD is paramount in service or process design because there are additional voices beyond VOC and VOB that must be captured, understood, and analyzed for any new service or process design work. These are the voices of the users of that service or process, which are often different from the final customer voices. Additionally, employees working in the appropriate service or process must be interviewed to capture their voices (VOE) if the new service is to be adopted readily. Without meeting employee needs, the processes will be inefficient and perhaps even ineffective for the customers. A good checklist for gathering user and employee environments and voices is the "6 M's." These are six broad variables that can create challenges for a new process or service if left unexplored:

- **Man**: people in and around the process or service
- **Machine**: any automatic mechanisms

---

4. "Inaccurate Work Order Blamed for LA Outage," Associated Press, September 16, 2005. Available online at HighBeam Research, www.highbeam.com/doc/1P1-113201581.html.

5. Charles Wolrich, "The Best Corporate Complaint Sites," Forbes.com (August 21, 2002), www.forbes.com/2002/08/21/0821hatesites.html.

- **Methods**: processes
- **Materials**
- **Measurements** or metrics, and
- **Mother Nature:** the environment

In addition to meeting the goals of efficiency and effectiveness, service processes must be flexible to keep customers satisfied. That can be achieved only if flexibility is a stated design goal. It is often an unstated need of service customers. Six Sigma Process Design (SSPD) is a series of steps designed to:

- **Define** the process or service challenges, determining key stakeholders and project outcomes, and developing and validating key customer needs
- **Measure** current process or service performance vs. defined needs and outcomes
- **Analyze** existing process or service data for improvement opportunities
- **Conceptualize** ways to meet customer and business needs with the right value propositions
- **Design** a process or service that meets customer, stakeholder, and business needs
- **Optimize** the process or service design to be effective, efficient, and flexible, with the right metrics, targets, and specifications to:
- **Control** the processes or service to always deliver the value expected

These seven steps are a hybrid of Operations and Process-focused Six Sigma and DFSS. QFD work may begin at either the Define or Conceptualize phase listed above, depending upon whether a process or service already exists that is to be modified and/or replaced. These phases are linked with a pair of key decisions and are dependent upon those decisions, as illustrated in Figure 3-6.

Designing a service involves several facets. In designing a service to augment a product, such as additional support or information distribution, it may take the needs of customers and the business directly from the HOQ. The HOQ will still need to be modified to include the needs of the users of the service and of employees working in the service processes.

However, in designing a service that significantly enhances the value proposition of the product or can provide its own unique value proposition(s), this is a different matter. Consider Apple iTunes and the company's iPod MP3-format music players. The combined revenues of these two products shattered Apple's previous profit and sales records. The iTunes service provides its own unique value proposition(s), which are

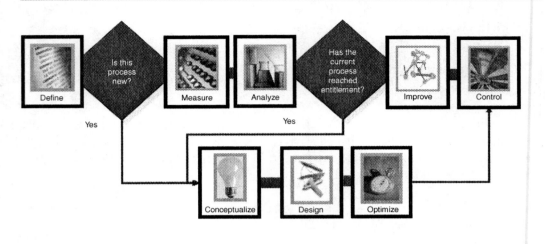

Figure 3-6   Six Sigma Process Design Phases

enhanced by the iPod product(s). In such a case, when designing a unique service, a new set of QFD matrices should be constructed, and a set of core processes will need to be designed. In this situation, VOC, VOB, and VOE must all be collected separate from the product QFD work which may already have been done due to the longer nature of product-development work. In collecting these voices, care must be taken to holistically capture enough voices to frame value. A good checklist of value to be voiced and collected looks like the following.

- Service Value, according to Quality, Cost, and Time:
  - Quality value added from
    - Features
    - Ease of Use
    - Performance
    - Benefits

  - Minus the Costs of
    - Acquisition/Installation
    - Maintenance
    - Disposal/Recycle

- ° Minus the (cost of) Time to
  - — Purchase or select
  - — Modify or customize
  - — Setup
  - — Install
  - — Use
  - — Recycle or dispose

Each customer values time differently, and some market segments may place more or less emphasis on the cost of time in the value equation. Large auto-repair franchises purchasing tire-installation equipment for multiple locations may place more emphasis on the setup time. Likewise, a consumer purchasing a laptop over the Internet may be concerned with reduced time to set up or reduced need to install software on the laptop.

Designing or redesigning a set of core processes takes real effort, and requires QFD to be imbued with the right set of voices. Creating a new service requires the same efforts multiplied by the number of processes to be created or affected. Some key tools beyond the QFD portion of SSPD include process simulation and functional analysis, which are beyond the scope here. The tools that follow QFD work in process design take their cues from the critical front end of QFD and the HOQ and matrices that follow. I have seen tremendous results in industry areas that are currently applying SSPD including HVAC, chemical, and automobile-engine manufacturing and servicing.

3.3 QFD AND MARKETING FOR SIX SIGMA

3.3.1 WHAT IS MARKETING FOR SIX SIGMA?

Marketing for Six Sigma (MFSS) has a critical goal of a successful launch of new products and services into key market segments, thereby increasing overall company market share. MFSS therefore must work hand-in-hand with Design for Six Sigma to make certain that the products and services being developed have the right value propositions for the targeted market segments. MFSS is focused upon the planning and implementation of new products and services into new markets or existing markets, or the launch of existing products into new market segments. The efforts to gather the Voice of the Market must also focus on how a company's current branding or reputation is unique. This is a key part of what distinguishes MFSS from DFSS. MFSS requires a distinct marketing strategy based upon searching out alignments between company strategic thrusts, or VOB, and the Voice of the Market (VOM).

One well-known example of strategy and market-alignment success is Apple Computer's Digital Hub Strategy and alignment to the MP3 music player market. In January, 2001, Apple unveiled its Digital Hub Strategy with the computer being the epicenter or hub for all things involved in the digital lifestyle, including music, video, personal scheduling, and information storage. The idea was to connect the Macintosh to all digital products, such as Personal Digital Assistants (PDAs), digital cameras, digital camcorders, and cellphones. The Macintosh Digital Hub Strategy was published by Jim Heid in 2002. By aligning the company's Digital Hub Strategy with the emerging market of MP3 music players, Apple was able to gain significant market share and garner staggering revenues by creating and delivering iPods and iTunes in an already-developed market. In Apple's typical style, the company observed the challenges existing customers faced with MP3 music players and services, and provided a product and a service to better meet those needs. Apple customers have long been accustomed to Apple products that seem to better anticipate their needs. This VOM, when connected with Apple's Digital Hub Strategy (the VOB), helped make the iPod product and the iTunes service great successes.

In summary, Marketing for Six Sigma uses science, statistics, and data to

- Gather data about current market trends, timing, company branding, and VOM
- Apply statistical tools in analyzing and forecasting markets and their segments
- Determine which value propositions the company will offer
- Understand customers' decision processes
- Help guide development of products and services with attention to customer voices, and then
- Launch the appropriate products and services successfully in the right market segments, with the right value communications, taking advantage of company branding

See Figure 3-7.

3.3.2 LINKING MFSS AND QFD

For those companies that are truly market focused, the linkage between MFSS and QFD comes about in two distinct areas. The first use of QFD is brought about in the efforts to summarize the VOM and plan deployment. Market trends and segmented needs are delineated and brought into the left side of the House of Quality, in place of customer voices. Across the top we place the potential product portfolios and services being considered and then evaluate them according to the typical QFD HOQ approach.

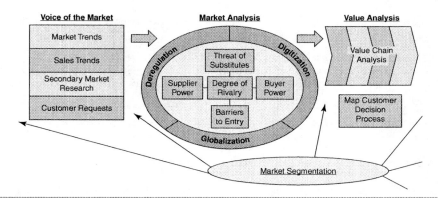

Figure 3-7 Front-End Portion of Marketing for Six Sigma (MFSS)

This approach lets portfolio development become market driven according to segment needs and competing alternative evaluations.

The second application of QFD within MFSS is more like the traditional QFD flow once products and services undergo development, as has been previously described. The only difference is that additional customer needs are provided based upon the MFSS work to understand customer decisions, satisfaction, and retention. This approach bene-fits by inclusion of business needs and direction per the VOB work typically done within MFSS or DFSS. Such VOB needs are usually expressed along the lines of resources needed, including labor and capital investments, plus financial measures like Net Present Value or Internal Rate of Return.

3.4 QFD AND TECHNOLOGY FOR SIX SIGMA

3.4.1 WHAT IS TECHNOLOGY FOR SIX SIGMA?

Technology for Six Sigma (TFSS) is similar to Design for Six Sigma in that both are focused on development. However, while the goal of DFSS is new-product development, the goals of TFSS are development and certification of a new *technology* and verification of its operating parameters and boundaries for use in current and future *product plat-forms and manufacturing efforts.* These can be new manufacturing technologies, product technologies, or even service technologies like bar-code readers or new IT services like instant messaging. Several incremental technologies may impact service or delivery methods, and therefore do not go into product design, but rather go straight to service-delivery methods. See Figure 3-8 for an illustration of this concept.

In TFSS, the Voice of Technology (VOT) or, equally, the Voice of (R&D) Employees needs to be considered. In this case, we are obtaining *expert voices* on new and emerging technologies and technology trends, in both relevant and adjacent technologies, and even at times in emerging technologies in unrelated fields, for use in current and future product efforts. Just as in Marketing for Six Sigma, we also need the Voice of the Market (VOM). Technology-development efforts can have enhanced chances for success when these voices are incorporated into development of new products, services, or manufacturing efforts.

3.4.2 TECHNOLOGY FOR SIX SIGMA: THE METHOD

The approach in TFSS is planned around developing technology-based product portfolios aimed at specific markets. TFSS developers provide technology platforms that address fundamental needs in the market with new and emerging technologies. In TFSS, we continuously seek out and evaluate new and potential technologies as well as new and existing markets. When technology timing and market timing are right, new products

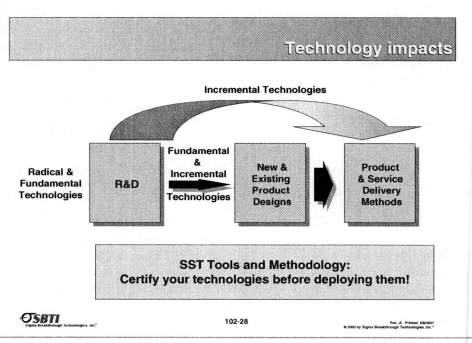

Figure 3-8 TFSS and Potential Technology Impact on Product and Service Delivery

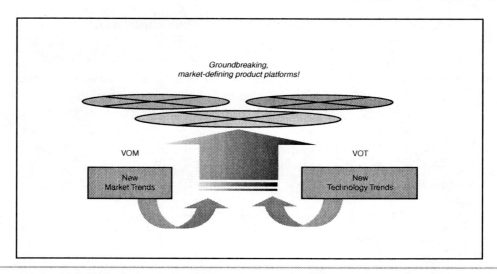

Figure 3-9 New Platform Creation by Blending Technology and Market Trends

may be introduced that can shift the market in major and fundamental ways. Figure 3-9 illustrates this combined timing effort.

The keys to evaluating these potential opportunities lie in proper capture of VOM and VOT or VOE (R&D expert employees), and in creation of a strategic technology roadmap. The voices are analyzed for underlying needs, and can be fed into the House of Quality to evaluate which potential product portfolios should include which technologies and be launched into appropriate market segments. With TFSS for applied research, we gather VOM market trends and technology trends, as summarized in the following outline.

- Technology must be assessed against Maturity, Value, and Likely Market Acceptance
  - Maturity
    - **Robustness:** How robust is this technology in its suitability when there are small variations in suppliers, materials, and processes?
    - **Tunability:** Can the technology be fine-tuned as it gets into product development?
    - **Supply-Chain Development:** Is there more than one source?
  - Value, according to the three of the directed evolution laws of TRIZ[6]
    - **Law of Ideality:** How well does this new technology move the system(s) involved towards an "ideal function"? Versus alternatives? The ideal function is that which

---

6. Victor R. Fey and Eugene I. Rivin, "Guided Technology Evolution (TRIZ Technology Forecasting), *TRIZ Journal* (Jan. 1999). Available online at www.triz-journal.com/archives/1999/01/c/.

accomplishes the primary system objective with virtually no resources and in virtually no time. Think of the contrast between on-demand movies at home versus going to the theater in terms of cost, time, and travel investments.

— **Law of Dynamization:** Does this technology make system elements more dynamic? Folding knives versus fixed-blade knives are a simple example of this, allowing for the elimination of a sheath and instead carrying the knife in one's pocket.

— **Law of Flexibility:** Does this technology create a more-flexible system overall? For example, technologies that allow creating an MP3 player inside a cell-phone allows for one product serving two functions (calls and music).

- Value, according to fundamentals (Value = Benefits versus Price)
  - Qualities & Benefits in
    - Performance
    - Features
    - Appearance or Aesthetics
    - Conformance
    - Reliability
    - Durability
    - Serviceability
    - Cost Savings
    - Time Savings
    - Perception (e.g., reputation, image)

  - Minus Cost(s) for
    - Acquisition
    - Installation
    - Use or Operation
    - Replenishment or Replacement
    - Removal
    - Disposal or Recycling

- o Minus Time to
  - – Purchase the new technology
  - – Unpack the new technology
  - – Set up the new technology
  - – Install the new technology
  - – Train or learn the new technology
  - – Use or apply the new technology
  - – Recycle or dispose of the new technology

- o Market Acceptance[7]
  - – Introductory value proposition message(s) about the new technology
  - – Allowing or facilitating existing customer behaviors
  - – Enabling technology for markets not previously open
  - – Disruptive technology in underserved markets
  - – Needs satisfaction for prospective customers of the new technology
  - – Reshaping the supply chain or business model, creating new advantages
  - – Acceptance of the new technology in first/early adopter market segments
  - – Definition of market volume segments of the new technology

These elements would go into the left-hand side of the HOQ, after first being organized and ranked by the technology-development team. In other words, they would replace the traditional VOC portion of a product QFD.

A simplified approach in evaluating synergy of market segments and various technologies is illustrated in the following two-step process. First, gather all technologies from your R&D experts that may be on the near and far horizons and rank them (VOT/VOE). Then, gather a list of all market segments in the near and far horizons from key marketing analysts (VOM). Simple scales for ranking these are illustrated below.

- • Rank all technologies on the horizons
  1. Potential disruptive technology[8]—not yet feasible
  2. Disruptive technology—feasible, not yet practical

---

7. Several of these points are covered in Clayton Christenson, *The Innovator's Dilemma: The Revolutionary Book That Will Change the Way You Do Business* (Cambridge, Mass.: Harvard Business School Press, 1997).
8. Christenson, *The Innovator's Dilemma.*

   3. Disruptive technology—feasible, practicality demonstrated, not yet in products

   4. Disruptive or other technology—feasible, practical, in emerging products

   5. Disruptive or other technology—mainstream and practiced in other products

- Rank all market trends on the horizons

   1. Potential emerging or new market trend—entry not feasible

   2. Potential emerging market—entry feasible but not practical

   3. Emerging, feasible market—entry is early but carries some risks

   4. New market—entry is warranted, risk of being late is on horizon

   5. Developed or developing market—competitors already have presence

To evaluate the synergy between market trends and technology trends using a matrix or a graph, multiply the ranks for each technology in each appropriate market segment, as illustrated in Figure 3-10.

These inputs provide the direction and kernels that TFSS Green Belt (GB) and Black Belt (BB) developers use to design and certify robust and tunable "proof-of-concept" models and platforms that product developers may then employ with confidence when developing new product designs. A roadmap for technology development can be seen in Figure 3-11. The essential elements are to:

- **Define** what's important in market, business, and technology trends
- **Invent/Innovate** along the lines of current and feasible trends and voices
- **Develop** the technologies that look most promising
- **Optimize** the technology for use in product portfolios, making it robust and tunable
- **Certify** the technology for use in products and manufacturing

Technology Idea & Target Market	Technology Rank	Market Rank	Combined Score
Tech 1, Market 26	4	4	16
Tech 1, Market 23	4	5	20
Tech 3, Market 26	3	4	12
Tech 3, Market 23	3	5	15
Tech 4, Market 21	2	4	8

Figure 3-10   Technology and Market Synergy Scoring

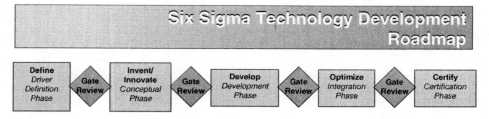

The 5 Phases in the D-I²-D-O-C Roadmap focus on the following Risk Management issues:

- D = Define (*Assess Drivers from Voice of the Market, Business, and Technology*)

- I² = Invent/Innovate (*Conceptual Feasibility & Modeling of new Technology Concepts*)

- D = Develop (*Base-line or nominal Performance of new Technologies*)

- O = Optimize (*Robust Performance of new Technologies under stressful conditions*)

- C = Certify (*Certification of new Technologies for maturity & safety for transfer into Product or Mfg. Process Design*)

Figure 3-11 Technology Development Roadmap

3.5 SUMMARY

In this chapter we have explored the differences between key Six Sigma methods, and how QFD plays a planning or managing role in each of the methods. In all cases, QFD must be utilized in one form or another to manage and deploy customer needs, serving as a "customer advocate" and also a "linkage manager" for the efforts. In Design for Six Sigma (DFSS), real customer voices have needs that must be prioritized, deployed into development efforts, and met. In Six Sigma Process Design (SSPD), both customer voices (VOC) and user voices need to be deployed when designing a new process or service. In Marketing for Six Sigma (MFSS), market trends, or Voice of the Market (VOM), must be deployed into new-product commercialization. Finally in Technology Design for Six Sigma (TDFSS), VOM and Voice of Technology (VOT) voices must be deployed when developing new technologies.

3.6 DISCUSSION QUESTIONS

- How does QFD tie into Design for Six Sigma? What voices need to be collected?
- What roles does QFD play during development of new products and services?
- How much does your marketing team participate in your QFD efforts? Key suppliers? Customers?
- How much analysis do you apply to customer voices before moving forward into product or service development?
- How much validation do you complete before moving forward?
- If you consider applying QFD into your development process, what is your plan? With DFSS or not?
- When designing a new service or process, which voices do you collect? How do you collect them?
- In new-service creation, how do you currently deploy customer and user needs in the design process?
- How do you obtain Voice of the Business (VOB), and how is it deployed in launching new products?
- When launching new products and services, how are the value propositions deployed and communicated?
- If you develop new technologies, what key market information is gathered, and how is it deployed?
- How do you release new technology into product-design areas as fit for use?

PART II

QFD at
Ground Level

Support Tools for QFD

4

With Steve Zinkgraf

To explain QFD and its uses in Six Sigma, we must first acquaint ourselves with some procedural building blocks. QFD utilizes selected tools from Total Quality Management (TQM) problem-solving and planning, specifically from a set called the **Seven Management and Planning Tools**. Six Sigma uses several fundamental statistical and organizational tools to identify sources of variability, quantify them, and reduce or eliminate their effects on products, services, and processes. This chapter introduces these tools and describes how they are used in QFD. These tools will come up over and over, just as a hammer or a saw is used repeatedly in building a physical house. If you have not seen these tools before, please take time to learn about them here. Even if you *have* seen them before, it might be helpful to read this chapter to find out how they are applied to QFD.

We are all accustomed to the concept of tools. A tool helps us to perform some function more easily than we could without the tool. For example, a hammer allows us to drive a nail into wood more easily than if we had no hammer. We might be able to drive a nail into wood without a hammer—for example, by using some heavy, hard object, such as a stone. But clearly the hammer is the preferred device for this purpose. Similarly, measuring to decide where to place the nail and marking its precise location involve other tools.

Once we have a hammer and have learned how to use it for its basic function, we can apply it to other uses, such as clearing glass slivers from a broken window. In other words, most tools have uses beyond those for which they were originally designed. In fact, we take for granted the way in which our familiarity with a particular tool allows us to perform a wide variety of tasks. Tools such as the hammer may be thought of as

mechanical tools in the sense that they make it easier to perform a task, assuming we know what we want to do in the first place.

From TQM, we use a variety of tools that assist us in tasks we perform in the workplace. TQM tools generally differ from mechanical tools in that they are **decision-making tools**. We use mechanical tools when we have already decided what we want to do, and we just want to do it quickly and with minimum effort. In contrast, decision-making tools help us decide what to do in the first place.

The TQM decision-making tools help us to organize ideas and data, to interpret the ideas and data, and to decide how to act on that interpretation. If mechanical tools are muscle enhancers, then decision-making tools are mind enhancers. This chapter introduces some of these decision-making tools and shows their relevance to QFD.

4.1 THE SEVEN MANAGEMENT AND PLANNING TOOLS

In the late 1970s a book appeared in Japan, published by the Japanese Union of Scientists and Engineers (JUSE), entitled *The Seven New Tools*.[1] These tools were intended to provide a level of problem-solving power in the conceptual domain equivalent to the power of the "Seven Basic Tools" in the process-improvement domain.

The Seven New Tools are usually called the Seven Management and Planning Tools in the United States. The list of tools and even the number of tools may vary a bit from one reference source to another, but most of the tools appear in all lists. The following tools are the mainstays of QFD:

- Affinity Diagram
- Tree Diagram
- Matrix Diagram
- Prioritization Matrix

Other tools often included among the Seven Management and Planning Tools are not directly required for QFD, but are helpful for some of the follow-up work. They include

- Interrelationship Diagram
- Process Decision Program Chart

---

1. Shigeru Mizuno, ed., *Management for Quality Improvement: The 7 New QC Tools* (New York: Productivity Press, 1988).

- Matrix Data Analysis
- Arrow Diagram

All of these tools are worth learning. Good sources[2] are available for learning them. Anyone involved in strategic planning of any type will benefit from the use of these tools. For our purposes, we will look closely at the Affinity Diagram, the Tree Diagram, and the Matrix Diagram. The Interrelationship Diagram (sometimes called the Interrelationship Digraph) is described briefly in Chapter 9. We recommend that you learn about the other tools from other sources.

4.2 AFFINITY DIAGRAM

The Affinity Diagram (Figure 4-1) is a powerful tool for organizing qualitative information. It provides for a hierarchical structuring of ideas. The hierarchy is built from the bottom up, and the relationships between the ideas are based on the intuition of the team creating the diagram.

The initial ideas in an Affinity Diagram can come from one of two main types of sources: *internal* or *external.*

Internal ideas are brainstormed by the team developing the diagram. Brainstormed ideas would be appropriate for a team that has no data to begin with. Team members may have some initial ideas about the problem they are beginning to solve, but they may not have begun collecting data, and they may not even know what data to collect.

The purpose of brainstorming is to develop a structure of ideas that describes the team's current understanding of some problem area. Having created this structured model of the problem area, the team could then decide which parts of the problem area to attack first, and which parts require gathering more data first.

External ideas are information the team has acquired, generally in the form of anecdotes that apply to the problem at hand. This is usually the situation for QFD teams that want to understand their customers' wants and needs.

To prepare for QFD, factual data is gathered that consists of phrases taken directly from the customer. While customer phrases may not necessarily refer to anything concrete, the customer phrases themselves *are* concrete, since they are direct evidence of customer wants and needs, and customers actually uttered the phrases.

---

2. Michael Brassard, *The Memory Jogger II: A Pocket Guide of Tools for Continuous Improvement and Effective Planning* (Salem, N.H.: GOAL/QPC Inc., 1994).

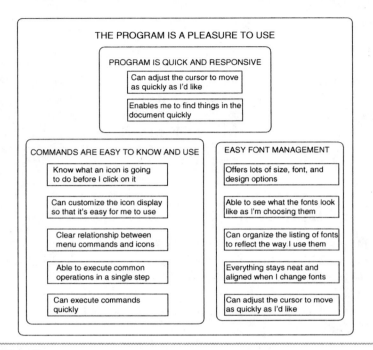

THE PROGRAM IS A PLEASURE TO USE

PROGRAM IS QUICK AND RESPONSIVE

Can adjust the cursor to move as quickly as I'd like

Enables me to find things in the document quickly

COMMANDS ARE EASY TO KNOW AND USE

Know what an icon is going to do before I click on it

Can customize the icon display so that it's easy for me to use

Clear relationship between menu commands and icons

Able to execute common operations in a single step

Can execute commands quickly

EASY FONT MANAGEMENT

Offers lots of size, font, and design options

Able to see what the fonts look like as I'm choosing them

Can organize the listing of fonts to reflect the way I use them

Everything stays neat and aligned when I change fonts

Can adjust the cursor to move as quickly as I'd like

Figure 4-1   Affinity Diagram

Customer phrases, sometimes called **verbatims**, are usually obtained from interviews that developers have conducted with customers in preparation for the QFD work. However, there are many other possible sources. (See a discussion of this in Chapter 16.)

To construct an Affinity Diagram, the team writes each phrase on a small card or Post-it Note. The cards are spread on a table, or the Post-it Notes are placed on a wall, where they can all be seen by the entire team. The team then moves the cards around to form clusters that intuitively go together. Intuition is an important element when using the Affinity Diagram. The use of intuition allows for discovering unexpected relationships, which in turn often spark breakthroughs in understanding the data.

Following is the general procedure for drawing Affinity Diagrams.

4.2.1 SCRUB THE DATA

- After writing ideas on cards, scrub the data. Scrubbing is the process in which each team member explains what he or she wrote on each card, to ensure that the team members all understand it the same way. Often, scrubbing helps the team to identify two cards with the same meaning, in which case they are replaced by a single card which best expresses the meaning. Scrubbing also helps reveal cards that have more than one idea on them. These are then replaced by multiple cards, each of which expresses a single idea.

For example:
The system offers lots of fonts, and I can see what each one looks like
would be replaced by two cards:

The system offers lots of fonts
and
I can see what each font looks like

4.2.2 Sort Cards into Clusters

- After all cards have been scrubbed, sort the cards into piles that "feel" as if they belong together. There should be no discussion during the sorting process.
- If a card continues to shuttle between two piles because one person wants it with one set of cards and another person wants it elsewhere, make a duplicate of it and put it in both piles. This rule is helpful during the early stages of sorting. It eliminates the possibility of a test of wills between two team members. In the long run, however, it may be confusing to have a card show up in two places in the hierarchy. Once the team begins assigning titles to the piles (next step), a duplicated card usually turns out to belong in only one pile.
- Teams often encounter difficulty in "leveling" during the affinity-diagram process. This is the problem of ensuring that the cards sorted into a pile are all at the same level of abstraction. For example, consider the following list of ideas:

  Can figure out how to turn on the windshield wipers
  Can set the car's clock without reading the driver's manual
  Can find and use the seat adjustments
  The car's adjustment controls are intuitive

All of these ideas could have come from interviews with automobile customers. The first three ideas deal with ease of understanding specific controls in the automobile. The fourth idea, "the car's adjustment controls are intuitive," deals with all of the controls. Since it's more general, it would make a good title for the other three ideas, and probably for other ideas that relate to individual controls.

Often the starting ideas in an Affinity Diagram are at many levels. The card-sorting and card-pile-naming processes are the places for noticing these level mismatches and for making the appropriate adjustments. Some cards move up in the hierarchy, some move down. There are no foolproof guidelines for determining the right level for each card; level determination appears to be an art today, not a science. Here are a few hints that might be useful:

- Keep the number of cards in one pile low. In this way, it's easier to compare each card with all the others in order to judge whether or not they are at the same level.
- Be on the watch for "world hunger" cards: phrases that are at such a high level of abstraction that they are not useful. Typical examples of such phrases might be

  A good product
  A world-class solution
  Best in its class

- Look for phrases that are at a higher level of detail or specificity than the others. These cards probably belong at a lower level in the hierarchy.

4.2.3 Create Titles for Each Cluster

- After the silent sorting process is complete, discussion begins again. The team now creates a title card for each pile. The name of the title card should express the common element in all the cards in the pile. The title card summarizes the data on the cards in the pile at a higher level of abstraction. Usually, the team discovers in this process that various adjustments must be made to the composition of cards in the piles. For example, the following customer needs for word-processing software might have been sorted together during the silent process:

  Offers lots of size, font, and design options
  Able to see what the fonts look like as I'm choosing them
  Can organize the listing of fonts to reflect the way I use them
  Everything stays neat and aligned when I change fonts

The team's choice for the title card might have been

  Easy Font Management

However, after discussion, the team might decide that "offers lots of size, font, and design options" is related to

  Can Alter the Appearance of My Material

but not to "Easy Font Management." In that case, "offers lots of size, font, and design options" would be moved out of this pile and placed in another pile, as shown in Figure 4-2.

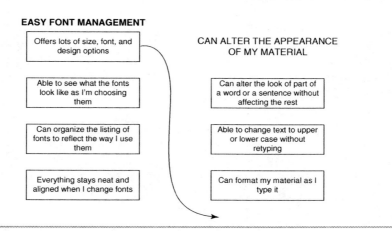

Figure 4-2 Moving a Card from One Pile to Another

4.2.4 THE FINAL GRAND CLUSTERING

- **Subdividing the Piles**: The most common reason for removing cards from piles is that there may be too many cards in a pile. Some teams adopt the rule that there can be no more than ten cards in a pile. If there are more than ten, they assume there must be more than one common theme, so the pile should be broken down into smaller piles. Some teams choose other maximum card counts, such as seven or even three.

- **Clustering the Clusters**: Having named the piles, the team now sorts the piles into clusters of piles, according to how they intuitively go together. One practical method for doing this is to first place all the subordinate cards in each pile underneath the title card, so they are hidden, and only the title cards can be seen. Then the team sorts the title cards (and all the hidden cards underneath) as if they were at the level of the first set of cards. The team then creates title cards for these new, higher-level piles, just as they created titles for the first level of cards. Each round of sorting and naming results in a set of titles that are summaries or abstractions of the sorted cards. The number of title or summary cards is about one-third to one-tenth as many as the sorted cards. This process of sorting, summarizing, and naming continues until the number of piles is lower than the maximum card count per pile.

4.2.5 AFFINITY DIAGRAMS COMPARED TO THE KJ METHOD

The Affinity Diagram method was popularized by quality training organizations such as GOAL/QPC in the early eighties. Later on, another similar method, called the KJ method, was introduced into the U.S.

The KJ method for sorting piles of cards, developed by Jiro Kawakita,[3] probably ante-dated the Affinity Diagram and may have provided the evolutionary basis for it. The KJ method stipulates very specific steps for affinity diagramming, including a color-coding scheme. Cards at the lowest level are written in black ink. Their title cards are written in red. When the red title cards are sorted and grouped, their title cards are written in another color, blue for instance.

There are other specific differences between the KJ method and the Affinity Diagram method. KJ proponents believe strongly in the value of these differences. The KJ method is robustly linked to a style of problem solving for continuous improvement known as the WV Method.[4] TQM problem solving in general is beyond the scope of this book, but I urge the reader to gain familiarity with the concept.

Typical dimensions for Affinity Diagrams are:

- 50 to 150 cards to begin with (tertiary level)
- 15 to 25 piles at the first summary level (secondary level)
- 5 to 10 piles at the second summary level (primary level)
- 3 levels in total

The highest level of abstraction is called the **primary** level, and the other levels are called **secondary, tertiary**, and **quaternary**. It is rare to see an Affinity Diagram deeper than four levels; most commonly, Affinity Diagrams are two or three levels deep. The more cards or ideas you begin with, the more levels you are likely to have. Most teams will have difficulties dealing with more than 150 or 200 cards. The author has worked with as many as 1,000 cards and does not recommend it.

Figure 4-3 is an example of an Affinity Diagram of customer needs for a word-processing program. The Affinity Diagram format shown in Figure 4-1 would be too awkward for the display of so many customer needs, so I have reformatted the needs into an equivalent indented-text format. Often, a team creates Affinity Diagrams by arranging Post-it Notes on a large wall chart. When the diagram is completed, it is transcribed to an indented text format so it can be printed and distributed to others in a more-compact form.

The text indented the least corresponds to the primary level; the next-most-indented titles represent the secondary level; the most-indented text is at the tertiary level.

---

3. Shoji Shiba, Alan Graham, and David Walden, *A New American TQM: Four Practical Revolutions in Management* (New York: Productivity Press/Center for Quality Management, 1993).

4. Shiba, *et al., A New American TQM.*

The customer needs in this example are incomplete, and yet there are a lot of them. The diagram is incomplete in another respect also: The number of customer phrases in some piles exceeds the recommended limit of ten. The hierarchy has been further refined in Figure 4-4.

Notice how the hierarchical structure, even in its somewhat-unrefined state, allows us to "zoom in" or "zoom out" on the data. The primary-level headings provide an overview. We can get a more-detailed understanding of a primary level by examining its secondary level. Each secondary level is more fully defined by its tertiary level, and so on.

You may not agree with the organization of the material. There is no right organization, nor are there right titles, because the relationships and titles are based on subjective judgment. In QFD, we attempt to organize the customer wants and needs the way customers would. Although many teams organize the customer wants and needs themselves, it is possible to enlist customers in the process. Including customers to organizing the data affords the team substantial additional insights. See Section 16.5 for a more detailed discussion of this point.

In other applications of Affinity Diagrams, where the phrases to be organized may come from the team's brainstorming, the right organization is whatever makes most sense to the team.

The Program Is a Pleasure to Use
No Surprises
Commands Are Easy to Know and Use

Know what an icon is going to do before I click on it
Can execute commands quickly
Can customize the icon display so that it s easy for me to use
Clear relationship between menu commands and icons
Able to execute common operations in a single step
Don't have to read the manual to figure out how to use the program
Program informs me about all its capabilities and features
The manual is easy to understand and use
"Help" function tells me how to do things, not just what things are
No complicated key strokes to memorize in order to do simple operations
Don't have to go into Help to understand how to do what I want to do

Program Is Quick and Responsive

Can adjust the cursor to move as quickly as I'd like
Enables me to find things in the document quickly

Easy Font Management

Offers lots of size, font, and design options
Able to see what the fonts look like as I'm choosing them
Can organize the listing of fonts to reflect the way I use them
Everything stays neat and aligned when I change fonts

Figure 4-3 Word Processor Wants and Needs (Partial)

No Surprises
What I See Is What I Get

Know what the document will look like when I print it
Able to see the whole page at once
Can see what I type as I type it
Can see all the pages in my document together, side by side
Able to see subtle spacings easily

Can Control the Shape of My Document
Can Work With Many Page Styles

Can create my own document templates
Can organize my text into tables and charts
Easy to set up, change, or eliminate headers and footers
Easy handling of material in multiple columns and rows
Can use different paper sizes and orientations
Easy envelope addressing
Offers me a variety of document types, e.g., letter, invoice, brochure
Simple to save settings as a default

Can Work With Text and Graphics

Able to mix text and graphics
Able to create charts and pictures in my document
Allows me to easily create presentation style material
Can add pictures/symbols to my document

Can Create and Manage Document Structure

Easy to create footnotes
Able to organize my document as an outline
Easy to create a table of contents
Easy to create an index for my document

Can Modify My Document Any Way I Want To
Can Change the Document Around Any Way I Want To

Easy to move things where I want them in the document

Can Alter the Appearance Of My Material

Can alter the look of part of a word or a sentence without affecting the rest
Able to change text to upper or lower case without retyping
Can format my material as I type it

Can Create Error-Free Documents

Able to easily undo any changes I make
Able to make global reformatting changes effortlessly
The program can proofread my text
Warns me if I'm about to do something wrong, like delete a file
Can check the spelling of words
Makes it easy to edit and correct my work
Won't lose my original text when I type into a section I have highlighted
Warns me if the software is going to bomb
Easy to save my work

Can Get My Ideas On Paper Easily
Can Mix Material From Many Documents

Able to view more than one document at a time
Can combine parts of different documents to form a new one
Enables me to know what is in a document before I open it
Able to move text from one application or document to another
When I bring in documents created in another program, they retain their original appearance
Able to convert document into other types of files (e.g., ASCII)
Can find/view information produced by other programs without exiting
Easy to retrieve and reuse work I created previously

Enhances My Creativity

Helps me quickly capture and save my ideas (e.g., when brainstorming)
Enables me to organize and reorganize my lists

Figure 4-3   Continued

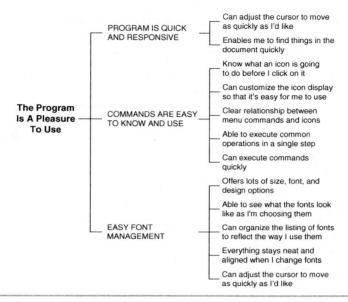

The Program Is A Pleasure To Use

PROGRAM IS QUICK AND RESPONSIVE
- Can adjust the cursor to move as quickly as I'd like
- Enables me to find things in the document quickly

COMMANDS ARE EASY TO KNOW AND USE
- Know what an icon is going to do before I click on it
- Can customize the icon display so that it's easy for me to use
- Clear relationship between menu commands and icons
- Able to execute common operations in a single step
- Can execute commands quickly

EASY FONT MANAGEMENT
- Offers lots of size, font, and design options
- Able to see what the fonts look like as I'm choosing them
- Can organize the listing of fonts to reflect the way I use them
- Everything stays neat and aligned when I change fonts
- Can adjust the cursor to move as quickly as I'd like

Figure 4-4   Tree Diagram

4.3 TREE DIAGRAM

The Tree Diagram, like the Affinity Diagram, is a hierarchical structure of ideas. In contrast to the Affinity Diagram, which is built from the bottom up based on an intuitive feeling for how the ideas go together, the Tree Diagram is built from the top down, and uses logic and analytical thought processes.

The Tree Diagram usually starts with some already-existing structure—for example, the hierarchy created by the Affinity Diagram process. The team then examines each level of the Tree Diagram, starting with the most abstract or highest level (the primary level), and analyzes that level for completeness and correctness.

> For example, the list of primaries in Diagram 4-3 is
> Can Control the Shape of My Document
> Can Modify My Document Any Way I Want To
> Can Get My Ideas On Paper Easily

Looking at this list analytically, a development team might conclude that the list is incomplete. For example, there is nothing in the list relating to installation, reliability, or interoperability with other software packages. In Figure 4-4 many needs are intentionally left out, just to make the example a manageable size for this chapter. In real-world projects, there could be real-world reasons why certain areas of customer need are missing. Perhaps these factors didn't show up because of the method used for collecting the VOC: limited time for interviewing, or unavailability of certain types of customers to interview, for example. Since the development team knows that these missing factors *ought* to have shown up, they may elect to add them after the Affinity Diagram process is complete.

Likewise, at lower levels, the team may make other additions and amendments, based on their expert knowledge of the subject matter. Working from the top level downwards, the team fills in and completes the hierarchy.

An important caution: When development teams overlay their knowledge on top of customer data, they are assuming some risk. The reason for listening to the customer in the first place is based on the notion that development teams can easily develop misimpressions of what customers' needs are.

In an ideal setting, the data received directly from the customer would be 100-percent complete and 100-percent correct. The Affinity Diagram would represent exactly the way customers view the relationships between the data. There would be no need to modify the data in any way, and the Tree Diagram process would not be necessary.

In real life, the VOC process is often incomplete or flawed in one way or another, and development teams often feel the need to correct the data. This must be done carefully and with humility, or the whole intent to listen to the Voice of the Customer will be undone.

Although the Tree Diagram practice of modifying the hierarchy as the team sees fit must be used cautiously, if at all, with customer data, it can very effectively be used to enhance data that comes from the development team itself—via brainstorming, for example. The Technical Response section of the HOQ is a good example of where this applies, and we'll be discussing it in Chapter 7. One way of generating the Technical Response (also called the Substitute Quality Characteristics, or SQCs) is for the development team to brainstorm ideas, create an Affinity Diagram, and then complete the Affinity Diagram by applying the Tree Diagram process to it.

When Tree Diagrams are completed, they are usually drawn in the form shown in Figure 4-4. Because Affinity Diagrams and Tree Diagrams are both hierarchical structures, if they are drawn in the same format there is no way to tell whether one is looking at one or the other type of diagram. The difference between them is the method of producing them, not their format. Affinity Diagrams start with the raw data and end with a hierarchical structure (bottom up); Tree Diagrams start with a presumed structure and end up with a detailed elaboration of that structure (top down). Both methods have value, and it is best that they be clearly labeled to avoid any confusion between Affinity Diagrams and Tree Diagrams.

4.4 THE MATRIX DIAGRAM

The **matrix** is a simple but powerful tool that lies at the heart of QFD. Its versatility is heavily exploited throughout QFD. Even though we see matrices in many forms in our everyday work, it's still worthwhile to review the basic concepts here.

A matrix is a rectangular diagram divided into horizontal **rows** and vertical **columns** (see Figure 4-5). Where a row and a column intersect, we have a **cell**. The cell is uniquely associated with one and only one row-column pair.

We list a range of comparable items along the left side of the matrix. By "comparable items," we mean items that are all attributes or facets of the same general topic. For example, green, red, blue, and yellow are all comparable in that they are all colors. Other comparable items are

- **Subsystems of an automobile**: suspension subsystem, chassis, steering subsystem, transmission, fuel and ignition subsystem
- **Customer needs**: easy to learn, easy to use, doesn't endanger me, easy font management

Each of these comparable items is associated with a row of the matrix. We list another range of items along the top and associate each of those with a column. We can use each cell in the matrix to record some relationship between the item associated with the row and the item associated with the column.

In Figure 4-6, the black circle represents a relationship between "C" and "2." The absence of black circles in other cells indicates that the kind of relationship existing between "C" and "2" does not exist anywhere else in this matrix. This type of entry, which indicates that a relationship exists or does not, is called a **binary** entry in the matrix.

In Figure 4-7 we have a matrix showing several binary relationships between various column and row items. For example, the relationship represented by the black circle

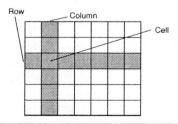

Figure 4-5   The Matrix Diagram

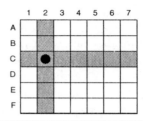

Figure 4-6   Binary Relationship

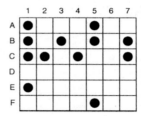

Figure 4-7   Multiple Binary Relationships

exists between items A and 1, between items F and 5, and between items B and 7, as well as between other pairs of column and row items.

Another way of viewing the matrix is to look at entire rows or columns, not just cells. At a glance, we see that the second row shows all the relationships that row item B has to any of column items 1 through 7. There are four circles in this row, so we can quickly see that item B relates to more of the numbered items than any other lettered item except item C, which also has four circles. Likewise, column 1 relates to more lettered items than does any other numbered item.

Other, more subtle patterns are also evident: Most of the relationships are in the top half of the matrix, suggesting that items A, B, and C are somehow different from D, E, and F. Looking from left to right, however, we don't see a very pronounced grouping of any of the numbered items, so in terms of the relationship symbolized by the black dot, there isn't much differentiation between the column items.

The *absence* of relationships is also important in this binary relationship matrix. The column corresponding to item 6 shows no relationship with any of the lettered items, and the absence of circles in the row corresponding to item D also shows that "D" has no relationship to any of the numbered items. As we'll see later on, this observation is very important in QFD.

4.5 THE PRIORITIZATION MATRIX

The Prioritization Matrix is an extension of the Matrix Diagram. It allows us to judge the relative importance of columns of entries.

We can put many different things into the cells of a matrix. In Figure 4-7, we used a form of binary data: The black circle or its absence indicated that a relationship existed or did not. Obviously, any other symbol could have been used. Check marks, X's, dashes, diagonal strokes, and open circles are all common representations of binary relationships.

Besides binary relationships, we can enter numbers or symbols representing those numbers into the cells. A common QFD practice is to enter numbers that express the strength or degree of the relationship between a column item and a row item. Traditional QFD practice from Japan uses the symbols ◎, ○, △ and blank to indicate "strong relationship," "moderate relationship," "slight or possible relationship," and "no relationship," respectively. Thus, in Figure 4-8:

- Column item 2 has a strong relationship with row item A
- Column item 7 has a moderate relationship with row item C
- Column item 2 has a slight or possible relationship with row item F
- Column item 5 has no relationship with row item D

Generally, we assign numerical values to these degrees of relationship, but many people believe graphical symbols are easier to see (when the matrix is drawn on a wall chart), and they carry more visual impact than do numbers. Some QFD practitioners prefer graphical symbols, others numbers. We'll use both in this book.

In Figure 4-9, we can easily see patterns in the relationships between column items and row items. Row item A has two strong relationships and one moderate relationship with column items; only Row item C also relates strongly with more than one column

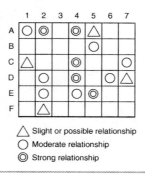

Figure 4-8 Multivalued Graphical Entries

Graphic symbol	Numerical values representing strengths of relationships
◎	9 (less common: 10, 7, 5, 3)
○	3 (less common: 2)
△	1
(blank)	0

Figure 4-9   Common Relationship Values

item. Likewise, we can see at a glance that of all the column items, column item 4 has the strongest relationships with the row items as a group.

The absence of strong relationships can also be observed easily. Column item 3 stands out because it has no entries. Not only has it the weakest set of relationships with the row items, it may be considered irrelevant. In the QFD House of Quality, where the row items are customer needs and column items are technical responses to those needs, an empty column item may indicate an unnecessary technical response, sometimes referred to as a "pet project."

In QFD, we assign numerical values to the graphical symbols representing these relationships. The most common numerical values are shown in Figure 4-9. If we redraw Figure 4-8, substituting these values for the graphical symbols, we get the matrix in Figure 4-10.

The numbers below the matrix are the sums of the strengths of the relationships in the columns. For example, the number 30 at the bottom of column 4 represents the aggregate strength of the relationships that column item 4 has with all the row items. It's the largest of the sums, and in QFD our interpretation is that column item 4 is the most important of the column items, in terms of its relationship with all the row items.

The practice of totaling the strengths of relationships, as in Figure 4-10, gives us a good estimate of the overall relationship of the column items with the row items, *assuming that the row items are equally important.* Usually, however, the row items are *not* equally important. In QFD, we usually associate numerical weights or priorities

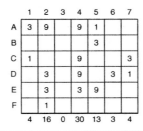

	1	2	3	4	5	6	7
A	3	9		9	1		
B				3			
C	1			9			3
D		3		9		3	1
E		3		3	9		
F		1					
	4	16	0	30	13	3	4

Figure 4-10   Multivalued Numerical Entries

with items in a list. When we enter such item weights in a matrix, we can combine those weights with the relationship values in the cells of the matrix in order to more realistically estimate the importance of the column items. Let's take a look at this in Figure 4-11.

In Figure 4-11, we have added a column of numbers representing the relative importance of each of the row items. (For now, let's not worry about how we arrived at those relative-importance figures.) We have divided every cell into two parts, separated by a diagonal line. Above the diagonal is the original number (from Figure 4-10) representing the strength of the relationship between the column item and the row item. In QFD this is usually called the **impact** of the column item on the row item. Below the diagonal is the product of the strength of the impact and the relative importance of the row item. In QFD terms this is usually called the **relationship** of the column item to the row item.

For example, the strength of the relationship (the impact) between column item 4 and row item E is 3, shown above the diagonal in the cell corresponding to column 4 and row E. The relative importance of item E is 4 (shown just to the right of item E). Therefore, the number below the diagonal in the cell is 3 times 4, or 12. This number combines the strength of the relationship and the importance of the row item into a single number (the relationship). By multiplying importance times strength, we are expressing the "importance of the strength." To estimate the importance of the column items, we add the "importance of the strengths" of the column items.

We can now look at Figure 4-10 in a new light. In that matrix, the numbers at the bottom of the matrix are simply the sums of the strengths of the relationships. They do not take the relative importance of the row items into account. Another way of viewing this is that we treated the row items as if they were all equally important. In Figure 4-11, if all the row items had been equally important, and the importance value had been set to 1,

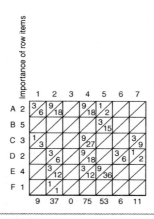

Figure 4-11 Prioritization Matrix: Weighted Rows and Weighted Cells

the totals at the bottom of the columns would have been the same as in Figure 4-10. In this sense, Figure 4-11 is a special case of Figure 4-11. Figure 4-11 is called a **Prioritization Matrix**. The matrix translates the priorities of the rows into priorities of the columns. It's the most common type of matrix in QFD.

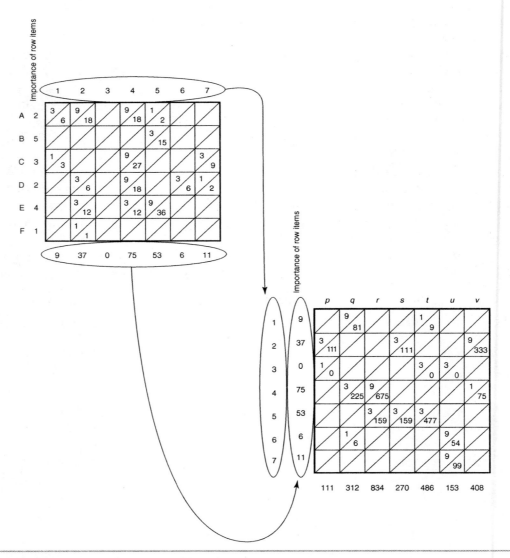

Figure 4-12   Multiple Prioritization Matrices

In QFD, we transfer column-prioritization information from one matrix to the rows of another matrix, as in Figure 4-12. Figure 4-11 provides a detailed look at how that happens. In the upper Prioritization Matrix, the row items, A through F, are the Whats. In the HOQ, the Whats would be the customer needs. The importance of the Whats in the HOQ is determined by the development team's work in filling the Planning Matrix (as described in Chapter 7). The result of this work is expressed in two Planning Matrix columns: the Raw Weights column (see Section 7.6), and the Normalized Raw Weights column (Section 7.7). The information in these two columns is equivalent— the Normalized Raw Weights are proportional to the Raw Weights—but expressed as percentages rather than large whole numbers. Either of these expressions of the Raw Weights is transferred to the columns of the second Prioritization Matrix (the lower one in Figure 4-11). They are labeled "Importance of row items" in the lower matrix. The development team generates a new set of Hows (denoted by the symbols P through v). The team then determines the impacts of P through v on the column items 1 through 7. These impacts along with the importance ratings of row items are used to prioritize the column items P through v.

4.6 ADDITIONAL TOOLS FOR QFD PRACTITIONERS FROM SIX SIGMA

Six Sigma tools most often associated with QFD work will be described next, but only in a brief overview of their relevance to QFD work, and in no particular order. The interested reader is encouraged to delve more deeply into either Ian Wedgwood's text *Lean Sigma, A Practitioner's Guide*[5] or *Commercializing Great Products with Design for Six Sigma* by Randy Perry and David Bacon.[6]

4.6.1 CRITICAL PARAMETER MANAGEMENT

Critical Parameter Management (CPM) was first described, to the best of my knowledge, by John Fox in his book *Quality Through Design*.[7] The intent of CPM is to manage the parameters that matter most in the customer's eyes, and the relationships between prod-

---

5. Ian Wedgwood, *Lean Sigma: A Practitioner's Guide* (Upper Saddle River, N.J.: Prentice-Hall, 2006).

6. Randy C. Perry and David W. Bacon, *Commercializing Great Products with Design for Six Sigma* (Upper Saddle River, N.J.: Prentice-Hall, 2006).

7. John Fox, *Quality Through Design: The Key to Successful Product Delivery* (New York: McGraw-Hill, 1993).

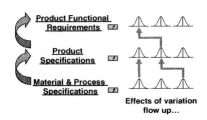

Figure 4-13   Parameter and Variation Diagram

ucts, subsystems, components, materials, and process parameters that spawn from designing products that meet these customer needs. CPM is a necessary tool that accompanies good QFD work. QFD does the planning to meet customer needs, and CPM keeps track of all the engineering relationships and performance numbers that are developed to meet those needs, along with the amount of variability that flows up from engineering simulation, manufacturing, and supplier capabilities. This is illustrated in Figures 4-13 and 4-14.

It is important to note that CPM is applied only to *critical* parameters, not to all parameters. The definition of what is critical may vary widely; however, **new, important**, or **difficult** customer needs are helpful guidelines in making the criticality determination. If it's a *new* customer need we are focused upon, clearly it may be critical to the product launch, or entry to a new market segment. If customers have told us this need is *important* in their business or their purchase decision, then it's critical. Lastly, if a customer need has been *difficult* to meet, it is therefore critical to differentiation in the marketplace. Further explanation of CPM may be found in the excellent book by Perry and Bacon.[8] CPM summarizes the important relationships and measures associated with the plan laid out by Quality Function Deployment (QFD) practices. The reader is advised that the focus in CPM is on parameters and relationships; that is the inherent value this tool provides to the QFD practitioner. The QFD practitioner must decide which customer needs will be the focus and thus be further addressed with CPM.

4.6.2 THE DESIGN SCORECARD

The Design Scorecard is used to assess the overall design quality against *all customer requirements* distinguishing it from CPM and some of the flow downward of parameters in QFD. The Design Scorecard does not keep track of relationships, which is another role

---

8. Perry and Bacon, *Commercializing Great Products.*

Customer Requirements:	Functional Product Requirements	Measurement Technique:	Target	USL	LSL	%P/T	Mean	Std Dew	Cp	Cpk	Transfer Function Model
Color stable candy wrappers	Increased thickness	On line micrometer	1 mil	1.55	0.456	28%	1.002	0.1279	1.46	1.46	Color Stability = 2.3 + 1.86* Thickness
Non-stick candy wrappers	Increased thickness	On line micrometer	1 mil	1.55	0.456	28%	1.002	0.1279	1.46	1.46	Stick Index = 0.31 + 2.01* Thickness
Films which stay wrapped on candy	Increased thickness	On line micrometer	1 mil	1.55	0.456	28%	1.002	0.1279	1.46	1.46	Wrap Stability = 0.0041 + 1.32* Thickness
Need to improve taste by 2% as measured by taste index	Increased thickness	On line micrometer	1 mil	1.55	0.456	28%	1.002	0.1279	1.46	1.46	Taste = 7.02 + .0034* Thickness
Need a film which processes without cutting	Increased thickness	On line micrometer	1 mil	1.55	0.456	28%	1.002	0.1279	1.46	1.46	Cut Resistance = 6.78 + 1.912* Thickness

Figure 4-14 Critical Parameter Scorecard Example

of CPM. What the Design Scorecard does accomplish, however, is to predict the probability of defect levels in the total product, in the areas of performance, parts, processes, and reliability or warranty. Additionally, if software is a concern, software-defect tracking is utilized to predict the level of errors or bugs likely to be found at product release.

The Design Scorecard does all these things by identifying

1. All customer requirements

2. All specified or purchased components, materials, and subassemblies, and

3. All internal manufacturing processes

Once all are listed, the capability or performance of each item is assessed against product or customer-specified requirements, and the defect quantities for each are predicted. These are totaled, and an equivalent Sigma level is calculated in each of these areas: performance, process, parts, and warranty or reliability. These estimates predict the likely quality of the product design and are then used to identify areas needing improvement as the design efforts progress. CPM is greatly utilized in Design for Six Sigma efforts. The value to QFD is now clear: The Design Scorecard measures *how well* we are meeting our QFD plan at the overall product level. See Figure 4-15.

4.6.3 MEASUREMENT SYSTEMS ANALYSIS

Sometimes there exist poor metrics or no metrics on what is important to customers, which can make Quality Function Deployment difficult if not impossible. Occasionally,

⊘SBTI			Top - Level Product Sigma		3.29
Sigma Breakthrough Technologies, Inc. Assembled Product Design Score Card					
Assembly Name	Turbochanger		Assembly Part Number	TS - 99	
Assembly	Parts DPU	Process DPU	Performance DPU	Predicted Field DPU	Total DPU
Turbochanger	0.007541	0.032438	0.101157	0.005829	0.147
Total DPU	0.007541	0.032438	0.101157	0.005829	0.146965
Opportunities	1	1	1	1	4
Top-level ST Sigma	3.93	3.35	2.77	4.02	3.289822

Figure 4-15   Sample Design Scorecard Showing Predicted Defects per Unit (DPU) and Product Sigma

what we are measuring does not make sense vis-à-vis the customer voices we hear during interviews. It is critical that what customers are measuring our products and services against is well defined and that the system is well developed, accurate, and understood if we are to improve customer satisfaction with our products and services. How customers measure lead times, service quality, product quality, and even total cost varies from customer to customer and from supplier to supplier. To improve customer satisfaction, these issues must be explored prior to designing or re-designing products and services. Certainly, all our QFD plans are dependent upon meeting customer needs, and if these are measured differently by the customers, we may miss our targets. There are frequently times when the measurement systems need evaluation and possibly correction or even replacement. The appropriate tool in each of these instances is Measurements Systems Analysis, sometimes called Gage Repeatability and Reproducibility Studies.

Measurement Systems Analysis (MSA) involves definition of the system itself, including all components, both physical and informational. Often the definition involves exploration of six broad areas of variability, sometimes called the Six M's: Man, Machine, Methods, Materials, Measurements, and Mother Nature. Each of these can represent a component of variation of the entire Measurement System. Once these factors are identified and designed, the study team works through separation of the random sources of variation from the non-random sources. The random sources are quantified, and the non-random sources are reduced or eliminated where possible. The amount of variation is checked against key guidelines and if unacceptable then steps are taken to reduce or eliminate the non-random variation, or to improve the system. When the subject is a measurement system involving operators and parts with repeated measurement possible, then these components are quantified first. The variability is then measured across all components in an active study to separate repeatability (same parts, same operators), reproducibility (same parts, different operators), and part-to-part variation. Initially, the goal is to separate and compare the amount of variation contributed by the measurement system itself versus the amount of variation that is being measured in the parts. These two groups are compared to the total variability to determine percent contribution, and also to specifications to determine percentage of tolerance. See Figure 4-16 for an illustration of key components leading to perceived variation.

A picture of one output from a Measurement System Analysis is illustrated in Figure 4-17, courtesy of Wedgwood, et al.[9] Note the Gage R&R percentage of about 33 percent, as illustrated by the green bar in the first grouping, making this system unsuitable for process-improvement work. In this figure, we see that the part-to-part variation is a large percentage contribution of the total system variation. However,

---

9. Wedgwood, *Lean Sigma.*

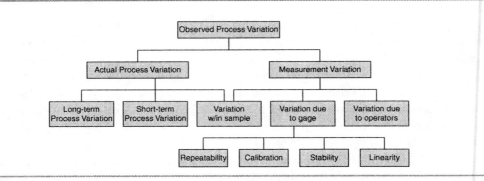

Figure 4-16 Observed Process Variation and Several Key Components

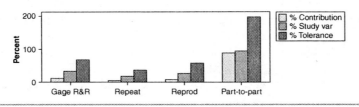

Figure 4-17 Example Gage R&R Components of Variation Plot
(Output from Minitab v14)

such a Measurement System is still unsuitable to detect improvement in products and services, and thus improvement is required before QFD work.

4.6.4 FAILURE MODE AND EFFECTS ANALYSIS, DESIGN EMPHASIS

Prevention of defects or non-working features in products is a key part of any quality design effort. If existing product failures are known, these must be designed out in new-generation products or replacement products if we are to win over new customers and continue satisfying existing customers. Each particular way that a product can fail to meet customer or performance expectations is called a **failure mode**. Each failure mode produces certain effects on the customer—some more severe, and some much less severe. Analyzing each possible failure mode for its effect on the customer, each potential cause, the likelihood of occurrence, the design controls in place, and whether they can detect or prevent the cause prior to customer receipt is called Design Failure Mode Effects and Analysis (DFMEA). DFMEA work is not to be undertaken lightly, and its best impacts are essentially risks avoided, and thus unseen. The use of DFMEA for improving products in aerospace and automotive product designs has generally increased overall product safety, prevented recalls, and increased customer satisfaction. Therefore, DFMEA usage can be a key element to improve customer satisfaction in our Quality Function Deployment planning. A sample DFMEA is shown in Figure 4-18.

4.6.5 DESIGN OF EXPERIMENTS

While first principles can sometimes give great insight, designed experiments are the only means by which many relationships between product performance and materials, components, and subassembly specifications can be determined. Design of Experiments (DOE) is used to verify and validate $Y=f(X)$, which is assumed as we relate the Whats to the Hows in the HOQ and the later matrices that follow. DOE is a hugely powerful tool set, and its potential and myriad complexities are far too comprehensive to be covered here. Suffice it to say that our QFD plans are only as good as the relationships we imply when we fill out our matrices. Hence, any assumed product-to-process or product-to-component relationships need to be developed, tested, and confirmed if our QFD plan is to have any value.

A DOE is a series of planned experiments to determine the relationships between independently varied parameters (Xs) and dependent parameters (Ys). In this process, product performance can be evaluated against differing materials, tolerances, specification ranges, and especially customer environmental variables. Often the experimental series generates profound discoveries and knowledge that would just not be possible using the "test only one variable at a time" approach sometimes found in product development. Interactions between independent parameters are often hidden by the latter approach, and usually they are discoverable only through DOE work. Testing of two-or-

Characteristic or Part	Potential failure Mode	Potential Failure Effects	S E V	Potential Causes	O C C	Current Design Evaluation or Control	D E T	R P N
What is the Characteristic or Part under evaluation?	In what ways does this characteristic lose its functionality?	What is the impact to the Customer (Internal or External)?		What causes the loss of function?	How often does causes FM occur?	What are the tests, methods, or techniques to discover the cause before design release?		
Coil resistance	Too high	Light emitted too law at specified current	8	Incorrect material	3	Material properties DOE	5	120
			8	Incorrect number of windings	2	Mechanical design simulation	2	32
			8	Wrong mateiral composition	2	Material specs	10	160

Figure 4-18   Example of Design Failure Mode Effects and Analysis (DFMEA)

C5	C6	C7	C8
Temp	Time	Chip	Rating
-1	-1	-1	60
1	-1	-1	72
-1	1	-1	54
1	1	-1	68
-1	-1	1	52
1	-1	1	83
-1	1	1	45
1	1	1	80

Figure 4-19   Sample DOE Experiment Matrix

more-variable interactions usually distinguishes the difference between providing a highly robust, reliable product to the customer and not. Our QFD work should identify where there are gaps in Y=f(X) knowledge needed to complete certain matrices, and plan for each DOE series around developing the required knowledge. A sample DOE matrix of three independent variables and their effects on one dependent variable (Rating) is shown in Figure 4-19.

4.6.6 KANO'S MODEL

Not all customer expectations have equal importance in terms of product advancement. Sometimes, improvement of certain features, or of product performance beyond a certain level, does not create customer excitement, or even greater customer satisfaction. Professor Noriaki Kano[10] taught us that customer expectations and experiences may be grouped into three broad categories, each of which has a different impact on customer satisfaction. Please refer to Figure 4-20 for an illustration of the Kano Model.

The first category is **Taken for Granted** or **Fitness to Standard** and in customer terms it means that beyond a certain minimum level of functionality, the feature's performance enhancement yields no greater customer satisfaction. Consider an automobile's electric windows as an example. Beyond a certain speed to move up or down, there isn't any noticeable customer satisfaction for the great majority of customers. If not meeting the minimum raise and lower speeds, only customer *dissatisfaction* can result. Often, product features in this area are considered minimum entry points into a customer's purchasing decision.

The second category is **The More the Better** or **Competitive,** which in customer terms means the more of these features or performance that is delivered by your product, the happier the customer. If we consider the automobile again, gas mileage would fit into this category. The higher the miles-per-gallon performance of a car in the economy-car segment, the

---

10. See Mizuno, ed., *Management for Quality Improvement.*

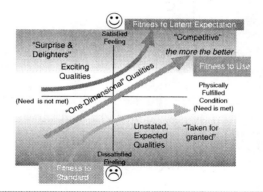

Figure 4-20   Kano's Model of Customer Satisfaction versus Functionality[11]

greater the customer satisfaction. Qualities or attributes in this category are often called competitive features because direct comparisons are meaningful to establish value in the customer's eyes, in particular when price is brought into the decision-making process.

The third category is called **Surprise and Delighters** or **Fitness to Latent Expectations** and in customer terms, satisfaction is greatly enhanced once this feature or quality is present. These types of product attributes or features are usually new, and differentiate one new product from those of its competitors, which often will then scramble to introduce competing or alternative options. By understanding unexpressed customer needs or anticipating new needs before they have been fully articulated or satisfied in the marketplace, great customer satisfaction and product demand may result. A recent example in the automobile industry is the inclusion of Bluetooth-enabled wireless coupling to a customer's mobile phone for hands-free operation.

The relationship to QFD of Kano's model is manyfold. Firstly, in determining in the HOQ which parameters we evaluate on competitiveness, we need to be certain that they are in the second or third category and not the first. Secondly, when collecting customer voices, we want to know which type of quality attribute a customer is voicing satisfaction or dissatisfaction about. If it is in the **Fitness To Standard** category, then these must be fixed as soon as possible, because the product and brand image may suffer due to being below standard. Lastly, when planning which features to address in our QFD work and develop further in new-product efforts, we want to know that these are in the second and, where possible, third categories. For more on this model, see the text by Perry and Bacon, *et al.*[12]

---

11. Shiba, *et al.*, *A New American TQM.*

12. Perry and Bacon, *Commercializing Great Products.*

4.6.7 PUGH CONCEPT SELECTION

Stuart Pugh created a method of looking at many design solutions and comparing them feature-by-feature in a matrix. In applying this technique, the team identifies which possible product designs have the most value and also what their weaknesses may be regarding certain features. The team then iterates the design concepts to form hybrid concepts, incorporating positive features from the remaining concepts into the top-rated concepts. By repeating this process in an iterative fashion, concepts may be developed which have team buy-in, and which incorporate the best possible feature set exceeding or meeting the greatest number of features identified by the team from customer voices.

In summary form, the way Pugh's Concept Matrix functions is illustrated in Figure 4-21. A **standard** or **datum** concept is selected first. This can be the current best concept available to customers, or some other standard of proven market and customer value. Customer needs and requirements must be listed along the left-hand side of the matrix. Listed across the top of the matrix are various concepts, the first being the standard or datum. Each concept is then compared to the datum for each customer requirement, and a judgment is made regarding whether each concept is the same, better, or worse at satisfying that customer requirement. A sample Pugh Matrix is shown in Figure 4-21, with +, -, and S representing **better than**, **worse than**, and **same as** the datum concept, respectively. In the figure, team member hybrid concepts are shown as having components selected from an **ideation** step (Figure 4-22) that has preceded the Concept Selection Matrix. In the ideation step, team members select aspects from prior concepts and "go shopping" to form their best concepts for comparative purposes.

Concept Attributes	Datum A1 B2 C1 D3	A5 B4 C1 D4	A6 B2 C1 D1	A3 B2 C5 D3					A2 B3 C1 D3
Req. 1	S	+	+	−					−
Req. 2	S	−	+	+					+
Req. 3	S	+	−	+					−
Req. 4	S	+	−	−					−
Req. 5	S	−	−	+					−
Σ+									
ΣS									
Σ-									
Rank									

Figure 4-21 Pugh Concept Selection Matrix

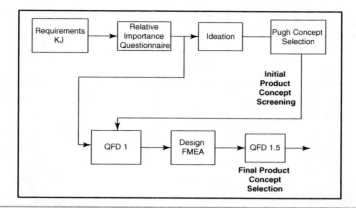

Figure 4-22 Product Development Flow[13]

In QFD work, we are planning to satisfy key customer requirements. A significant step in this planning is to arrive at a concept that is focused upon requirements based on customer needs. The Pugh Concept Selection Method insures that requirements based upon customer needs are used as the guideposts when developing the best concept possible. In Figure 4-22, from Perry and Bacon, one possible path is illustrated for using some key tools mentioned in this section along with the House of Quality.

In summary, the Pugh Concept Selection approach offers three distinct and valuable outcomes:

- Concepts have been evaluated, selected, and optimized around customer needs.
- The possible solution space has been explored in depth, arriving at a conceptual approach very likely to be the optimum value for listed customer needs.
- The team has great buy-in to the final concept, as it is usually a hybrid of the best features from many initial concepts.

4.6.8 PROCESS MAPPING

Process maps in all their various forms and formats are valuable tools for describing any process from many different perspectives. Some describe how work flows across various departments or functions, while others describe how value is added (or not) throughout a supply chain or core process. The most often used form is an input and output vari-

---

13. Perry and Bacon, *Commercializing Great Products*, Ch. 6.

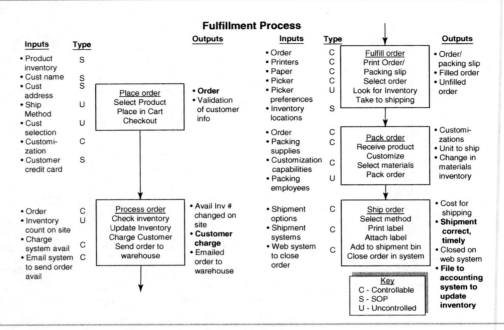

Figure 4-23   Illustration of Process Variables Map[14]

ables map[15], which describes the key steps of a work process, and the independent variables and dependent variables for each step. Such a process map is sometimes called a Process Variables Map. The last part of completing a process map is to classify the independent variables as controlled or uncontrolled, and identify whether the uncontrolled variables are **noise variables**. Noise variables are uncontrollable, and can be a nuisance or a downright cause of process instability.

The process-map tool is a foundational method for all process-improvement work. Six Sigma Black Belts begin with process mapping when starting process-improvement projects.[16] From a QFD perspective, when we are planning to launch new products and services, it is of the utmost importance to understand which processes will be affected by the launch and assess how these processes may need to change. See Figure 4-23.

---

14. Wedgwood, *Lean Sigma*.

15. Doug Sanders, Bill Ross, and Jim Coleman, "The Process Map," *Quality Engineering* 11:4 (July 1999): 555.

16. Sanders, *et al*, "The Process Map."

4.7 SUMMARY

In updating this second edition, we have added Six Sigma tools in this chapter to the original tools of Total Quality Management. These Total Quality Management and Six Sigma tools are usually described as "tool kits," Six Sigma Product Design and Process Improvement Tools, the Seven Basic Tools, and the Seven Management and Planning Tools. While the use of any of the tools is limited only by the imagination and creativity of the user, the Six Sigma Process Improvement Tools and Seven Basic Tools, which are mostly numerical tools, are primarily aimed at problem solving and continuous improvement, while the Seven Management and Planning Tools are intended, as their title indicates, to deal with the types of qualitative information usually associated with planning.

In this chapter, we have looked at key tools most applicable in QFD: the Affinity Diagram, the Tree Diagram, the Matrix Diagram, the Prioritization Matrix, Critical Parameter Management, The Design Scorecard, Measurement Systems Analysis (MSA), Design Failure Mode Effects and Analysis (DFMEA), Design of Experiments (DOE), Kano's Model, Pugh Concept Selection, and the Process Map.

Now that we have familiarized ourselves with these tools, it's time to use them in building the House of Quality. Chapters 5 through 12 will take us on a section-by-section, or "room-by-room," tour of the House of Quality.

4.8 DISCUSSION QUESTIONS

If you have never used the tools described here, take some time to do so before you attempt your first QFD. Additional research with the texts of Perry and Bacon[17] and Wedgwood[18] is strongly advised.

Affinity Diagram: What types of concepts or ideas in your work could be organized by means of an Affinity Diagram? Tasks in a project? Elements of a plan? Customer wants and needs? Pick an area of importance to your current work, brainstorm all of its elements, and "affinitize" them. First do this by yourself, then try it with a small team that has a stake in the subject.

- What were the differences between doing it by yourself and doing it in a team?
- Were there any surprises? If so, what were they, and how did the Affinity Diagram process contribute to them? If there were none, explain to yourself how using a process you've never used before could give you no new insights.

---

17. Perry and Bacon, *Commercializing Great Products.*

18. Wedgwood, *Lean Sigma.*

Tree Diagram: After completing an Affinity Diagram, use the Tree Diagram process to complete the hierarchy. Try it by yourself, then with a team, as with the Affinity Diagram process above. Ask yourself the same questions as with the Affinity Diagram process.

Matrix Diagram: Take the most recent plan that you had any part in creating. List the objectives of the project on the left side of a matrix, and list the actions identified in the plan along the top. Fill in the matrix, showing the strength of the relationship between each objective and each action. Based on your examination of the completed matrix, can you think of any way to improve the plan?

Critical Parameter Management (CPM): Find one of the most critical parameters to your customer of one key product. Trace back the path down to the lowest levels of materials, components, and process parameters. How many relationships did you find? How many branches, and are they all well known and understood? What capabilities are there for measurement data and comparisons to requirements along the way? Without doing this once on your own, how will you get an appreciation for the value of CPM?

The Design Scorecard: List all the product performance tests done on a final product evaluation by you and your customer. What specific items are tested, and are there good records overall for the mean values and the standard deviations for each performance parameter? If you can do this for all test parameters, you have begun the first page of a Design Scorecard. Z-scoring each parameter and calculating expected defects will allow an overall performance defect prediction, which may be used to calculate the performance sigma of the product design. While not simple, it is a very effective means of predicting long-term yields at final test. Who in your organization would own the performance scorecard section? The process scorecard? And lastly, the parts or materials scorecard?

Measurement Systems Analysis: What measures do your customers use to accept your products? Is the amount of measurement error for those acceptance tests well known and quantified? What about acceptance testing within your business? Do you have good measurement error estimates? If not, how are you certain that you are not failing good products and passing bad products?

Design Failure Modes and Effects Analysis (DFMEA): Take a look at your design function within your organization and check into whether DFMEA is used regularly, once in a while, or not at all. Chances are your company requires its usage. Ask the engineers how much this adds to the chances for overall product success. If the answers are all positive, then this tool is part of your culture. If not, when will you take the opportunity to begin discussions about it with key engineering managers?

Design of Experiments (DOE): How are relationships between process and product parameters quantified in your current product-development activities? How are experiments designed and conducted to optimize product performance? Robustness? DOE is among the most powerful tools ever created to obtain product and process knowledge. If it is not being used extensively, what barriers exist to prevent extensive use?

Kano's Model: Not all requirements or customer needs are created equal. Can your organization tell from customer-voice collections which ones are Standard, which are Competitive, and which are Delighters? Kano's model is equally important in planning, collecting, and interpreting customer needs. How will you use the model to modify these three key areas of evaluating VOC?

Pugh Concept Selection: How does a team arrive at the best final concept for products? For services? How many alternatives are truly explored? Is the team fully satisfied when the final product concept is chosen? Are compromises being made, or is the concept being enhanced by the options discussed? Finally, how are customer needs utilized to make choices about which concepts get passed over and which get moved forward?

Process Mapping: All work is a process. Which processes will be affected with a new product or service offering? Since the processes are what create the value according to design instructions, drawings, and specifications, how will those changes be communicated and recorded? Seek out the parts of your New Product Development that link to key informational and manufacturing processes and determine whether they require process maps.

Overview of the House of Quality

With Steve Zinkgraf

We'll now expand the very brief overview of the House of Quality (HOQ) that was given in Chapter 1. While still an overview, the description in this chapter is considerably more detailed than the previous one. In this chapter, we introduce each of the steps of the House of Quality, and we discuss what information each step produces and how the information is used. Once we've completed this overview, we look at each step individually.

The House of Quality is the central construct of QFD. Most of Parts II and IV of this book deal with the House of Quality. Other parts of the book, especially Part V, describe QFD as a process that links the various phases of the development process together, from customer to delivery.

There are two reasons we have chosen to focus so much on the HOQ. First, it contains many of the features we will see in other parts of QFD, so once we have studied it, the remaining QFD matrices and charts will be easy to understand. Second, just about everyone using QFD for the first time starts with the House of Quality. For better or worse, many teams choose not to use the rest of QFD. Therefore, a basic introduction to QFD *must* include the HOQ. Once it is mastered, the HOQ leads fairly naturally to various extensions.

If you choose to stop with the House of Quality, you owe it to yourself to know what you have chosen not to do. Please be sure to familiarize yourself with the organizational implications of QFD (in Part III), and with the possible follow-on steps after the HOQ (in Part V). In most cases, the first matrix a QFD team puts together is the House of Quality. Many teams never create any other matrices (although the use of additional matrices is increasing). Since we must put first things first, let's turn our attention to the House of Quality.

In this chapter we'll present the "textbook" House of Quality. It contains just about everything that anyone has ever put into the HOQ. As you read this chapter, and all of the rest of the book, bear in mind that every aspect of QFD is a candidate for modification or omission. New QFD elements can and should be added according to the needs of the development team. In almost every application, the HOQ as described here is modified to better help a team solve its particular problem. As we explore more of the details of the HOQ throughout this book, we'll consider the most common variations, and we'll look at the benefits and drawbacks of each.

5.1 TOUR OF THE HOUSE OF QUALITY

The House of Quality is a very complex matrix in the sense that it consists of several matrices attached to each other.

Referring to Figure 5-1, let's a take a quick tour of the House of Quality.

5.1.1 THE FIRST STEP

The first step (or "room") of the HOQ to be constructed will (almost always) be the Customer Needs and Benefits step. The customer's wants and needs are normally derived

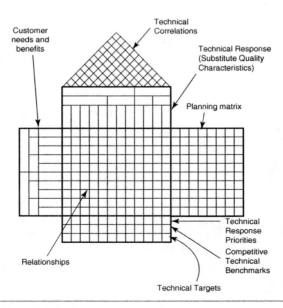

Figure 5-1   The QFD House of Quality

from the actual words of the customer, by any of several methods. We'll be covering many of those methods in detail in Chapter 17. Once gathered, the customer phrases are developed into a hierarchy by means of the Affinity Diagram process, with the most-detailed needs at the lowest level and the more-abstract needs at higher levels of the hierarchy, as we saw in Chapter 4.

Most development teams collect the Voice of the Customer from interviews, and then create the hierarchy of wants and needs themselves. It is possible—and desirable—to have customers strongly influence or completely determine the structure, or at least to validate the structure before it is used. Failure to maximize customer involvement in this process often leads to important misunderstandings of the customers by the development team. When product development teams don't understand their customers' needs well, the product-planning activity often gets mired in confusion. This results in slow product planning and costly midcourse corrections, not to mention noncompetitive products.

5.1.2 THE SECOND STEP

The Planning Matrix (sometimes called the Preplanning Matrix) is often the second step to be constructed. Traditionally, this step is shown on the right side of the HOQ, as in Figure 5-1, although as a practical matter it's more convenient to draw it immediately between the Customer Needs and Benefits and the Relationships. In this matrix, the development team records its answers to a variety of marketing and product-planning questions.

The Planning Matrix calls for high-level product goal setting based on the team's interpretation of the market research data. The goal setting has the effect of combining the company's business priorities with the customer's priorities. It is therefore a crucial step in product planning.

The specific information in the Planning Matrix, for each customer want/need, is

- How important is the need to the customer? (Generally answered by market research)
- How well does the team's current most-similar product or service offering meet customers' needs? (Generally answered by market research)
- How well does the competition's current most similar product or service offering meet customers' needs? (Generally answered by market research)
- How well does the team want to meet customers' needs for the product or service being planned? (Determined by the team)
- To what extent could meeting a need well be used as a sales point? (Determined by the team)

The answers to these questions combine to create a prioritization or rank ordering of the Customer Needs and Benefits.

One reason to complete the Planning Matrix immediately after the Customer Needs and Benefits are completed is because once customer needs are prioritized, the QFD team may choose to restrict its analysis to only the highest-ranking customer needs. This considerably reduces the time required to complete the QFD process. If the Planning Matrix were postponed until another time, say after the Relationships step was filled in, the development team would not be able to restrict its analysis, since it would not know which customer needs were most important.

Some practitioners generate the Technical Response Priorities and even determine the Relationships before developing the Planning Matrix. An advantage to this sequence is that team members will be required to become extremely familiar with the customer needs in order to generate the Technical Response Priorities. Hence, they will be that much better prepared to do the goal setting and high-level analysis in the Planning Matrix when they get to it.

5.1.3 THE THIRD STEP

The third step in completing the HOQ to complete is Technical Response Priorities. This can be thought of as a set of product or process requirements, stated in the organization's internal language. Sometimes they are called **corporate expectations**, to distinguish them from the **customer expectations**. A variety of different types of information may be placed here. The most common alternatives are

- Top-level solution-independent measurements or metrics
- Product or service requirements
- Product or service features or capabilities

Whichever type of information is chosen, we call it Substitute Quality Characteristics (SQCs). Just as the Customer Needs and Benefits represent the Voice of the Customer, the SQCs represent the Voice of the Developer. By placing these two voices on the left and top of the matrix, we will be able to systematically evaluate the relationships between them.

Whether teams use measurements, requirements, or features as their SQCs depends on the design methodology of their organization. Some organizations prefer a structured design process in which the design is first expressed very abstractly or independently of the ultimate design. It is then progressively (more concretely) developed in a series of steps in a process sometimes referred to as **stepwise refinement**. This approach is particularly favored by software engineers. Other organizations are more comfortable creating a concrete design of the product or service at the earliest possible moment. In fact, many organizations conceive of the design before they assess customer or market needs. These differences in style are all compatible with QFD and can be handled by "Static/Dynamic" analysis, as described in Chapter 13, Section 13.2.1.

Regardless of which type of SQC is placed along the top, the amount of detail may need to be managed in much the same way that the amount of customer detail needs to be managed. When there is a great deal of detail, the SQCs can be arranged hierarchically by means of the Affinity Diagram process, followed by the Tree Diagram process, as in Figure 5-2. The hierarchy gives the QFD team some freedom to conduct the analysis at a high or low level of detail by choosing the Primary, Secondary, or Tertiary level of the hierarchy. The higher the level, the smaller the Relationships step; the lower the level, the more detailed the analysis.

5.1.4 THE FOURTH STEP

The fourth step is to complete the Relationships step of the House of Quality. This is the largest step of the matrix, and therefore represents the largest volume of work. Various shortcuts are possible, which we'll discuss later in the book. This step uses the Prioritization Matrix method. For each cell in the Relationships step, the team enters a value that reflects the extent to which the SQC (at the head of the column) contributes to meeting the customer need (to the left of the row). This value, along with the prioritization of the Customer Needs and Benefits, conprises the contribution of the SQC to overall customer satisfaction. We'll discuss the details of computing the contribution in Chapter 8.

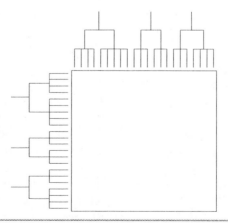

Figure 5-2 Tree Diagrams on the Left and on the Top

Once the contributions of all SQCs have been computed, the SQCs are essentially prioritized. Those with greatest overall impact on customer satisfaction are most important. This fundamental result is one of the most important outcomes of using QFD.

5.1.5 THE FIFTH AND SIXTH STEPS

Some QFD teams, especially those working in less-disciplined environments, abandon QFD after the Relationships step, and use the priorities of the SQC to plan later stages of the development project.

Other teams next use the prioritization of the SQCs to provide guidance about further HOQ product-planning activities. These activities are Competitive Benchmarking and Target Setting, the fifth and sixth steps in completing the HOQ. They occupy the bottom two lines of the HOQ, labeled Competitive Technical Benchmarks and Technical Targets in Figure 5-1. Competitive Technical Benchmarks and Technical Targets are normally expressed in language compatible with the language of the SQCs.

5.1.6 THE SEVENTH STEP

The seventh and usually final step in completing the House of Quality is to fill in the Technical Correlations matrix (the "roof"). This matrix is used to record the way in which SQCs either support or impede each other. This information helps QFD teams to identify design bottlenecks, and it helps them to identify key communication paths among designers.

This completes our overview of the House of Quality. In Chapter 6, we'll begin looking at each of the steps more closely.

5.2 SUMMARY

We've just completed a tour of the House of Quality. The HOQ matrix acts as a repository of marketing and product-planning information. The key inputs to the matrix are customer wants and needs, product strategy information, and Substitute Quality Characteristics (a high-level formulation of product or service requirements). Other information that can be placed in the HOQ includes product benchmarking data and target values.

The HOQ contains many steps or "rooms," each of which can and should be customized by the development team to meet its needs. While various sequences for working on the steps each have advantages, the team must consciously choose a sequence, and plan its work accordingly.

The most strategic judgments the team must make are in the Planning Matrix. Here the team sets customer-satisfaction goals, which have the effect of combining the company's business objectives with the customer's priorities.

Another step of the HOQ, the Technical Response Priorities, contains high-level product or service requirements. The Relationships step entails mapping how the elements of the Technical Response Priorities relate to meeting customer needs, and hence to overall customer satisfaction.

The Technical Correlations step ("the roof") of the HOQ is used to analyze relationships among elements of the Technical Response Priorities.

The step below Relationships contains the priorities of the technical-response elements, along with a roadmap for competitive benchmarking and target setting.

We're now ready to drop down to another level of detail and discuss the steps of the House of Quality one by one. The first will be Customer Needs and Benefits.

5.3 DISCUSSION QUESTIONS

- Identify which information normally used for the HOQ was actually used in a recent product planning activity that you participated in.
- Did you use information beyond the data required for the HOQ?
- How could you customize the HOQ to include this extra information?

Customer Needs and Benefits Section

This chapter describes how QFD represents the Voice of the Customer (VOC). As mentioned in Chapter 1, there are several different groups from whom to collect voices for various aspects of product, process, and technology design; however, the voice collection and processing methods remain the same. We will treat them all as VOC for the purposes of this chapter. As a reminder to the reader, customer satisfaction is of paramount importance. To that end, only true customer needs may be satisfied, but not really voices. Hence the reader is reminded to fully analyze each customer focus segment VOC to discern the true customer needs that will be satisfied and the priority order, or hierarchy, of those needs. The ultimate metric is customer satisfaction, which fundamentally determines the longevity of every business venture.

In QFD, one of the hardest aspects is to properly relate the Voice of the Developer (VOD) to the VOC. In this chapter we will explore Substitute Quality Characteristics (SQCs). SQCs are those measures the developers manage during development of the products and services. SQCs in essence represent the customer by proxy from their relationships to the VOC developed by analyzing customer needs and completing the HOQ. In order to understand the VOC representation, we'll discuss how the VOC is collected and analyzed. The main steps are:

- Prepare for and conduct customer visits. Listen to the customer and capture the customer's unstructured words and images that are of use in the customer's environment.
- An image of a customer struggling with packages (see Figure 6-1) suggests a need, whether voiced or unvoiced. Describing an image like this, of use in the customer's environment, is what customer images are all about (see Figure 6-2).

Figure 6-1   Customer Need: Struggling with Packages[1]

Figure 6-2   An Image of Need that Can Be Tied to Voices[2]

1. Courtesy of iStockphoto.com, Glenda Powers.

2. Courtesy of iStockphoto.com.

- Sort out the different types of comments the customer has made into various categories, and relate the comments to relevant images for context.
- Take each of the categories, the true customer needs, and create a hierarchical structure of those needs that allows them to be worked with at various levels. Validate the hierarchy of needs with a followup survey or questionnaire.

This is a very important step in QFD, for the obvious reason that the VOC is one of the main inputs into the QFD process. If you believe that "garbage in" produces "garbage out," then you know how important it is to do this phase of QFD right. By the time we've finished this chapter, you'll have a good idea what the VOC should look like.

This chapter deals with VOC basics, which include a single customer segment or category and methods for distinguishing between customer needs and solutions. The chapter also deals with the structuring of customer needs using the Affinity Diagram. Chapter 17 covers more-advanced issues, including dealing with differing types of customer needs and various methods for interviewing customers, some of which are aimed at developing breakthrough product concepts.

6.1 GATHER THE VOICE OF THE CUSTOMER

The Customer Requirements section of the House of Quality (see Chapter 1, Figure 1-2, Item 1) contains a structured list of needs customers have for the product or service being planned. This section is usually derived from the VOC—literally, statements or fragments of statements made by customers or potential customers.

The usual steps in creating the Customer Requirements section are

1. Gather the Voice of the Customer
   - Prepare for and visit customers
   - Gather all customer voices, positive and negative
2. Sort the Voice of the Customer into major categories, including
   - Needs and Benefits
   - Substitute Quality Characteristics
   - Reliability Requirements
   - Other
3. Structure the needs in an Affinity Diagram
4. Arrange the needs in the Customer Requirements section

Before gathering the Voice of the Customer (VOC), the team must decide who the customer is. If, as is often the case, there are more than one category (or market segment) of customer, the team must decide on the relative importance of the various customer categories and treat them appropriately. All of this is covered in Chapter 16, Section 16.3. At this point, we'll simplify our discussion by dealing with the treatment of a single customer category.

6.1.1 PREPARE FOR AND VISIT CUSTOMERS

The Voice of the Customer may be gathered by a variety of methods, all of them aimed at asking the customer to describe his or her needs for a product or service of the type being planned. Some developers implement this step by conducting a survey in which, for the most part, customers are asked their opinions on a series of predetermined topics. This is a big mistake, because the survey designers have no basis for determining the topics into which to inquire. Many times, this flawed approach yields ambiguous statements without context that may be interpreted several ways. The wrong interpretation of needs can become a serious error upon moving forward into the design stage of product development.

A much better approach is to identify customer needs by visiting them with discussion guides developed around open-ended questions. Interviewing and analysis techniques are described in Chapter 17. Significant training is warranted for those on the interview teams so that they understand how to

- Conduct themselves properly during visits with customers
- Ask open-ended questions seeking facts and objective measures of need
- Probe for facts by digging deeper when only opinions or conclusions are offered
- Understand Kano's model when probing to help identify types of needs

The idea is to let customers speak for themselves as much as is practicable. The admonition "do no harm" is a guideline for anyone visiting with customers in the context of listening only, and not trying to solve any problems that may be discussed or observed. In addition, the interview team is encouraged to watch, if possible, the customers using existing and/or competing products and services that the new products or services are intended to displace. By watching and capturing descriptive images of unmet needs or problems faced, team members may add context to the voices, providing more meaning to the design team later.

In addition, a significant benefit in image descriptive capture is to help identify any *unstated needs*. These may be the kind described in the Kano model discussed in Chapter 2.

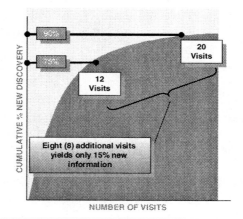

Figure 6-3   Number of Visits versus Percentage of New Information Obtained

Determining how many visits to make is not an easy task. Figure 6-3 illustrates some work by Griffin and Hauser[3] that shows that, *in general,* the point of diminishing returns for new-issue discovery is often reached after 12 visits. Depending on the complexity and number of needs, the exact number of visits appropriate to a given case is difficult to predetermine, as this is a discovery process.

The result of the interviews is a set of customer phrases and image descriptions representing the customers' wants and needs. Because customers as a rule don't structure their thoughts about product needs, the customer phrases will initially be a mixture of true needs, most-favorite and least-favorite product features, complaints, suggestions, and other types of comments. All of these comments have potential value during the development process, but only the customers' true needs are required for the HOQ. Section 6.2 will discuss the sorting of customer phrases.

6.1.2 GATHER CUSTOMER COMPLAINTS

Another source of customer needs, in addition to interviews, is customer complaints. Most organizations have special organizations and processes for handling complaints, since they represent a major nightmare to any company: the nightmare of customer dissatisfaction. From our discussion of the Kano model in Chapter 2, we recall that these are often related to Fitness to Standard expectations that have not been met. These must be addressed to maintain credibility in the eyes of the customer.

---

3. Abbie Griffin and John R. Hauser, "The Voice of the Customer," *Marketing Science* 12:1 (Winter 1993): 1.

Too often, companies regard complaint management as their quality-control mechanism. As we have learned from the Kano model, this strategy is not enough to make a company competitive. However, removing dissatisfiers from a product is certainly a necessary, if not a sufficient, step to competitiveness. Hence, it can be useful to include customer complaints in the complete VOC.

Typically, companies will maintain databases of customer complaints. These databases can be quite large, and their organization will not normally be convenient for merging into a customer needs Affinity Diagram. Most complaints are classified by the severity of the complaint or by the type of response required to deal with an unhappy customer.

The following is a suggested method for extracting useful VOC information from the complaints:

1. Randomly sample a manageable number of complaints from the database, say 200 to 400 complaints. Sample from *all* complaints. There is a tendency for customer-service departments to discount or ignore certain types of complaints—for example, misuses of the product, requests that are beyond the control of customer-service representatives, and requests for new features. However, these types of comments from customers are extremely important to the developers of new products.

2. Using the expertise of developers and customer-service personnel, translate the complaints into positive phrases or concepts that represent the underlying customer needs expressed by the complaints.

3. Winnow the resulting customer-needs phrases by removing duplicates. Maintain an indication with each phrase that it was derived from complaints.

4. Merge the resulting phrases with the phrases acquired by interviews.

5. Develop the Affinity Diagram of customer needs.

Consider restructuring the complaints database so that it can provide more readily available customer input in the future.

6.2 SORT THE VOICE OF THE CUSTOMER INTO MAJOR CATEGORIES

Because customers often ask for solutions without revealing the underlying need, and because customers' words are not constrained by any particular discipline, the phrases must be sorted before the customer needs can be structured. The standard QFD construct

I.D.	Customer demographic	Customer need	Use									
			What		When		Where		Why		How	
			I/E	Data	I/E	Data	I/E	Data	I/E	Data	I/E	Data

Figure 6-4 Voice of the Customer Table, Part 1[4]

for doing this sort is called the Voice of the Customer Table (VOCT).[5] The VOCT traditionally has two parts.

Part 1 of the VOCT (Figure 6-4) is aimed at capturing the context of customer needs, so that the widest possible range of customer needs is identified and can be understood at a glance. The VOCT is not a matrix; it is a columnar list of customer phrases. The columns are used to provide quick visual clues as to the nature of the data.

The **I.D.** column identifies the source of the customer phrase. For example, it could contain an interview number, page number, line number, or date of interview. Its purpose is to provide a link back to the source of the phrase in case further information about the phrase is required.

The Customer Demographic column stores information such as age, income bracket, occupation, and location of the person who provided the data.

The Customer Need column contains the want or need as it appears or will appear in the Affinity Diagram. Generally, the same need will be expressed in slightly different ways by different customers. The development team will settle on one wording of this need that best represents all the variants.

The Use section holds information that describes what customers do. Such information has implications for the design of the product or service. For example, if the product being designed is a flashlight, the designers will be interested in how customers use flashlights, in order to design a competitive product. Do customers use the flashlight to provide light while changing a car tire in the dark? Do customers use the flashlight as a reading light in bed? These uses probably require very different designs. The VOCT provides a way of listing the range of uses so the developers can make the right design decisions.

---

4. GOAL/QPC, used by permission.

5. This diagram appears in many GOAL documents, but in particular: Satoshi "Cha" Nakui, "Comprehensive QFD System," paper presented at the Third Symposium on QFD (Methuen, Massachusetts: GOAL/QPC, 1991).

The Use section is broken down into several subsections in order to provide a structure for understanding the context of use. A phrase such as

"I carry the flashlight in my car in case I need to change a flat tire at night"

could result in two or three data entries: "in my car" would go in the Where column, and "to change a flat tire at night" would go in the Why column. The phrase

"I only use the flashlight for reading"

could yield "beside the bed" in the Where column, and "for reading" in the Why column.

The headers What, When, Where, Why, and How cover the usual categories of general questions that help interviewers and data analyzers uncover as many aspects of a - situation as possible.

Beside each Data column is an I/E column (Internal/External) to indicate whether the data was generated by a developer or company employee (I) or came from a customer (E). The idea behind the I/E column is this: If a customer mentions a specific context for use of the product, the developers may be able to generalize from this specific context to a wider range of contexts. For example, "change a flat tire at night" might suggest:

Take rubbish out at night

Fix a leaky roof at night

Patch a broken window at night

Complete an urgent outdoor chore at night

These additional internally generated contexts provide a broader picture of what may be required of the product or service, and they make the requirements more vivid to the developers. However, they should be labeled (I) to indicate that the phrases did not come directly from customers.

Part 2 of the VOCT (Figure 6-5) sorts the data a different way.

In Part 2, the customer phrases are placed in one list or another depending on whether the phrase is a true customer need, a suggested or requested product function, or any of the other categories the development team may be interested in. Here are some example customer phrases that might be sorted into VOCT Part 2 categories. These phrases all relate to the use of icons in a word processor.

Customer need:

"I don't want to click the icon in order to find out what it does"

(converted into:)

Figure 6-5 Voice of the Customer Table, Part 2

"Know what an icon is going to do before I click on it"

Substitute Quality Characteristic:

Percentage of correct identifications of icon meaning by users in a panel test (more is better)

Function:

Icon, when clicked, causes operation to be executed

Icon, when cursor passes over it, causes explanatory message to be displayed

Reliability:

Whenever explanatory message is displayed, it is the correct message

Everything still works well no matter when or how often the user clicks the icon

Target Value:

96 percent or better correct identifications of icon function by test-panel participants

Not everyone uses the VOCT—certainly not both parts of it. Nevertheless, the VOCT is worth understanding, as it was developed to deal with some common problems of interpreting the Voice of the Customer. Everyone processing the VOC must do at least some of the analysis implied by Part 2 of the VOCT.

Some general guidelines for distinguishing between the main categories in VOCT Part 2 follow.

6.2.1 CUSTOMER NEEDS

A Customer Need is a statement, in the customer's words, of a benefit that a customer gets, or could get, or might get, from a product or service. Needs should be issues that are fairly stable and do not often change. For example: "I need a reliable means to get to and from work." Ideally, the need or benefit should be stated in a way that is independent of the product or service being proposed to meet the need. The word **benefit** is often used in conjunction with **need** to characterize the type of information appropriate for a customer need. **Benefit** shifts the focus toward the customer and away from the Technical Response.

This seemingly simple guideline is much more difficult to apply than you may think. Consider customer needs for an automobile, for instance. Some needs, such as

Safe transportation
Comfortable ride

are reasonably independent of the concept of a vehicle with four wheels, a motor, seats inside a metal body, and the other general characteristics that define an automobile. However, other needs seem to have the automobile as we know it built into the need:

Steering wheel that is easy to control
Responsive to my foot pressure on the accelerator pedal

The interviewer can attempt to uncover more general needs by probing:

"If it was responsive to foot pressure on the accelerator pedal, how would you benefit from that?"

Sometimes the underlying benefit is too general to be useful in making product decisions. Unfortunately, we have no hard-and-fast rules that can guide us as to how general or how specific Customer Needs should be. Suffice it to say they should be general enough to allow for multiple solutions, and specific enough to be clearly applicable to the problem at hand.

6.2.2 SUBSTITUTE QUALITY CHARACTERISTICS

Substitute Quality Characteristics (SQCs) represent an abstract description of the product or service in a company's or developer's technical language. These are only as useful as the VOC collected in proper context. A poorly-defined VOC, like "**I want a car that goes faster**," can lead a developer to SQCs that are misleading, like "**Top Speed**." The customer may have needed faster acceleration from a stop, or greater acceleration when passing at highway speeds, or even faster capability around hairpin turns. Top Speed would be the wrong SQC for any of these three customer needs. We'll be looking at various types of SQCs and various methods for generating them in Chapter 8. Just as customer needs should be completely defined by the customer, SQCs should be aspects of the product or service that are under the control of the development team.

An SQC can be **solution-independent,** or it can define the solution by listing its elements. The best product and service developers know that solution-independent descriptions up front provide the best chance for creative solutions later on. Obviously, SQCs that define the solution up front provide no chance for creative solutions.

Most SQCs are generated by developers in response to customer needs. However, customers sometimes suggest SQCs when interviewed for their wants and needs, and we can use Part 2 of the VOCT to sort these SQCs into the appropriate columns. In fact, SQCs can come from anywhere; the alert developer will always be receptive to suggested SQCs from whatever source.

The VOCT Part 2 column for SQCs could be thought of as a resource into which SQCs can be stored, regardless of their source, and from which SQCs can be drawn later on in the QFD process.

In any case, we cannot control what types of SQCs customers may articulate; we must be watchful for any type. In Chapter 8, we'll discuss how the team evaluates the importance of the SQCs.

6.2.3 FUNCTIONS

Functions are descriptions of ways in which the product or service operates. A Function contains an action verb and a noun, and typically adds value by transforming information or energy or moving mass from one area to another. Examples of functions of an automobile are

Open window
Speed up
Slow down

Basic functions of a word processor would be

Edit text
Save document to disk
Check spelling of text

Functions are often made up of subfunctions, which would also be classified in Part 2 of the VOCT as functions. Examples are

Open window

Stop window from opening too far

Hold window in place when open

Move cursor

Select text

Deselect text

Insert characters

Delete characters

When we are sorting customers' phrases, we will often encounter mentions of functions and sub-functions. Later on in QFD, we can deal with the varying levels of these descriptions by using the Affinity Diagram method. This is described further in Chapter 8.

6.2.4 RELIABILITY REQUIREMENTS

Reliability relates to a sense of confidence that customers have, or would like to have, in the service or product in their operating environments over time. David Garvin, in his study on air conditioners,[6] found that the use of reliability engineering correlated to customer perceived quality. Long-term customer satisfaction is usually keyed with highly reliable products. Lack of reliability can show up in negative phrases:

Warns me if the software is going to bomb

I don't have to take it (automobile) for servicing very often

It can also show up in positive phrases:

Keeps working no matter what happens

My data is preserved no matter what happens (word processor)

In the Kano model, reliability needs are classified as Dissatisfiers. Customers are very dissatisfied when products or services cannot be relied upon, but are not highly satisfied just because they can be relied upon.

---

6. David Garvin, "Competing on the Eight Dimensions of Quality," *Harvard Business Review* (Nov. 1, 1987).

These statements give the developers valuable clues as to how customers view reliability, and therefore how to plan and test for the reliability of the product or service. Any developer knows that there is an infinite number of ways a product or service can fail, and it is impossible to design or test for them all. The Voice of the Customer can help us to avoid those failure modes that matter most to the customer.

6.2.5 Target Values

Target values are indications as to how much of some technical characteristic a customer wants. Any time customers mention numbers in the articulation of their needs, there's a good chance they are providing target values. Care should be taken by the interviewer to explore these further to understand if these are ideal, minimum, or maximum expected values. An additional reason why it is important to get the units of measurement correct is that human nature sometimes leads to misstatements. Quite often, people will substitute minutes for hours or yards for feet, or even mix up metrics and English units. Examples of target values are:

Have as many as ten documents open at once

0 to 60 mph in ten seconds

Won't wait for telephone service more than one minute

Target values supplied by customers must be considered cautiously. There are many technical issues surrounding target values of which customers may not be aware. The most important issues are methods of measurement, appropriateness of metric, and appropriateness of value.

Methods of Measurement

When a customer requests

Have as many as ten documents open at once

for a word processor, the developer must determine how to count the documents. Users may or may not be aware that even when one of *their* documents is open, the word processing-software may have other documents open, such as temporary files which may be needed to undo user actions (restore the document to its state prior to the last several user actions), a file with user formatting information in it, and a dictionary file.

When a customer requests

0 to 60 mph in ten seconds

Figure 6-6 Image of Acceleration Need Context[7]

for an automobile, the developer must understand what the customer's assumptions are. What type of gasoline is the car using? How many passengers is the car carrying? What is the condition of the tires? If it is a standard-transmission automobile, how is the gear shifting to be performed? In this area, it is especially useful to understand the *context* of the targets; image descriptions can be quite helpful when attached to a voiced target image, as in Figure 6-6.

When a customer specifies

Won't wait for telephone service more than one minute

for a hotline call, is the customer including dial-up time and connect time in the computation of one minute? Does the minute start when the initial menu of choices starts, or after the customer has made a choice?

These considerations are not likely to be articulated by customers, yet customers may have very definite assumptions about how to count open documents, how to drive a car, and how to measure one minute. At the same time, developers are likely to be accustomed to standard, internally defined methods of measuring, which could carry with them very different assumptions.

In short, customer target values should be probed during the interview in order to arrive at the underlying customer need. They should not be taken too literally at any time. We'll discuss techniques for interviewing customers in Chapter 17.

---

7. Courtesy of iStockphoto.com, Michael Pettigrew.

Appropriateness of Metric

Let's consider again the target value statement

> Have as many as ten documents open at once

Here the customer has an underlying need, which we can only guess at without further probing. If the developers knew what the customer's underlying need for "ten documents open at once" was, they might be able to meet that need by a solution that did not relate to numbers of open documents at all. For example, suppose the underlying need was

> Able to view more than one document at a time

The customer's solution to meeting this need might be

> Have as many as ten documents open at once

because that is the only method the customer knows for meeting the need. However, other solutions might be available to the developer, such as storing key information (the first line of every paragraph, for instance) from many files in a single compressed file that the user can peruse. Perusing a single compressed file may have disadvantages for some users, but the balance of advantages and disadvantages of this solution might be just right for other users. Determination of whether this solution really meets the user need would depend on deeper analysis of the need. Perhaps some type of contextual analysis (see Chapter 17) would be required.

In this case, the metric, "number of documents open at once," probably is not the right measurement: It presumes one solution, precludes other solutions, and obscures the real need.

While the developer must take the customer's proposed metrics seriously, the proposals must be probed to uncover the real needs. Without knowing these needs, the metrics could lead the developer on a fruitless journey.

Appropriateness of Value

Given the likelihood that the customer may have proposed an inappropriate metric, we should not be surprised if the proposed target value for that metric requires careful scrutiny also.

In general, metrics fall into three categories: The Larger the Better (LTB), The Smaller the Better (STB), and Nominal the Best (NB).[8]

---

8. An excellent discussion of these terms can be found in Chapter 2 of Madhav S. Phadke, *Quality Engineering Using Robust Design* (Englewood Cliffs, N.J.: Prentice Hall, 1989).

LTB refers to metrics for which the worst value is zero, and the best value is arbitrarily large. Examples of LTB metrics are mean time between failure, strength of permanent bond adhesives, shelf life of consumer products, and support strength of architectural columns.

STB refers to metrics for which the best value is zero, and the worst value is arbitrarily large. Examples of STB metrics are audiotape noise (hiss) level, heat loss through an insulating layer, and time to resolve a customer problem.

NB refers to metrics for which the best value (called the **nominal**) is a specific value determined by the situation, and the worst value is arbitrarily greater or arbitrarily smaller than the nominal. Examples of NB metrics are clothing size, voltage output of a power supply, and temperature of food. Care must be taken with NB value setting relative to the demographic variations of customer segments. What is nominal for one group may not be the target for another group.

When the customer says

Won't wait for telephone service more than one minute

the developer must ask which of the metrics categories is being proposed: LTB, STB, or NB. This is helpful in two situations.

First, there is the possibility of exceeding the customer's expectations. The customer's limit of one minute is likely to be what the customer will tolerate, not what the customer really wants. Rather than accept the target of one minute, the developer has the opportunity of exceeding the customer's expectations by generalizing the target value to a direction and providing a product or service that does even better than the customer asked.

Second, the customer-defined target may be unrealistic. In this case, the developer may be able to produce a competitive product by approaching the customer's target and exceeding the competition's performance on this metric, but without actually meeting the customer's target.

The terms LTB, STB, and NB are like most other QFD terms: They have not been standardized. Some confusion can arise if the development team is not careful. Various alternative terms are displayed in Figure 6-7.

LTB (Larger the Better)	LTB (Less the Better)	TB (Target Best)
MTB (More the Better)	STB (Smaller the Better)	NB (Nominal Best)
MB (Maximum Best)	MB (Minimum Best)	
GTB (Greater the Better)		

Figure 6-7 Variations on LTB, STB, and NB

Notice the duplication of abbreviations LTB and MB in the figure. It pays to keep the terminology consistent.

6.3 STRUCTURE THE NEEDS

After collecting customer needs from many sources—and even after sorting out Substitute Quality Characteristics and many other items that are not truly needs—there will still be a large, unmanageable list to deal with. In the QFD process, the needs are arranged into an Affinity Diagram, which is then completed and refined by using the Tree Diagram process.

Most commonly, the Tree Diagram is three levels deep. If it has more levels, the lower levels are used as definitions and clarifications of the higher levels. For example, in Figure 6-8, the Tree Diagram is four levels deep in some places (Intuitive Controls is at the tertiary level, and its subordinates, beginning with "Know what an icon is going to do before I click on it," are at the quaternary level).

If the tertiary level has been chosen for analysis, then the quaternary level, where it exists, can be used to define more fully the associated tertiary terms. The development team chooses one level for analysis, placing that level against the left edge of the Relationships section of the HOQ (see Figure 6-9).

Notice that by arranging the customer needs in a hierarchical tree structure, we preserve all detail. The hierarchy also allows us to manage the information by choosing to work at a particular level.

The Program Is a Pleasure to Use
Commands Are Easy to Know and Use
Intuitive Controls
    Know what an icon is going to do before I click on it
    Clear relationship between menu commands and icons
    Don't have to read the manual to figure out how to use the program
    Don't have to go into Help to understand how to do what I want to do
Controls under my fingertips
    Can execute commands quickly
    No complicated keystrokes to memorize in order to do simple operations
    Able to execute common operations in a single step
Can customize to suit my working style
Can customize the icon display so that it's easy for me to use
Easy to get the information I need
    Program informs me about all its capabilities and features
    The manual is easy to understand and use
    "Help" function tells me how to do things, not just what things are
Program Is Quick and Responsive
    Can adjust the cursor to move as quickly as I'd like
    Enables me to find things in the document quickly

Figure 6-8 Customer Needs and Benefits

Easy Font Management
   Offers lots of size, font, and design options
   Able to see what the fonts look like as I'm choosing them
   Can organize the listing of fonts to reflect the way I use them
   Everything stays neat and aligned when I change fonts
No Surprises
What I See Is What I Get
   Know what the document will look like when I print it
   Able to see the whole page at once
   Can see what I type as I type it
   Can see all the pages in my document together, side by side
   Able to see subtle spacings easily
Can Control the Shape of My Document
Can Work With Many Page Styles
   The page styles I need at my fingertips
   Can create my own document templates
   Simple to save settings as a default
   Offers me a variety of document types, e.g., letter, invoice, brochure
   Handles all my document organizational needs
   Can organize my text into tables and charts
   Easy to set up, change, or eliminate headers and footers
   Easy handling of material in multiple columns and rows
   Can use different paper sizes and orientations
   Easy envelope addressing
Can Work With Text and Graphics
   Able to mix text and graphics
   Able to create charts and pictures in my document
   Allows me to easily create presentation style material
   Can add pictures/symbols to my document
Can Create and Manage Document Structure
   Easy to create footnotes
   Able to organize my document as an outline
   Easy to create a table of contents
   Easy to create an index for my document
Can Modify My Document Any Way I Want To
Can Change the Document Around Any Way I Want To
   Easy to move things where I want them in the document
   Able to make global reformatting changes effortlessly
Can Alter the Appearance of My Material
   Can alter the look of part of a word or a sentence without affecting the rest
   Able to change text to upper or lower case without retyping
   Can format my material as I type it
Can Create Error-Free Documents
Protects Me Against My Mistakes
   Able to easily undo any changes I make
   The program can proofread my text
   Warns me if I'm about to do something wrong, like delete a file
   Can check the spelling of words
   Makes it easy to edit and correct my work
   Won't lose my original text when I type into a section I have highlighted
Protects Me Against Its Mistakes
   Warns me if the software is going to bomb
   Easy to save my work
Can Get My Ideas On Paper Easily
Can Mix Material From Many Documents
Can find material in other documents easily
   Able to view more than one document at a time
   Enables me to know what is in a document before I open it
   Can find/view information produced by other programs without exiting

Figure 6-8   Continued

Can use material in other documents easily
    Can combine parts of different documents to form a new one
    Able to move text from one application or document to another
    When I bring in documents created in another program, they retain their original appearance
    Able to convert document into other types of files (e.g., ASCII)
    Easy to retrieve and reuse work I created previously
Enhances My Creativity
    Helps me quickly capture and save my ideas (e.g., when brainstorming)
    Enables me to organize and reorganize my lists

Figure 6-8   Continued

The Program Is A Pleasure To Use	Commands are Easy to Know and Use	Intuitive Controls	
		Controls under my fingertips	
		Can customize to suit my working style	
		Easy to get the information I need	
	Program Is Quick and Responsive	Can adjust the cursor to move as quickly as Id like	
		Enables me to find things in the document quickly	
	Easy Font Management	Offers lots of size, font, and design options	
		Able to see what the fonts look like as I'm choosing them	
No Surprises	What I See Is What I Get	Know what the document will look like when I print it	
		Able to see the whole page at once	
		Can see what I type as I type it	
		Can see all the pages in my document together, side by side	
		Able to see subtle spacings easily	
⋮	⋮	••••	
		••••	
		••••	

Figure 6-9   Wants and Needs in the House of Quality

6.4 SUMMARY

We've covered a lot of material in this chapter.

The Customer Needs and Benefits section of the House of Quality is the starting place for all QFD activities. The most common source of customer phrases representing their wants and needs is the customer interview, preferably at their point of use to obtain context and image descriptors of needs. Examination of customer complaint data is also useful. Customers' language is not, however, as structured as is desirable for QFD. The developer must be aware that customers will provide data of many types, which the developer must sort, classify, and structure in order to make useful the needs identified from customer visits and interviews.

The most common categories used for classifying the Voice of the Customer are Customer Needs, Substitute Quality Characteristics, Functions, Reliability, and Target Values. To make best use of the VOC, the developers must record the customer phrases and sort them into categories such as these.

Customer Needs are the most important category for the early stages of QFD. Because there are usually so many needs, we use the Affinity Diagram method and the Tree Diagram method to arrange them in a tree-structured, hierarchical format. The most important parts of this structure are usually its primary, secondary, and tertiary levels. The structure allows the developers to "zoom in" or focus on the customers' needs at different levels of detail, depending on the appropriate level of analysis. Once the needs are structured, one level—usually the tertiary level—is used for subsequent QFD analysis, and is placed on the left of the House of Quality.

Having collected, sorted, and structured the VOC, development teams are justified in celebrating! This is an immense, mind-opening task. The number of "aha"s experienced by most teams by the end of this stage is generally dramatic. If "50 percent of problem solution lies in the statement of the problem," then the development team has gotten halfway to a world-class product or service by reaching this point.

The VOC we've seen in this chapter is *qualitative data.* In other words, it describes *what* the customer wants. In QFD, we'd also like to have *quantitative* descriptions of the VOC. These tell us *how important* each of the needs is to the customer, and also *how well* we and the competition are doing in meeting those needs. This quantitative data, along with how to collect it and how to use it to set product strategy, will be described in the next chapter.

6.5 DISCUSSION QUESTIONS

- What data do you have that represents the Voice of your Customer? How was it acquired? How has it been structured?
- Can you represent it in a tree-structured format?
- Which customer phrases fit into the categories of Customer Need, Function, and Reliability?
- Are there any target values? How do your customers measure these values?
- What customer needs are the metrics associated with?

The Product Planning Matrix

With Joe Kasabula, Randy Perry

This chapter shows how QFD helps a team to conduct strategic planning for its product. Just as the Customer Needs and Benefits section is a repository of *qualitative* customer data, the Planning Matrix is the repository for important *quantitative* data about each customer need. The development team will use this data to decide which aspects of the planned product or service will be emphasized during the development project.

The customer-data and Planning Matrix portions of the QFD are developed through VOC efforts. These may be based on customer interviews, KJ analysis (see Chapter 4, Section 4.2.5), and follow-up customer surveys, for example.[1] After completion of customer interviews, the development team will identify the most important customer requirements using KJ analysis. The KJ analysis results will be used to populate the left-hand side of the QFD as shown in Figure 7-1 below. Based on the KJ analysis results, the team will develop a follow-up customer validation survey that will provide additional quantitative ranking information and develop the Planning Matrix portion of the QFD as described in this chapter. A sample Planning Matrix linked to the KJ Diagram in Figure 7-1 is shown in Figure 7-2.

This is a long chapter, because the standard, "textbook" Planning Matrix consists of seven very different types of data, each of which must be described separately. The actual number and nature of these seven columns of data is the subject of considerable customization and variation from one QFD to the next. We try to indicate what those

---

1. Randy C. Perry and David W. Bacon, *Commercializing Great Products with Design for Six Sigma* (Upper Saddle River, N.J.: Prentice-Hall, 2006).

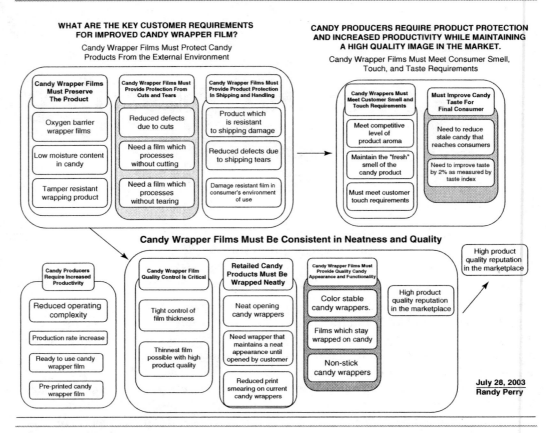

WHAT ARE THE KEY CUSTOMER REQUIREMENTS FOR IMPROVED CANDY WRAPPER FILM?

Candy Wrapper Films Must Protect Candy Products From the External Environment

CANDY PRODUCERS REQUIRE PRODUCT PROTECTION AND INCREASED PRODUCTIVITY WHILE MAINTAINING A HIGH QUALITY IMAGE IN THE MARKET.

Candy Wrapper Films Must Meet Consumer Smell, Touch, and Taste Requirements

Candy Wrapper Films Must Be Consistent in Neatness and Quality

July 28, 2003
Randy Perry

Figure 7-1   Example Requirements KJ

variations are. We generally indicate the most common usage for each Planning Matrix column first, with variations afterwards.

The Planning Matrix is the tool that helps the development team to prioritize customer needs. Figure 1-2 Section #3 and Figure 7-3 are schematic representations of the Planning Matrix. Figure 7-26 is a worked-out numerical example.

The Planning Matrix provides a systematic method for the development team to

- Compare the performance of the current product or service in meeting customers' needs to the competition's performance
- Develop a strategy for customer satisfaction that optimizes the organization's ability to both sell the product (short-term customer satisfaction) and keep the customer satisfied (long-term satisfaction)

	Importance Rating	Function		
Maximize, minimize, or target				
Customer Needs				
Meet competitive level of product aroma	8			
Maintain the "fresh" smell of the candy product	8			
Must meet customer touch requirements	8			
Need to reduce stale candy that reaches consumers	4			
Need to improve taste by 2% as measured by taste index	4			

Figure 7-2　Planning Matrix Example Tied to Requirements KJ Diagram

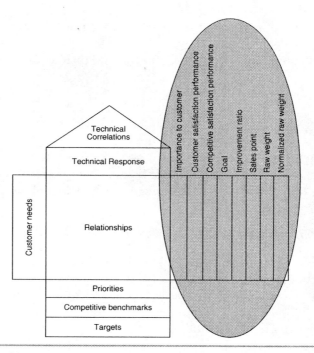

Figure 7-3　Planning Matrix

The goal setting that the development team does in the Planning Matrix will set the tone for the rest of the project.

The Planning Matrix contains a series of columns that represent key strategic product planning information and questions. Sometimes these questions have been called "embarrassing" questions because many organizations either don't know the answers or cannot agree on them. Because QFD systematically forces team members to ask all the questions, their level of knowledge about their own product and the competition can become very obvious. The data placed in the Planning Matrix allows the development team to make strategic decisions about the product or service it is planning. Therefore, the Planning Matrix is extremely important.

The Planning Matrix asks the following key questions for each customer need:

- How important is this need to the customer?
- How well are we doing in meeting this need today?
- How well is the competition doing in meeting this need today?
- How well do we want to do in meeting this need with the product or service being developed?
- If we meet this need well, can we use that fact to help sell the product?

Let's take a look at the Planning Matrix columns and their associated questions one by one.

7.1 IMPORTANCE TO THE CUSTOMER

The Importance to Customer column (Figure 7-4) is the place to record how important each need or benefit is to the customer. Three types of data are commonly used in this column: **Absolute Importance, Relative Importance,** and **Ordinal Importance.**

7.1.1 ABSOLUTE IMPORTANCE

The Absolute Importance entries are usually chosen from a scaled selection of importance. The number of points on such a scale has been known to range from three to ten. We'll base our examples on a five-point scale, where the values 1 to 5 may be defined as

1. Not at all important to the customer
2. Of minor importance to the customer
3. Of moderate importance to the customer

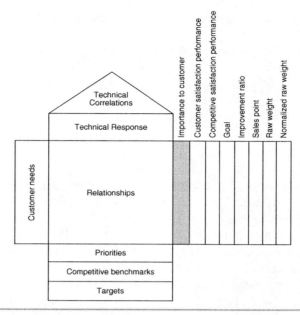

Figure 7-4   Planning Matrix: Importance to the Customer

4. Very important to the customer

5. Of highest importance to the customer

A five-point scale originated by Dr. Rensis Likert,[2] commonly used in surveys today, is minimally sufficient for our purposes. This type of scaling is subject to **central tendency bias**, as some respondents may avoid the extreme responses (1 and 5). Knowing this issue makes data from these scales fairly useful. Odd-numbered scales are useful in determining difference from a central value statement in a survey—for example, in measuring agreement or disagreement along with the degree of agreement/disagreement. Appropriately, the central value statement is the one anchored at the 3 value on the five-point Likert Scale.

The Hedonic Index is a seven-point scale that has found wide application as well. It is currently used in areas ranging from taste-testing to quality surveys involving the Consumer Price Index. A seven-point scale has the advantage of finer gradation. This must be weighed against the dangers of allowing user bias (upward or downward) to

2. Rensis Likert, "A Technique for the Measurement of Attitudes," *Archives of Psychology* 140 (1932): 1.

creep in if there is no central statement anchored at the center value of 4 in a Hedonic Index survey .

Absolute Importance values are usually obtained by a survey in which respondents are asked to rate the importance of each need on a scale provided by the interviewer (or described in the survey form). Surveys of this type are often designed and implemented by the development team, without the help of professional market research firms. In some very-low-budget QFDs, the product-development team members will estimate these ratings themselves, or with the help of a few customers—a risky undertaking, but not unheard of.

Even assuming that accurate and representative data in an absolute scale is available, there is still a problem using Absolute Importance: Customers tend to rate almost *everything* as being important (as in Figure 7-5). While everything *is* important to the customer, the development team is still forced to make trade-offs, because of constrained resources. In addition, customers can select products that do not meet their needs equally well and still be satisfied overall with them. If customers can clearly differentiate the importance of different needs, the QFD process can help the developer translate those differences into prioritized technical responses. If the absolute-weighting data tends to be bunched near the highest possible scores, it does not contribute strongly to helping developers prioritize.

A better method for measuring Importance to the Customer uses Relative Importance, which unfortunately requires skills not usually found in development teams.

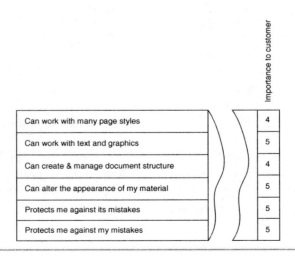

Figure 7-5 Absolute Importance

7.1.2 RELATIVE IMPORTANCE

The Relative Importance entries reflect the significance of various needs in comparison to each other. If one need is twice as important as another to the customer, then the importance score of the more-important need would be twice the score of the less-important need. It is critical that customers grouped together are from the same customer segment. Mixing different-segment customers together will lessen the value of Relative Importance ratings.

Relative Importance values are typically placed on either a 10 point, or a 100-point scale or on a percentage scale. The highest number indicates the highest possible importance to the customer. Not every customer in a survey will assign the same weight to each customer need, so it is unlikely that any customer need will ever be scored at the maximum value, typical ranges of Relative Importance scores are from about 40 to 85 on the 100 point scale.

Figure 7-6 shows typical Relative Importance scores

Relative Importance (sometimes called **ratio-scale importance**) is measured by asking customers to compare the attributes to each other and indicate importance. There are many methods.[3] One widely used technique presents the customer attributes to respondents in pairs, and asks the respondent to indicate how much more important one member of each pair is compared to the other. Only one pair of choices is presented at a time, and the respondent selects a set of percentages of preference (0/100, 10/90, 20/80, 30/70 and so on, up to 100/0) for each pair of choices. This method is called "constant sum paired comparisons." Each attribute is compared to every other attribute in such pair-wise comparisons. Information reflecting these choices is stored in cells in a matrix . The matrix is processed in a manner similar to the Analytical Hierarchy Process method (see Chapter 16, Section 16.3.3). The analysis results in weights for each customer attribute that indicate the relative importance of the needs.

Any such process of pair-wise comparison carries with it the risk of inconsistent judgments. That is, there is nothing that would stop a respondent from stating that

A is more important than B

B is more important than C

C is more important than A

This type of circular reasoning cannot be easily avoided in surveys, although it can be tested for after the fact. The possibility of this outcome of inconsistent judgment is

---

3. For some examples, see Glen Urban and John Hauser, *Design and Marketing of New Products,* 2nd ed. (Englewood Cliffs, N.J.: Prentice-Hall, 1993).

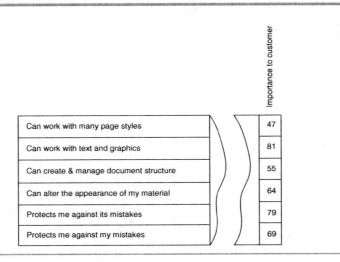

	Importance to customer
Can work with many page styles	47
Can work with text and graphics	81
Can create & manage document structure	55
Can alter the appearance of my material	64
Protects me against its mistakes	79
Protects me against my mistakes	69

Figure 7-6   Relative Importance

in fact a key reason why customer interviews are a preferred method for gathering customer-importance data. Nevertheless, if a Relative Importance survey process can be constructed in a way that guarantees that inconsistent judgments won't be made, there is no reason why the resulting weights would not reasonably represent the way people feel about the relative importance of their choices. Again, this is work best done by professionals.

Another widely used technique presents the customer with a complete list of possibilities and asks the respondent to arrange these in ascending or descending order of importance. Optionally, the respondent may be asked to assign numerical values representing degree of importance in the sorted list. (The Vocalyst process, described in Section 17.5, uses this technique.) This process has the advantage over pair-wise comparisons of assuring consistency. A disadvantage of this process is that certain methods of collecting data are impractical. For example, if a telephone survey is conducted, the respondent will probably not be able to visualize more than about seven attributes. Again, on-site customer interviews to gather this information are preferred.

Every method for measuring and computing importance is nothing more than a mathematical model of how numbers of people feel. I find Relative Importance the most useful measure of importance for QFD.

When considering multiple markets, each with diverse market segments and many different customers, Relative Importance models can be complex enough to justify the use of professional market researchers. A variety of statistical and data-acquisition decisions must be made during the design of this type of research, and it should not be attempted by amateurs. For many combined design and marketing teams, on the other hand, these efforts can be highly focused on key markets and targeted segments, and will not necessarily require professional market researchers. The business risks and complexity of the model needed should drive the decision as to whether to hire market-research professionals or not.

7.1.3 ORDINAL IMPORTANCE

Ordinal Importance, like Relative Importance, is an indication of order of importance. Unlike Relative Importance, which indicates *how much* more or less important one attribute is compared to another attribute, Ordinal Importance indicates only *that* one attribute is more or less important than another. Typical Ordinal Importance scores might look like those in Figure 7-7.

The highest number in the Ordinal Importance column ("Importance to customer") indicates the attribute that is most important to the customer. Most people use Ordinal Importance numbers just the opposite way: the number "1" indicates most important, and higher numbers indicate lower importance. However, QFD arithmetic always treats higher numbers as more important.

Typical methods for measuring Ordinal Importance involve surveying customers and asking them to rank-order the customer attributes, or to assign importance numbers to the attributes as with Absolute Importance surveys. Various forms of averaging of the responses can be used to arrive at Ordinal Importance. For example, if customers are asked to rank-order importance:

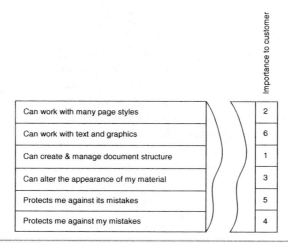

Figure 7-7   Ordinal Importance

- Assign the number 1 to the lowest-ranked customer need on each survey response. Assign ascending numbers to each higher-ranked customer need on each response. The highest number assigned (to the customer need judged most important in a survey) would be equal to the total number of customer needs ranked by the customer.
- Add the assigned numbers for each attribute in each survey response.
- The customer need with the highest score will be most important. The customer need with the next-highest score will be next most important, and so on.
- The scores *are not* proportional or weighted importance numbers, since the customers were not asked to provide any information about Relative Importance.

The resulting Ordinal Importance numbers are reasonable estimates of the way the customers surveyed felt. Even though the development team cannot be very confident of *how much* more important one customer need is than another, they can be confident that items at the top of the list are more important overall than items at the bottom of the list. The validation survey mentioned at the beginning of this chapter is one way many Design for Six Sigma teams resolve this uncertainty.

To compute Raw Weights (Section 7.6), the importance value will be multiplied by other values in the Planning Matrix. There are two issues to be aware of in this regard:

- It is not strictly valid to multiply an Ordinal Importance value by the proportional values used elsewhere in the Planning Matrix. Nevertheless, this practice is fairly common in QFD, and succeeds in giving the development team a fair indication of what the priorities are in their project.
- The range of ordinal numbers used as multiplier values is extremely wide compared to the ranges we see for Absolute Importance and Relative Importance values. For example, with 15 customer needs, the ratio of largest to smallest ordinal numbers will be 15 to 1. The ratio of largest to smallest Absolute Importance numbers is theoretically 5 to 1, but in practice it is usually about 5 to 3, or 1.6 to 1. The ratio of largest to smallest Relative Importance numbers is theoretically 100 to 1, but in practice it is usually about 85 to 40, or 2.1 to 1.

Thus, the Ordinal Importance scale, when multiplied by other values in the Planning Matrix, tends to make the highest Raw Weights much larger than the lowest Raw Weights, emphasizing the difference between the most-important customer needs and the least-important ones.

7.2 CUSTOMER SATISFACTION PERFORMANCE

Customer Satisfaction Performance (Figure 7-8) is the customer's perception of how well the *current* product or service is meeting the customer's needs. By "current product," we mean the product or service currently being offered or delivered that most closely resembles the product or service we plan to develop. The information is placed in the column titled "Customer Satisfaction Performance."

The usual method for estimating this value is by asking the customer, via survey, how well he or she feels the company's product or service has met each customer need. This satisfaction level is usually expressed as a **grade** or **performance level.** Grades are usually given on a four-, five-, or six-point scale, although sometimes scales up to ten points are used. Often, customers are asked to supply letter grades (A through F) because of their similarity to grades given in U.S. schools. The grades would be converted to numbers, where the highest performance would be translated into the highest number. Typical survey questions for measuring satisfaction levels are shown in Figure 7-9.

The respondent is expected to answer each question by checking or circling one of the responses. Allowance is often made for respondents who don't know the answer, or for whom the question does not apply.

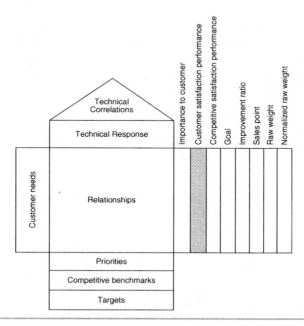

Figure 7-8   Planning Matrix: Customer Satisfaction Performance

	Very poorly	Poorly	Neutral	Well	Very well	Does not apply
How well has the service met your need for courteous telephone operators?						
How well has the word processor met your need for control over appearance of your document on the computer screen?						
How well has the product met your need for minimizing the time it takes to recharge the battery?						

Figure 7-9   Satisfaction Survey

One method of using data from surveys like this in QFD is to assign numerical values to the possible responses and then compute weighted averages.

Figure 7-10 illustrates a filled-out example.

The weighted average in this case is Weighted Average Performance:

$$\text{WAP} = \{\Sigma_i[(\text{Count of respondents at performance value } i) \times i]\} \div (\text{total number of respondents})$$

The Weighted Average Performance score for a particular question could be the value we use in the House of Quality (HOQ) Planning Matrix. If almost all customers were to answer a question with the same value (as in Figure 7-11), then the reaction of the customer base would be homogeneous with respect to this customer need. However, there is the possibility that the answers may not cluster around a single value. Take a look at Figures 7-11, 7-12, and 7-13. For all three cases, the Weighted Average Performance is

Customer attribute: "Offers lots of size, font, and design options"	Performance/grade	Number of respondents	Performance weight [(Number of respondents) • (Performance)]
Very poorly	1	50	50
Poorly	2	157	314
Neutral	3	626	1878
Well	4	180	720
Very well	5	40	200
Totals		1053	3162
Weighted Average Performance Score (3162 ÷ 1053)			3.0028

Figure 7-10   Weighted Average Performance

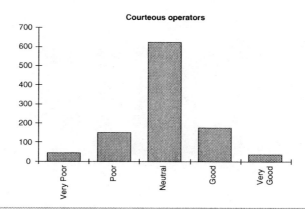

Figure 7-11 Homogeneous Customer Satisfaction Performance (By Far, Most Customers Responded with "Neutral")

very close to 3, but in Figures 7-12 and 7-13, a large percentage of the customers' performance ratings are well above and below 3. In other words, the Weighted Average Performance value is not representative of many of the customers and should be used with caution.

Charts like Figures 7-12 and 7-13 may indicate a segmentation of the customer base. By segmentation, we mean that the needs of, or selling opportunities to, a substantial proportion of the customers are different from those of the other customers. They are not being satisfied completely by the existing product. If the developers want to satisfy most of the customers, they may have to develop a technical solution different from the current one.

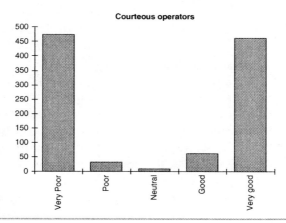

Figure 7-12 Segmentation of Customer Satisfaction Performance (As Many Customers Responded with "Very Poor" As Responded with "Very Good")

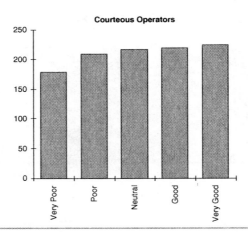

Courteous Operators

Figure 7-13   Segmentation of Customer Satisfaction Performance (About the Same Number of Customers Responded with Each of the Five Possible Responses)

The main point here is that the distribution of Customer Satisfaction Performance responses to a survey question must be understood before the QFD team blindly represents the performance level of all the customers by a single number, such as the Weighted Average Performance.

7.3 COMPETITIVE SATISFACTION PERFORMANCE

In order to be competitive, the development team must understand the competition. This may sound simplistic, but many development teams do not study their competition very carefully. Because it is usually much harder to reach the competition's customers than their own customers, development teams often operate in the dark with regard to their competition's strengths and weaknesses.

Development teams sometimes try to rely on trade journals' evaluation reports for comparisons of their products or services with the competing alternatives. Since the trade journals' criteria for comparison are unlikely to match the customer attributes that a development team has created from its own customer interviews, the trade journals' evaluations are very difficult to use in these side-by-side comparisons.

QFD provides a method by which the development team can record the competition's strengths and weaknesses alongside its own. The comparison can be shown at two important levels: first, in terms of customer needs, and second, in terms of technical response (Substitute Quality Characteristics). In the Planning Matrix, the development team has the opportunity to compare, side-by-side, how well its current product and the competition's are meeting customer needs (Figure 7-14).

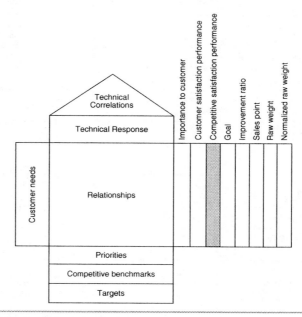

Figure 7-14 Planning Matrix: Competition's Rating

Any survey that asks customers how satisfied they are with your product can also be sent to your competition's customers. Most companies have ready access to their customers (or can get such access through their distributors). Access to the competition's customers may require more resourcefulness, such as making use of commercially available mailing lists, or surveying people at street corners, shopping malls, or trade conventions. Most market-research firms can provide effective advice in this regard.

As for benchmarking the competition's technical performance, this topic is generally beyond the scope of this book, although Chapter 11 covers it in a general fashion.

Benchmarking allows them to set Customer Satisfaction Performance goals strategically. The team can choose to aim for high Customer Satisfaction Performance where the competition is weak, or for high Customer Satisfaction Performance where the competition is strong. In comparing Importance to Customer, Customer Satisfaction Performance, and Competitive Satisfaction Performance, a number of possible strategic choices become apparent. In Figure 7-15, the comparison of our current rating ("Customer satisfaction performance") and the competition's rating ("Competitive satisfaction performance") shows, at a glance, some important gaps.

"Controls under my fingertips" and "Enables me to find things in the document quickly" are areas in which our development team's product is substantially behind the competition's. "Can customize to suit my working style" and "Offers lots of size,

Customer Needs	Importance to Customer	Customer satisfaction performance	Competitive satisfaction performance	Goal
Intuitive controls	84	2.9	2.8	
Controls under my fingertips	83	3.1	4.4	
Can customize to suit my working style	81	4.6	3.8	
Easy to get the information I need	80	4.7	4.6	
Can adjust the cursor to move as quickly as I'd like	49	2.9	2.8	
Enables me to find things in the document quickly	48	3.1	4.4	
Offers lots of size, font, and design options	45	4.6	3.8	
Able to see what the fonts look like as I'm choosing them	42	4.7	4.6	

Figure 7-15 Planning Matrix: Strategic Comparisons

font, and design options" are areas where our development team is doing significantly better.

An alternate method of displaying ratings is by graphics. Many HOQ diagrams in the literature show performance levels as points on a graph, connected by lines. The points may be represented by circles, squares, or triangles to distinguish the developer's product or service from the competition's. Figures 7-16 and 7-17 illustrate this representation, using the same performance values that appear in Figure 7-15. Most QFD software provides the option for either representation. I find graphical representation, as in Figure 7-16, difficult to read. The use of a radar chart as shown in Figure 7-17 may be more illustrative for the team.

I worked with one team whose members attempted to fill in the Customer Satisfaction Performance section of the Planning Matrix only to discover that they knew almost nothing about their competition. The team guessed at the competition's strengths, and where they could not agree, they entered question marks. Their Planning Matrix looked like Figure 7-18.

When their management discovered all the question marks instead of concrete data on the competition, they made a decision to launch a competitive benchmarking project. One of the benefits of QFD is that it creates a structure that leads developers to ask the right product-planning questions. In this case, the questions that still needed answers were painfully obvious, and the team knew what it had to do before it could complete its planning.

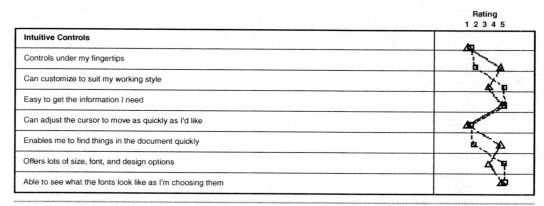

	Rating 1 2 3 4 5
Intuitive Controls	
Controls under my fingertips	
Can customize to suit my working style	
Easy to get the information I need	
Can adjust the cursor to move as quickly as I'd like	
Enables me to find things in the document quickly	
Offers lots of size, font, and design options	
Able to see what the fonts look like as I'm choosing them	

Figure 7-16 Planning Matrix: Line-Graph Style Graphic Representation of Customer Satisfaction Performance

Competitive Satisfaction Performance data should appear in a QFD project in the same form as Customer Satisfaction Performance data. In the most typical market-research scenarios, surveys are designed to capture Competitive Satisfaction Performance in the same way as Customer Satisfaction Performance, so that the resulting quantitative data will be comparable.

Often, however, Competitive Satisfaction Performance data is not as neatly wrapped as we would like it to be. Most QFD teams *do the best they can* with such data. My suggestion in such a case is to gather together all available Competitive Satisfaction Performance data and present it in a single document. Sometimes the Affinity Diagram method is very helpful for sorting and organizing such data.

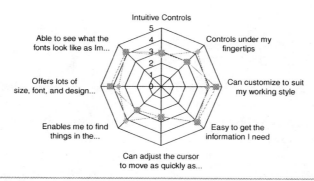

Figure 7-17 Planning Matrix: Rader-Chart Style Graphic Representation of Customer Satisfaction Performance

Customer Needs	Importance to Customer	Customer satisfaction performance	Competitive satisfaction performance	Goal
Intuitive controls	84	2.9	?	
Controls under my fingertips	83	3.1	5	
Can customize to suit my working style	81	4.6	?	
Easy to get the information I need	80	4.7	5	
Can adjust the cursor to move as quickly as I'd like	49	2.9	?	
Enables me to find things in the document quickly	48	3.1	?	
Offers lots of size, font, and design options	45	4.6	?	
Able to see what the fonts look like as I'm choosing them	42	4.7	3	

Figure 7-18   Planning Matrix: Unknown and Guessed Competitive Satisfaction Performance Levels

Having gathered Competitive Satisfaction Performance data, which may not have been collected in a form that matches the tree diagram of customer needs, the next step is to relate it to the customer needs. In cases where the data clearly lines up and fits the structure of customer needs already established for the QFD, simply enter the data. There will probably be gaps where the Competitive Satisfaction Performance data is not available. This should be indicated in the Planning Matrix.

One may be tempted to create a market research project to fill in all gaps in Competitive Satisfaction Performance data. This is probably not necessary. I recommend getting new data only for those customer needs that emerge as very high priority after a first pass through the Planning Matrix.

7.4 GOAL AND IMPROVEMENT RATIO

In the Goal column of the Planning Matrix (Figure 7-19), the team decides what level of Customer Satisfaction Performance to aim for in meeting each customer need—the Goal. The performance goals are normally expressed in the same numerical scale as performance levels. The Goal is compared with our current rating to set the Improvement Ratio. The Improvement Ratio is one of the most important multipliers of Importance to Customer; therefore, setting the Goal is a crucial strategic step in QFD.

Often, the questions are asked: "Why set goals at all? Why not set all goals as high as possible? Don't we want to excel in all areas?" These questions relate to one of the fundamental

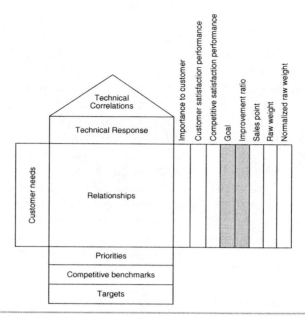

Figure 7-19 Planning Matrix: Goal and Improvement Ratio

benefits of QFD. If we had unlimited resources, we could indeed aim to have our product or service be perfect in all aspects. But no projects ever *do* have unlimited resources. Development teams must always make choices regarding where we will place special emphasis or extra resources, and where we won't. Trade-offs must be made on all projects. Most teams will be prioritizing goals according to product or company branding practices. If you are Apple Inc., your customers tend to have an expectation of innovation and ease of use. These goals will usually be set higher than others.

From the point of view of limited resources, it is a strategic necessity to choose which aspects of a product or service will excel, and which won't. Thus, goal setting in QFD involves comparing ourselves to the competition, and noticing which customer needs are most important.

Setting performance goals in the Planning Matrix of the House of Quality generally has far-reaching effects on priorities throughout the development project. This is because the goals, combined with our current rating, determine the Improvement Ratio column, a measure of effort required to alter Customer Satisfaction Performance for a customer attribute. If the company has a product or service in place, with a Customer Satisfaction Performance level of 3, say, for one customer attribute, then it will normally take about the same effort to achieve that same level of performance in a new version of the product or service as it did previously.

If the goal is higher than the current level—for example, 4, as compared to 3 for the previous product or service—then one may infer that *something special* will have to be done to have a positive effect on Customer Satisfaction Performance. "Something special" could mean an innovative redesign of at least part of the product or service, or it could mean a radical change in the way the product or service is packaged or delivered. Such changes often require corresponding multi-departmental changes. These types of changes are never easy to accomplish; therefore, the development team must not take goal setting for Customer Satisfaction Performance lightly.

The arithmetic in the Planning Matrix is set up to reflect the difficulties of these changes, although the traditional QFD arithmetic as described in early QFD articles and books is not perfect. Current Satisfaction Performance and Goal are combined arithmetically to produce a value called the Improvement Ratio. The Improvement Ratio is a multiplication factor that effectively scales the Importance to Customer, and thus reorders the importance of the customer needs. The most common method for determining the Improvement Ratio is to divide the Goal by Current Satisfaction Performance:

$$\text{Improvement Ratio} = \text{Goal} \div (\text{Customer Satisfaction Performance})$$

The more aggressive the Goal compared to Current Satisfaction Performance, the larger the Improvement Ratio, and hence the more important the customer need. However, this simple ratio may not provide the appropriate Improvement Ratio for many cases.

For instance, when Current Satisfaction Performance is very low (1, 2, or 3)—in other words, when the denominator in the Improvement Ratio formula is very low—then the Improvement Ratio itself will be quite large, even for modest improvement goals. On the other hand, when Current Satisfaction Performance is high (4 or 5), and therefore the denominator in the Improvement Ratio formula is high, then the Improvement Ratio itself will not influence the overall importance of the customer need very much.

Let's look at a few examples (Figure 7-20). In cases 1, 2, and 3, the Goal has been set at 2 points better than Current Satisfaction Performance. The Improvement Ratios, however, range from a high of 3 down to a low of 1.67, differing by almost a factor of 2. This difference may or may not reflect the relative difficulty of achieving new levels of Customer Satisfaction Performance.

When Customer Satisfaction Performance is quite low, conventional wisdom suggests that it is usually relatively easy to make modest improvements. This same wisdom suggests that it becomes increasingly difficult to approach perfection. For example, in semiconductor manufacturing, yields of new products are traditionally quite low, perhaps as low as 20 percent. The production problems are generally fairly obvious, and yields can

Case	Current Satisfaction Performance	Goal	Improvement Ratio
1	1	3	3
2	2	4	2
3	3	5	1.67
4	1	2	2
5	4	5	1.25

Figure 7-20 Improvement Ratios

be brought up to 70 or 80 percent without great difficulty. As yields approach perfection, improvements become increasingly difficult.

This phenomenon is sometimes called the **low-hanging fruit** factor. In the beginning of most quality-improvement activities, many problems affecting quality are apparent to everyone and can be identified and dealt with easily. These correspond to the low-hanging fruit in the orchard: easy to see and easy to pluck. Once all the low-hanging fruit has been removed, the fruit higher in the tree remains. This fruit is harder to see, and when spotted is harder to pluck, because it cannot be reached by an orchard worker standing on the ground.

In the same way, when Customer Satisfaction Performance is low, problems often abound that are easily identified and easily fixed. When these obvious problems are out of the way, Customer Satisfaction Performance will have improved, but still may not measure up to world-class competition. More-sophisticated problem analysis and more-elusive solutions may be required.

Referring back to the Improvement Ratio, one may now infer that it is easier to move Customer Satisfaction Performance from a 1 to a 2 than it is to move it from a 4 to a 5. Yet the Improvement Ratio arithmetic implies just the opposite: The Improvement Ratio of 5/4 is much smaller than the Improvement Ratio of 2/1. There are two alternative arithmetic approaches to the Improvement Ratio that can better reflect the low-hanging fruit phenomenon. The first is to substitute an Improvement Difference for the Improvement Ratio, and the second is to use a Degree of Difficulty judgment directly.

The Improvement Difference is defined as

$$\text{Improvement Difference} = 1 + (\text{Goal} - \text{Current Satisfaction Performance})$$

This formula has the characteristic that all improvement increments—whether starting from a low or a high level of Customer Satisfaction Performance—have the same impact on overall importance (raw weight) of a customer attribute. If the Goal is the same as Current Satisfaction Performance, a very common case, the difference is zero, which obviously should not multiply the Importance to Customer. Hence, the Improvement Difference formula includes the "1+" term.

The formula has two disadvantages. First, in the rare case that the Goal is *less than* the Current Satisfaction Performance, the Improvement Difference will be negative or zero, thus making it an inconvenient multiplier of Importance to Customer. Second, the low-hanging fruit theory suggests that it is much more difficult to improve Customer Satisfaction Performance when it is already high, as compared to when it is low. An even more realistic formula would express this sense of increasing difficulty.

The most straightforward (and oft-recommended) way of dealing with this problem is to sidestep the formula completely, and directly specify the degree of difficulty:

1   No change

1.2 Moderately difficult improvement

1.5 Difficult improvement

Some QFD teams use Current Satisfaction Performance and Competitive Satisfaction Performance as reference values. They omit the step of setting an explicit Goal and enter one of these values into the Improvement Ratio column directly.

7.5 SALES POINT

The Sales Point column (see Figure 7-21) contains information characterizing the ability to sell the product or service, based on how well each customer need is met. For example, for an automobile, a customer need might be for fuel efficiency. If the automobile could be designed to meet this need well, efforts to sell the product could capitalize on this capability.

The most common values assigned for Sales Points are

1   No Sales Point

1.2 Medium Sales Point

1.5 Strong Sales Point

Sales Points do not carry as much weight as other factors in the Planning Matrix, such as Importance to Customer or Satisfaction Performance Goal. This is because in the version of QFD that came from Japan, the ability to sell a product was not considered to be as important as the ability to increase customer satisfaction.

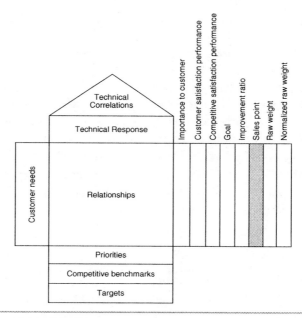

Figure 7-21 Planning Matrix: Sales Points

Obviously, performing very well on a customer need can also make it easier to sell the product. In this case, the Importance to Customer value might be high, the Goal value might be high, *and* the Sales Point value might be high. Some experts argue that a form of "double accounting" may be occurring, since the same general advantages are being expressed in all three values. While I acknowledge the possibility of such double accounting, I am not unduly troubled by it. The only harm in the possible double accounting is that the customer need's priority (raw weight) may be too high. With all these values high, performing well on this particular customer need should clearly be a priority in any case. QFD is not an exact science; the results should not be taken literally and blindly. The numerical manipulations give us a general idea of what's important, but must always be interpreted with common sense.

Not all customer needs represent sales opportunities. For example, depending on the product, fulfilling a need for safety or for compliance with long-established regulatory standards would not likely create customer interest that would justify a sales campaign.

Electronic devices such as audio-video receivers always conform to Underwriters Laboratories (UL) standards in the United States. These standards aim at eliminating the possibility of electric shocks to users during normal operation.

It would be pointless, even counterproductive, for a company to advertise that its audio-video receiver does not shock customers. First, this safety characteristic is shared by all similar products, so it is not a differentiator. Second, the suggestion that products already *assumed* to be safe are indeed safe is likely to raise questions in the minds of potential customers about the safety claims.

In general, products whose characteristics meet *expected* needs (in the sense of the Kano model or the Klein model—see Section 17.7.5) such as "will not shock the customer" are not likely to be candidates for high values in the Sales Point column. Products or services that meet needs not met by the competition or by previous offerings, or that meet needs better than the competition or previous offerings, *are* candidates for high values in the Sales Point column.

So, strong Sales Points might be

- High fuel efficiency (automobile)
- Long life (light bulb)
- No need for a second coat (house paint)
- Getting clothes white and bright (laundry detergent)

How strong these Sales Points are depends on how they compare to the competition's, and on how important it might be to the customer for the product to perform exceptionally well on these attributes.

At the time the Sales Point column is being filled in, the development team may have no idea what the design will be, or how specific customer needs will be met. One way of harnessing QFD's power is to set aggressive goals in the Goal column of the Planning Matrix in customer-need areas that could lead to competitive advantage, and then link the corresponding Sales Point values to those aggressive goals. This will allow the QFD process to highlight which parts of the design require breakthrough thinking in order to realize the advantage.

Guidance in deciding where to be aggressive can come from Kano analysis (Section 4.6.6) or Klein Grid analysis (Section 17.7.5). The Klein Grid model can help identify "hidden" customer needs which, if strongly met, could create disproportionate levels of customer satisfaction. The Kano model can help identify technical responses ("delighters") to these hidden needs.

7.6 RAW WEIGHT

The Raw Weight column (Figure 7-22) contains a computed value from the data and decisions made in Planning Matrix columns to the left. It models the *overall importance*

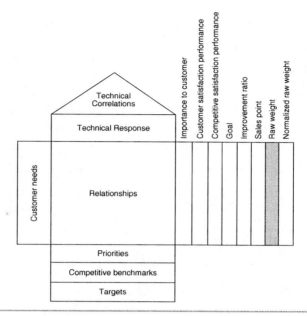

Figure 7-22 Planning Matrix: Raw Weight

to the development team of each customer need, based on its Importance to the Customer, the Improvement Ratio set by the development team, and the Sales Point value determined by the development team. The value of the Raw Weight for each customer need is

Raw Weight = (Importance to Customer) × (Improvement Ratio) × (Sales Point)

The most conventional formula for the Improvement Ratio is

Improvement Ratio = (Goal) ÷ (Customer Satisfaction Performance)

The higher the Raw Weight is, the more important the corresponding customer need is to the development team. The Raw Weight is a single number embodying Customer Satisfaction Performance, implementation effort, and sales potential. Hence, it provides an overall strategic business perspective on the importance of the Customer Needs to the success of the product or service being planned.

One of the attractions of QFD to many development teams is that the Planning Matrix, along with the Raw Weight, provides a mechanism for funneling the priorities and concerns of interested parties across the organization into an analysis process that

Factor	Minimum Value	Maximum Value
Importance to Customer (Absolute)	1	5
Importance to Customer (Weighted)	1	100
Customer Satisfaction Performance	1	5
Competitive Satisfaction Performance	1	5
Goal	1	5
Improvement Ratio, assuming: $$\text{Improvement Ratio} = \frac{\text{Goal}}{\text{Our Current Rating}}$$	0.2	5
Sales Point	1	1.5
Raw Weight (with Absolute Importance)	0.2	37.5
Raw Weight (with Relative Importance)	0.2	750

Figure 7-23   Weight Ranges in the Planning Matrix

takes all concerns into account. It also helps the team to attach weights to the various concerns. These weights are determined by the value ranges used for each of the terms in the Raw Weight formula.

Notice that with the ranges of factors as given in Figure 7-23, the Sales Point factor does not influence the Raw Weight as much as other factors, such as Importance to Customer. This reflects the philosophy that Customer Satisfaction Performance should be treated as more important than sales potential. Any QFD team is, of course, free to adjust the ranges of any of the factors to reflect the team's attitudes about which factors should most affect the planning process.

7.7 NORMALIZED RAW WEIGHT

The Normalized Raw Weight column (Figure 7-24) contains the Raw Weight values, scaled to a range of 0 to 1 or expressed as a percentage.

To calculate the Normalized Raw Weight, first sum the Raw Weights to compute the Raw Weight Total:

$$\text{Raw Weight Total} = \Sigma \ \text{Raw Weight}$$

The Normalized Raw Weight for each customer need is then the Raw Weight for the customer need divided by the Raw Weight Total:

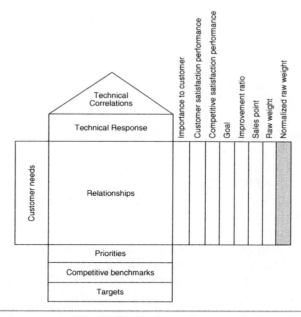

Figure 7-24 Planning Matrix: Normalized Raw Weight

Normalized Raw Weight = (Raw Weight) ÷ (Raw Weight Total)

The Normalized Raw Weight will be a decimal value between 0 and 1. Some people prefer expressing the Normalized Raw Weight as a percentage rather than as a decimal value. To express it as a percentage, multiply the Normalized Raw Weight decimal value by 100. Whether expressed as a decimal value or as a percentage, the values are equivalent, and we won't distinguish between them in this book. In Figure 7-26, the worked-out example of the Planning Matrix, the Normalized Raw Weights are expressed as decimal values.

Since the Raw Weight will be used in QFD as a proportional value, the Normalized Raw Weight carries the same information as the Raw Weight. In other words, if the Raw Weight for Customer Need A is double the Raw Weight for Customer Need B, the same ratio will apply to the Normalized Raw Weights of these two customer needs.

It's convenient to convert the Raw Weight to a Normalized Raw Weight for subsequent calculations in QFD. The Raw Weight, which is often in the range of 15 or higher, will be multiplied by other values in the Relationships section of the HOQ (see Chapter 9). These multiplied values will be added, and the resulting sums can be above 1000. If the Normalized Raw Weight is used instead, the resulting sums will be much smaller, and therefore generally easier to manage and display. As we have seen elsewhere, these sums will be used as weights to be transferred to other matrices, where they will in turn be multiplied by other numbers and added to create a new set of weights. Such weights tend to get large, and most QFD practitioners reduce them again by converting them to Normalized Weights as described here.

7.8 CUMULATIVE NORMALIZED RAW WEIGHT

The Cumulative Normalized Raw Weight, when used, is normally placed last, to the extreme right of the Planning Matrix, as in Figure 7-25. After the Raw Weights and Normalized Raw Weights have been computed, and the customer needs have thereby been prioritized, it is sometimes useful to view the customer need raw weights in terms of their overall importance. To do this, the team sorts the customer needs by Normalized Raw Weight in descending order (with the highest weight first), as in Figure 7-26.

Once the customer needs have been rearranged this way, it's possible to identify the most important customer needs as a group. The Cumulative Normalized Raw Weight column is based on the values in the Normalized Raw Weight column. The Cumulative

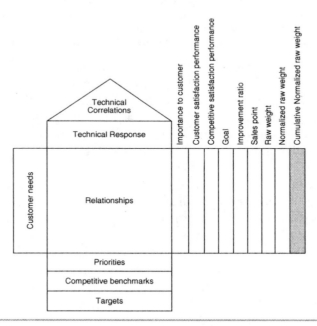

Figure 7-25   Planning Matrix: Cumulative Normalized Raw Weight

	Weighted Importance to customer	Customer satisfaction performance	Competitive satisfaction importance	Goal	Improvement Ratio	Sales Point	Raw Weight	Normalised Raw Weight	Cumulative Normalized Raw Weight
Can customize to suit my working style	81	4.6	3.8	4.6	1.00	1.5	559	0.19	0.19
Easy to get the information I need	80	4.7	4.6	4.7	1.00	1.2	451	0.16	0.35
Control under my fingertips	83	3.1	4.4	4.4	1.42	1.2	438	0.15	0.50
Intuitive controls	84	2.9	2.8	3.3	1.14	1.5	416	0.14	0.65
Enables me to find things in the document quickly	48	3.1	4.4	4.5	1.45	1.5	324	0.11	0.76
Offers lots of size, font, and design options	45	4.6	3.8	4.6	1.00	1.5	311	0.11	0.87
Able to see what the fonts look like as I'm choosing them	42	4.7	4.6	4.7	1.00	1.2	237	0.08	0.95
Can adjust the cursor to move as quickly as I'd like	49	2.9	2.8	2.9	1.00	1.0	142	0.05	1.00
Totals							2878	1.00	

Figure 7-26 Completed Planning Matrix

Normalized Raw Weight shows how much of the total raw weight can be attributed to the most important customer need, the two most important customer needs, the three most important, and so on.

Each Cumulative Normalized Raw Weight value in a row is formed from the sum of the Normalized Raw Weight of the customer need of the row, and the Normalized Raw Weights of all the customer needs more important (higher) than this need. By displaying the cumulative weights this way, we can easily see which customer needs contribute to various fractions of the total Raw Weight. For example, the most important needs, which as a group contribute one-half of the total Raw Weight, are found by scanning the Cumulative Normalized Raw Weight column from the top downward to the first value that is equal to or greater than 0.50. In Figure 7-26, we see that the first three customer needs contribute one-half of the total Raw Weight. Frequently, a small subset of customer needs accounts for most of the Raw Weight. This information can be useful in saving time during the QFD process (see Section 16.7.2), and, more importantly, in focusing the team on the most important project goals.

7.9 SUMMARY

The Planning Matrix is the portion of the House of Quality that contains strategic marketing information and planning decisions. Completing this section of the House of Quality is a major step in the QFD process. It prioritizes customer needs based on their importance to the customer *and* their importance to the development team's organization. This set of priorities will have a major impact on all future planning and development activities.

The quantitative information in the Planning Matrix normally is the result of market research. The team makes its strategic decisions based on its understanding of the customer needs, and its own and the competition's performance in meeting those needs. The most important and far-reaching strategic judgments are the setting of Customer Satisfaction Performance goals and the identification of selling opportunities for the product or service being planned.

The information that feeds the decision making in the Planning Matrix is

- Importance to the customers of each customer attribute
- Current Customer Satisfaction Performance, by customer need, with the development team's product or service most similar to the one being planned
- Current Customer Satisfaction Performance, by customer need, with the competitor's product or service most similar to the one being planned

The strategic determinations the development team makes and records in the Planning Matrix are

- Customer Satisfaction Performance Goal for each customer attribute
- The sales potential of performance on each customer attribute

The computations the development team makes in the Planning Matrix are

$$\text{Improvement Ratio} = \text{Goal} \div (\text{Customer Satisfaction Performance})$$

and

$$\text{Raw Weight} = (\text{Importance to Customer}) \times (\text{Improvement Ratio}) \times (\text{Sales Point})$$

The Raw Weight and the Normalized Raw Weight convey equivalent information, but the Normalized Raw Weight is more convenient when carried into later phases of QFD. Either calculation provides a rank ordering of customer attributes that will influence decision making throughout the planning process.

We've now covered QFD's handling of the Voice of the Customer both qualitatively and quantitatively. It's time now to consider the organization's response to these customer needs. Up to now, we've listened to the customer speaking in the customer's own language. In the next chapter, we'll look at the organization's language, and how QFD helps the organization speak its language.

7.10 DISCUSSION QUESTIONS

In your organization, how do you determine the importance to the customer of the various customer attributes? What do you like about the method? How could it be improved?

How do you set Customer Satisfaction Performance goals? Do you know Customer Satisfaction Performance levels for each customer attribute? If not, how could you gather such information? Do you know how well the competition is doing?

What strategic planning decisions do your developers make, and how do they compare to the decisions identified by the columns of the Planning Matrix? If you wanted to record your organization's planning decisions in a structure like the Planning Matrix, which columns would you add, delete, or change?

Substitute Quality Characteristics (Technical Response)

This chapter describes the way in which QFD deals with the development team's technical response to the customer's needs. Just as the Voice of the Customer (VOC) has qualitative and quantitative components (entered into the Customer Needs and Benefits Section and the Planning Matrix), so does the translation of the VOC into the Voice of the Developer. That translation, which we'll usually call Substitute Quality Characteristics (SQCs), will be placed in qualitative form on the top of the Relationships Matrix, and in quantitative form at the bottom (as Target Values and Competitive Technical Benchmarks). It is critical that any translation into an SQC be validated before proceeding too far along the path of product development, to verify that, in the final product, it will in fact meet the needs of intended customers.

In this chapter we'll look at a few alternative formulations for SQCs, along with some methods for generating them and some ways to verify SQCs.

Substitute Quality Characteristics is the term used for the internal, technical language an organization uses to describe its product or service. In QFD parlance, we use the term Quality Characteristics to denote the customer's needs (the VOC). The translation into technical terms is called SQCs because it represents a *substitution* of the organization's technical language for the customer's language. Any translation should be checked with the customer for accuracy early in the development process. These technical terms (see Figure 1-2 Section 4, in Chapter 1; and Figure 8-1) may describe the product from any of a variety of points of view. Most commonly, developers call this language the Product Requirements or the Design Requirements.

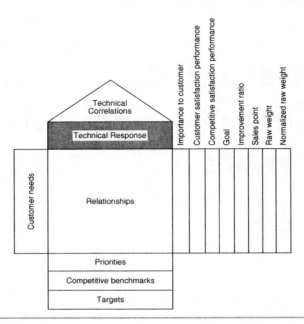

Figure 8-1   Technical Response

The nature of Product Requirements varies widely from group to group, and from industry to industry. Many organizations describe products and services in more than one set of terms. They distinguish between these various languages by giving them such names as Customer Requirements, Market Requirements, Top Level Specifications, Detailed Specifications, or Technical Specifications, to name but a few.

Some generic formulations of SQCs exist, notably one by Stuart Pugh (see Section 13.2, Enhanced QFD and Concept Selection). Such formulations can be used as "starter kits" to get a set of SQCs established rapidly, and to aid teams in arriving at a complete set. The generic formulations must of course be modified to meet the needs of the team's specific project. In particular, the language of the generic formulations is not likely to correspond to the development team's product-description terminology.

There is often much confusion about the boundaries between these various product description languages, and there is often little or no standardization of the vocabulary of each language. For example, in very hierarchical organizations, the most workable distinction between Top Level Specifications and Detailed Specifications may be defined by which part of the organization is responsible for writing the associated documents.

The ideal relationship between various product or service description languages is one in which the languages are defined and ordered according to how abstract, or solution-independent, each language is. If one product description allows for many possible implementations, it is more abstract than another that clearly describes or implies one and only one implementation.

In software, for example, a product requirement for "easy command selection" is more abstract than "commands available by pull-down menus." "Easy command selection" allows for at least the three most popular technical solutions—pull-down menus, iconic push-buttons, and keyboard shortcuts. "Commands available by pull-down menus" allows for only one of these possibilities, and is therefore more concrete, or less abstract.

Any development organization would do well to define its highest-level product- or service-description language by providing a vocabulary of terms that are included in the language, along with examples. Lower-level languages could be similarly defined. Most important in these definitions would be to show how a designer translates a statement from a higher-level, more-abstract language to one or more statements in a lower-level, more-concrete language.

One way QFD practitioners describe these various levels of abstraction is with the Whats/Hows metaphor (Figure 8-2). The language that appears on the left side of a QFD matrix represents What is desired. The language at the top of the matrix describes How the developers will respond to the Whats. To get the most out of QFD, the language of the Whats should be distinctly more abstract than the language of the Hows.

The Hows may still be abstract compared to other language available to the developers and to be used later in the development process. In later phases of development, and later phases of QFD, the Hows can be placed on the left side of another matrix so that they become the Whats at a more-detailed level (Figure 8-3). The development team then uses more-specific language along the top of the new matrix to represent the Hows of that matrix.

Given a term that describes some aspect of a product, it is not always easy to decide what language it belongs to. Section 17.3.1 illustrates the common situation of a customer expressing a need as a solution. The customer asks for "tinted glass" in an automobile instead of asking for "privacy." Because "tinted glass" is technically the Voice of the Customer, should we include "tinted glass" with the Whats and place it on the left side of the House of Quality? On the other hand, because "tinted glass" is a technical

Figure 8-2   What versus How

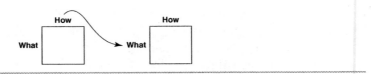

Figure 8-3   Deploying Hows to Whats

response to the customer need for privacy, should we place it along the top of the House of Quality?

A practical method for dealing with the problem of placing product requirements in the right place in QFD is to use the Voice of the Customer Table (VOCT) described in Chapters 6 and 17. Even if the developers lack preexisting definitions for their technical description language(s), the VOCT provides a concrete method for sorting out terms and for deciding how many levels of technical description languages they want to work with.

8.1 TOP LEVEL PERFORMANCE MEASUREMENTS

Probably the most valuable language for Substitute Quality Characteristics is the language of Performance Measurements. These are measurements that the development team derives directly from customer needs. They should be general enough to be applied to a product or service regardless of the specific implementation. This allows them to provide ideal measurements for benchmarking competitive products or services and for providing a solution-independent starting point for developing new concepts.

The standard method for developing Performance Measurements is to begin with the customer attributes. For each customer attribute:

1. Define measures
2. Define measurements

8.1.1 DEFINE MEASURES

Defining measures is the process by which the development team establishes the relevance and the relationship between its measurements and customers' perceptions. The word **Deployment** in Quality Function Deployment most readily applies to this process of defining measures. In a nutshell, the team translates, or deploys, each customer need into a technical-performance measure.

For each customer attribute, define one or a few technical-performance measures. These measures should be validated by the customer before proceeding too far with them. It is fairly easy to misunderstand a customer attribute and come up with measures that are not truly what the customer wants. For example, a customer may want a "less noisy" vacuum cleaner. The context is important: the customer may have meant less noisy *when vacuuming bare floors.* Arbitrary engineering measurements of motor noise or vacuum-exhaust noise might lead the design team in the wrong direction, expending unnecessary time and money in an area not well correlated to product success in the customer's eyes. The steps in Figure 3-1 (Chapter 3) and the KJ work described in Figure 17-1 (Chapter 17) should help here. Write these key measurements along the top of the House of Quality, as in Figure 8-4. Some examples of performance measures and their relationship to customer needs are also shown in Figure 2-9 (Chapter 2).

For each measure, be sure of the following:

- That it can be measured while the product or service is being developed, before it is shipped or deployed; in other words, that it can be used as a predictor rather than a lagging indicator of Customer Satisfaction Performance.

- That it can be controlled by the development team. The team should be able to make decisions that effectively would adjust the measurement up or down. A good way to think of the measure is as a dial the development team can rotate clockwise or counterclockwise. As the team conceptually rotates the dial, Customer Satisfaction Performance may conceptually be affected either positively or negatively.

- That it has the proper context in usage by customers, and that the measurements have been validated as having meaning in the eyes of customers. Contextual usage is captured with image KJ and requirements KJ. Validation may be captured through validation surveys and/or additional customer visits once the team has its performance measurements.

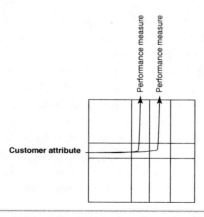

Figure 8-4 Customer Attribute Deployed to Performance Measurement(s)

To be properly defined, each performance measure should be characterized in a few ways. First, the *units* of the measure should be defined. Examples would be

- Voltage in volts
- Time in minutes
- Process complexity in number of steps
- Accuracy in Defects Per Million Opportunities (or transactions), DPMO

Second, the *direction of goodness* should be defined. Remember that goodness, or value to customers, means your new product may be able to perform its function better (effectiveness), in new ways (flexibility), faster (speed), with fewer steps (efficiency), or using less energy or materials (cost)—and, of course, any combinations of these goodness traits. There are three possible directions of goodness:

- *The More the Better.* The implied target is infinity. Examples are the following:
  ○ Reliability as measured by mean time between failures (MTBF)
  ○ Fuel efficiency as measured by miles traveled per gallon of fuel used
  ○ Bonding strength as measured by pounds supported per square inch of adhesive area
  ○ Number of instances of treating a customer respectfully

Some development teams confuse the implied target of infinity under the More the Better with an acceptable value (a **tolerance**) for the metric. This is the difference in Kano's model between Fitness to Standard and Fitness to Compete. Many developers express consternation when confronted with targets such as infinity. The idea behind using such language is to recognize that most technical objectives are points on a sliding scale. If we view them as such, we may see possibilities for increasing performance that we have not seen before.

On the other hand, arbitrarily high numerical objectives are usually impossible or very impractical to achieve. Teams need both the *aggressive* objective (infinity) and an *acceptable* value for the measure. While the market may *accept* a product with reliability of 20,000 hours MTBF, clearly a product with much higher MTBF would be more attractive, if all other factors—including price— were the same. It matters most that the product achieves noticeably more of the desired measure than the competing alternatives available at time of product deployment into the marketplace in the eyes of the customers. Otherwise, there is little to no product differentiation.

- *The Less the Better.* The implied target is nominally zero. Examples are the following:
  - Quality of service as measured by number of defects per transaction. Defects could be determined by monitoring all or a sample of transactions. If the transactions are interpersonal—for example, hotline service center calls—defects could be defined as instances of treating a customer disrespectfully, or of giving a customer incorrect information.
  - Simplicity of process, versus process complexity as measured by number of steps
  - Speed of startup as measured by time to launch a software application

There may be rare cases where a target is listed as minus infinity, but we have never seen such a case that made practical sense. If there was such a case where the target is minus infinity, the measurement scale could be shifted or changed so that zero becomes the lowest possible value. In a refrigeration example, absolute zero would be selected as the ideal value, as opposed to minus infinity for the temperature.

Additionally, arbitrarily low numerical objectives may be impossible or very impractical to achieve. As in the More the Better, teams need both the *aggressive* objective (zero) and an *acceptable* value for the measure. While the market may *accept* an airline baggage service with only seven mishandled bags per 1,000 passengers, clearly a service with zero mishandled bags would be more attractive, if all other factors—including price—were the same. It matters most that the product or service achieves noticeably less of the desired measure than the competing alternatives available at time of product or service deployment into the marketplace in the eyes of the customers. Otherwise there is little to no differentiation.

- *Target Is Best.* The target is as close as possible to a nominal value with as little variation around that value as possible. Examples are the following:
  - Exactness of fit as measured by the diameter of a steel rod intended to fit into a cylindrical sleeve. If the diameter is too large, the rod will not fit into the sleeve; if the diameter is too small, the rod will wobble in the sleeve.
  - Constancy of ideal temperature within a food freezer container. The ideal temperature is 4°F. Colder temperatures cause the food's flavor to deteriorate; warmer temperatures reduce the shelf life of the food.

In service applications, the majority of Top Level Performance Measurements will be of the More the Better and Less the Better types. Target Is Best is not common for service applications.

8.1.2 DEFINE MEASUREMENTS

Defining measurements is the process of describing *how* each measurement will be performed. The team must document all assumptions and comments about each type of measurement. A Measurement Systems Analysis (MSA) should be performed for every measurement, as described in Section 4.6.3. An MSA should be mandatory for any metric related to a customer attribute that is new, important, or difficult. The MSA work should be performed as soon as is practical in the development process.

Describing how each measurement will be performed is a step that escapes many developers. They may feel that the measurement method is self-evident and doesn't need explicit description. The omission of this step leads to much lost time during planning, and later during development.

This step *operationalizes* the definition of the measurement. W. Edwards Deming has pointed out that definition of the process of measurement is a key factor in defining the measure.[1] He says that operationally defined measures are measures "one can do business with." Conversely, measures that are not operationally defined are measures one *cannot* do business with. Such measures cause confusion, because one person will inevitably have in mind a different measurement process than another.

Consider the following measure: "Speed of startup as measured by time to launch a software application." One developer may assume the method of measurement involves

1. No other applications running on the computer
2. Operating system version 3.1, as shipped by the operating system supplier, installed with supplier-standard defaults
3. Maximum RAM configuration used for measurements
4. Clock starts when the "Open application" command is executed at the operating-system level
5. Clock stops when the application is ready to receive a user command
6. Application is launched five times, and the average speed of the five launches is taken

Another developer may have a very different set of assumptions:

---

1. W. Edwards Deming, *Out of the Crisis* (Cambridge, Mass.: Massachusetts Institute of Technology, Center for Advanced Engineering Study, 1986), Chapter 9. See also, W. Edwards Deming, *Quality, Productivity, and Competitive Position* (Cambridge, Mass.: Massachusetts Institute of Technology, Center for Advanced Engineering Study), Ch. 15.

1. Other applications are running on the computer, in particular applications A, B, C, and D, which many customers expect to use in conjunction with the target application

2. Application launches under operating-system versions 2.9, 3.0, and 3.1 must all be measured. Each operating system is used as shipped by the operating system supplier and installed with supplier-standard defaults.

3. Minimum and maximum RAM configurations are used for measurements

4. Clock starts when the "Open application" command is executed at the operating-system level

5. Clock stops when the application is ready to receive a user command

6. Sixteen configurations are identified using different combinations of applications A through D, operating system versions 2.9 through 3.1, and minimum and maximum RAM configurations. Target application is launched once with each configuration, and the average speed of the sixteen launches is taken.

Developers working to the first definition would optimize application launch for the single configuration specified. Such an application might do very poorly under minimum RAM, or with operating system 2.9. The detailed description of the measurement guides developers about what to optimize. It makes all discussions involving the measurement clearer and more efficient.

8.2 PRODUCT FUNCTIONS

A completely different approach to defining Substitute Quality Characteristics is to place *product* or *process functions* along the top of the House of Quality. It can be appropriate to use functions instead of performance measures under the following circumstances:

- The product or service concept has already been established (breakthrough ideas, at least at the strategic level, are not needed or are already incorporated into the concept). Many times, a successful version of a product or service is already in the field, and QFD is being used to define a "midlife kicker" or an upgrade to the previous offering. In such a case there may be a list of possible extensions—already expressed as features—that need to be prioritized.

- The development team lacks either the time or the resources at the moment to develop and prioritize performance measures. Since prioritization of performance measures does not define a product's or service's features, the QFD process must be used at least once more to translate prioritized performance measures into prioritized features. This extra step is time-consuming and may not always be worth the effort.

In addition, some development teams, especially software-development teams, are unaccustomed to using performance measures in their product-definition process. For such teams, these measures may only get in the way of what they see as their work. While translating customer needs directly into functions lowers the chances for breakthrough ideas, teams that don't normally use performance measures may be better doing just such a translation.

Many products and services have large numbers of capabilities or functions. It may be bewildering to place them all at the top of a House of Quality. Depending on the level of detail the developers use to describe the functions, the House of Quality could be correspondingly small or large.

The developers can use either the Requirements KJ or an Affinity Diagram to decide what level of functional detail they want to work at. The hierarchy of product or process function will have several levels (as do all KJs and Affinity Diagrams). Analyzing at the higher levels (with fewer elements) will present the advantage of quicker analysis. The corresponding disadvantage of working with less detail is less depth in the analysis.

Generally, for strategic analysis, a quicker analysis phase at a less-detailed level is appropriate. The House of Quality at the strategic-analysis level will indicate which few critical functional areas require more-detailed planning. These areas can then be singled out, and the development team can analyze only those areas in a subsequent House of Quality (see Figure 8-5).

8.2.1 FUNCTION TREES

In the previous section we described a brainstorming approach for creating an Affinity Diagram of the functions of a product or service. As is true of all Affinity Diagrams, functions are organized from the bottom up.

Another approach, the Function Tree method, uses the Tree Diagram method and organizes the functions from the top down. This method has been described by Don Clausing.[2]

In this approach, the primary functions of the product or service are identified first. Each primary function is then broken down into subfunctions. Each subfunction is elaborated into finer detail until the development team has reached the level of detail it needs. This top-down approach creates a *function tree* that helps the development team to focus on the most-important functions of the product or service.

The resulting Tree Diagram can have multiple levels of increasing detail, just like an Affinity Diagram. The development team can choose the level on which to perform the QFD analysis.

---

2. Dr. Don Clausing, *Total Quality Development* (New York: ASME Press, 1994).

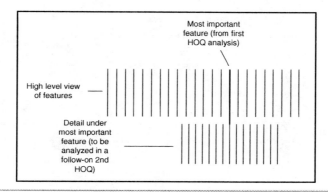

Figure 8-5 Explosion of Most Important Function

A Function Tree may be indistinguishable from an Affinity Diagram of functions, once it has been completed. The method of creating it and the associated point of view are different—just as affinity diagramming and tree diagramming are different. To get the best of both methods, the team may consider first brainstorming functions and affinitizing them, then completing the structure with the Tree Diagram method by working from the top down.

8.2.2 CRITICAL PARAMETER MANAGEMENT

As previously described in Section 4.6.1, Critical Parameter Management (CPM) is a tool to flow down and track key customer needs through the development of product systems, subsystems, process, and components. CPM helps answer How, Why, and How Well. CPM does this by linking key aspects of the design through the many $Y=f(x)$ equations that arise as a team flows down customer needs into the design process. In this manner, how we satisfy customer needs is documented, and why key design decisions were made is also captured. In addition, the supply-chain capabilities are entered into CPM scorecards to compare requirements to capabilities.

Modern, complex products often have many flow-down paths, and the picture can get complex very fast, as illustrated in Figure 8-6. Often, keeping track of so many paths and multiple relationships becomes burdensome to a team. Fortunately, there is software that can help. An excellent example of such software is provided by Cognition Corporation, in the form of its Cognition Cockpit package.[3] Figure 8-7 shows one view provided by the software for a Critical Parameter for a surgical stapler.

---

3. David Cronin, Cognition Corporation, Bedford, MA, www.cognition.us, personal communication, Autumn 2008.

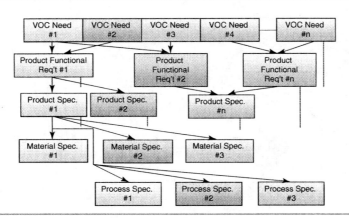

Figure 8-6   Many Functional Paths

The value of CPM and associated software in QFD work is that the Voice of the Customer flows down, while the Voice of the Processes flows up. This allows the design team to make trade-offs as the design takes shape, and to make data-based decisions regarding design features and associated costs to meet key customer requirements. By using software, the team can collaborate with design members who are not on-site, and also with key suppliers and manufacturers. This way, as requirements flow down, associated capabilities may be entered and compared to requirements to begin answering the questions, "How Well?" and "How Much?"

Modern software shortens the time-consuming tasks of Critical Parameter Management development and capture. By using CPM software, the team can both accelerate the QFD efforts and simultaneously improve the quality and amount of knowledge captured in the development process. Utilizing CPM within a QFD effort creates a more-detailed capture of engineering knowledge and functional relationships compared with some of the approaches mentioned earlier in this section.

8.3 PRODUCT SUBSYSTEMS

While the design elements of a product are not normally chosen to be the Substitute Quality Characteristics, there are occasions where the choice is appropriate. Most commonly, the QFD process translates from

1. The Voice of the Customer, to
2. Performance Measures, to

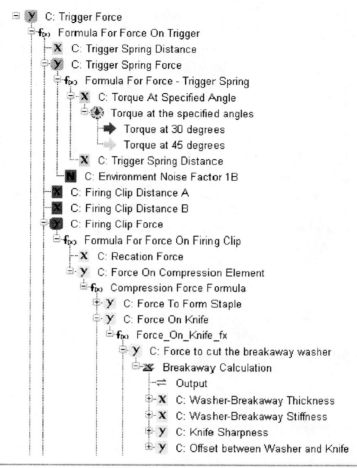

Figure 8-7   Cognition Cockpit[4] Software Sample View of Surgical Device Critical Parameter Management

3. Functions, to

4. Product Design

Each successive pair of topics (1 to 2, 2 to 3, 3 to 4) represents the left and top, respectively, of a new matrix. We have already discussed the possibility of a Voice of the Customer to Functions matrix (1 to 3). Such analysis would be appropriate if the product concept were **static**. We'll be discussing static/dynamic analysis later (in Section 13.2.1).

---

4. Cognition Cockpit software, www.cognition.us.

Development teams have also created Voice of the Customer to Product Design matrices (1 to 4). Such a matrix shows the development team how various components of the product design influence various customer-satisfaction attributes.

The most common method for representing the design is by use of the Tree Diagram. First, identify the primary subsystems of the product. For example, with a camera, the primary subsystems would be

Imaging subsystem

Film management subsystem

Viewfinding subsystem

Exposure time management subsystem

Every part of the camera should be identified within one and only one of these subsystems. The imaging subsystem might contain:

Lens

Film plane

Lightproof compartment between lens and film

The film management subsystem might contain:

Take-up spool

Film advance subsystem

Film plane subsystem (holds the film flat for accurate focusing)

Film supply subsystem

Each of these subsystems and components can be described at several levels of additional detail, thus providing the development team with the multiple-level Tree Diagram or CPM it needs for QFD analysis.

8.4 PROCESS STEPS

When developing new processes and services, design teams may have some advantage with regard to Substitute Quality Characteristics over product-design efforts. If properly planned and executed, the Voice of the Employees (or other users) in that process may be obtained to help validate the SQCs. This is a voice-capture effort mentioned in Section 3.2,

QFD and Six Sigma Process Design (Chapter 3). By interviewing process operators for some of the voices captured, possible SQCs may be identified, to be validated at some later point with a series of return interviews. In addition, if there are current processes or competing processes to evaluate, process mapping as described in Section 4.13 (Chapter 4) may yield insight into SQCs.

For teams developing new processes or services, the following choices for SQCs are as applicable as for products:

- Performance measures
- Functions

Performance measures for services are much the same as for physical products: their values are generally under the control of the team that designs or lays out the processes underlying the service. Each process will have one or more clearly defined inputs and outputs. Development teams can define performance measures for these processes, based on time, cost, or quality of result. Examples of such measures are the following:

Average time to process an input and produce the output (cycle time)

Average number of inputs processed per unit time (throughput)

Number of errors per standard number of transactions (quality; defect rate)

Average processing cost per input (cost)

In QFD, the service development team can evaluate these and many other process-performance measures to determine which ones drive Customer Satisfaction Performance most strongly.

While product subsystem analysis may not apply to services, a closely related method of analyzing services and processes does apply. Services are delivered by processes. These processes may be conceived of at various levels of abstraction. For example, Telephone Customer Support for a financial institution, such as a bank, could be conceived of—at a very abstract level—as a simple process, as in Figure 8-8:

1. Route incoming call to Customer Service Associate
2. Classify customer request
3. Respond to customer request

Anyone who has analyzed telephone customer support will know that beneath this simple process lies an enormously complex structure for handling customer requests.

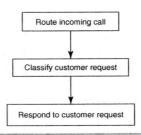

Figure 8-8   High-Level View of Process

This structure includes many decision points and subprocesses at each of the three main steps listed here. It also includes tools and technology to support these decision points and steps.

The decisions relate to managing the flow of incoming calls and balancing the volume of these calls against the available Customer Service Associates; identifying the customer, the customer account, and the type of customer request; deciding on the appropriate response; and acting accordingly. The tools and technology to support these steps include elaborate telephone equipment to queue and route incoming calls, as well as computers and associated databases to provide up-to-the-minute information about each customer to the Customer Service Associates and to update this information based on the nature of the customer call.

Each of these enumerated service components represents a subsystem or sub-service of the overall service. Each is a candidate to be an SQC in the House of Quality. These service components can be evaluated for their impacts on meeting customer needs.

The relative advantages and disadvantages of using process steps as compared to performance measures are shown in Figure 8-9.

Performance measures	Process steps
Advantages:	Advantages:
1. Generally solution-free, providing stronger likelihood of creative solutions.	1. Concrete, easily envisioned.
2. Measurements can be used to manage the processes.	2. Can be used in the HOQ at the level of detail appropriate to the team s needs.
Disadvantages:	Disadvantages:
1. Difficult to understand or to create in organizations where measurement is not the norm.	1. Incomplete definitions of process steps can lead to confusion during QFD.
2. Expensive to implement (cost-benefit issues aside).	2. Focus on concrete process steps too early reduces the chances for breakthrough solutions.

Figure 8-9   Performance Measures versus Process Steps

8.5 SUMMARY

Substitute Quality Characteristics represent the organization's internal, technical language. The development team uses SQCs to describe its product or service in an abstract manner. The QFD team must choose which of its possibly many technical formulations it wishes to use in QFD. More-abstract, solution-independent formulations provide the team with more breakthrough opportunities at the expense of more QFD steps to bridge the gap between customer wants and needs and action.

The process of defining a language for SQCs can be assisted by referring to the "Whats versus Hows" model. Learning to separate problems from solutions is an early benefit frequently cited by QFD teams.

The most commonly used SQCs are Top Level Performance Measurements.

Other SQCs include Product Functions, Product Subsystems, and Process Steps. In QFD applications where the SQCs are Top Level Performance Measurements, the team generally develops a subsequent QFD matrix that puts the rank-ordered Top Level Performance Measurements on the left, and one of these more specific forms on the top.

At this point, we know the Voice of the Customer and the Voice of the Developer. The next question QFD asks, and that the development team must answer, is, "How are these two voices linked up?" If we developed the SQCs from the customer needs, we have confidence that each SQC is linked strongly to at least one customer need. But that may be the tip of the iceberg. Figure 8-10 illustrates the expanding nature of this key aspect. There may be many other important linkages between SQCs and customer needs. We need to study them in order to decide which of the SQCs are power-

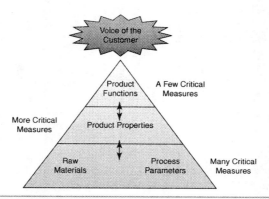

Figure 8-10   SQC Expansion from Customer Voices

ful drivers of customer satisfaction. In the next chapter, we'll see how these linkages are evaluated.

8.6 DISCUSSION QUESTIONS

In your development process, compare your description of Customer Needs and Benefits with your description of your Technical Response. Can you separate the Whats from the Hows? Do your colleagues agree with your analysis?

What language do you use for your Technical Response? List some typical nouns and verbs. Is your language solution-independent? What would be more solution-independent language? What would be more solution-specific language?

Impacts, Relationships, and Priorities

This chapter shows how the relationships between Substitute Quality Characteristics and customer needs are modeled in QFD. One of the particularly brilliant ideas embedded in QFD is to use a matrix to study these relationships. Prior to QFD use, the relationship between Technical Response and customer need was expressed by simple lists, and the complex many-to-many relationships remained only intuitively comprehended.

We will see how the Relationships matrix makes it possible to represent and visualize the patterns of these relationships much more easily than could be done otherwise.

In this chapter we'll see how QFD handles strong and weak relationships, and even negative relationships.

At the point when the development team is ready to fill in the Relationships, the team will have already determined a substantial amount of information about the marketplace. The customer needs will have been determined, the strategic market research information and decisions in the Planning Matrix will usually have been determined, and the Technical Responses or SQCs will have been decided upon. Very often, the initial product or service concepts will have been developed after the VOC work has been completed. If this had been done prior to QFD, the HOQ would have been easier and generally faster to complete. Within a DFSS context, this is less risky, because VOC would have been utilized as an input to the Pugh Concept Selection process. Outside of a DFSS context, completing the HOQ matrix prior to concept selection is advised to reduce the risk of incorrect concept selection.

The team's next task is to fill in the Relationships section. This chapter will show how the development team does this. We will also see how the team determines each SQC's relative contribution to overall customer satisfaction, and therefore that SQC's priority.

9.1 AMOUNT OF IMPACT

The Relationships section (Figure 9-1) provides a mapping between Substitute Quality Characteristics on the one hand, and Customer Needs and Benefits on the other.

Each Relationship cell represents a judgment, made by the development team, of the strength of the linkage between *one* SQC and *one* customer need. We call the strength of linkage the *impact* of the SQC on the customer need. The entire Relationship section of the HOQ contains cells for storing these impacts about *each* SQC-customer need pair.

We now come to the reason why the analogy of the SQC as a "dial" is useful (see Section 8.1.1, Chapter 8). As the SQC "dial" is conceptually rotated in the direction of goodness (or in the opposite direction), what will happen to Customer Satisfaction Performance with respect to a particular customer need? In QFD we recognize four possibilities:

1. Customer Satisfaction Performance with respect to the need is *not linked* to the SQC. In other words, for changes of any sort, large or small, in the amount or degree of the SQC, no noticeable change in Customer Satisfaction Performance of that need is predicted by the development team.

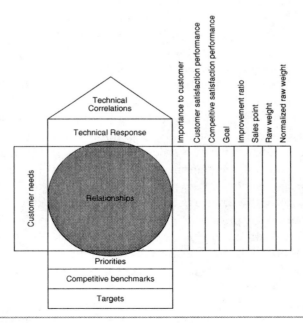

Figure 9-1   Relationships

2. Customer Satisfaction Performance with respect to the need is *possibly linked* to the SQC. For relatively large changes in the amount of the SQC, little or no change in Customer Satisfaction Performance of that need is predicted by the development team.

3. Customer Satisfaction Performance with respect to the need is *moderately linked* to the SQC. For relatively large changes in the amount of the SQC, noticeable but not major changes in Customer Satisfaction Performance of that need are predicted by the development team.

4. Customer Satisfaction Performance with respect to the need is *strongly linked* to the SQC. For relatively small changes in the amount of the SQC, significant changes in Customer Satisfaction Performance on that need are predicted by the development team.

For most QFD activities, the linkage is considered to be positive; that is, if the SQC is moved in the direction of goodness, customer satisfaction is assumed to increase. Negative linkage is possible, but it complicates the QFD process. Whenever possible, the development team should try to convert such negative linkage to positive linkage by redefining the SQC. Unfortunately, this is not always possible. Section 9.4 provides a detailed discussion of the special considerations for handling negative linkages.

Certain symbols are customarily used in QFD to denote these four possible impacts. The symbols, their meanings, and their numerical equivalents are as shown in Figure 9-2.

The symbol meaning "Not linked" is a blank. Occasionally this causes a bit of confusion in QFD, because it is not possible to distinguish a matrix cell that has been evaluated as "Not linked" from a cell that has not been evaluated.

Some QFD facilitators use a symbol such as a check mark or a lowercase "b" in place of the blank in order to avoid this confusion.

In Figure 9-3, one can see at a glance that Technical Response X makes a much greater contribution to Customer Satisfaction Performance than does Technical Response Y. The figure also suggests that Need A is being addressed more directly than Need B: Need A has a ◎ and a Δ to its credit, while Need B has only a single ○.

Symbol	Meaning	Most Common Numerical Value	Other Values
	Not linked	0	
Δ	Possibly linked	1	
○	Moderately linked	3	
◎	Strongly linked	9	10, 7, 5

Figure 9-2 Impact Symbols

9.1.1 JUDGING IMPACT OF PERFORMANCE MEASURES

The Performance Measure is the ideal Substitute Quality Characteristic for making impact judgments. The Performance Measure can be thought of as a continuous variable, and Customer Satisfaction Performance for any need is normally represented as a continuous variable (a simplification of the statistical basis for measuring satisfaction performance). Thus, the impact represents the mathematical relationship between the two variables (Figure 9-3).

$$\text{Customer Satisfaction}_A = f(\text{Performance Measure}_x)$$

In this context, the impacts ◎, ○, and Δ correspond to steep slope, moderate slope, and almost no slope. A blank impact corresponds to no slope.

This discussion assumes a *monotonic* relationship between Performance Measure$_x$ and Customer Satisfaction$_A$. In other words, as the Performance Measure moves in the direction of goodness, Customer Satisfaction Performance continues to improve. Engineers are aware that the relationships may be more complex than that. To make product planning simpler and cleaner, it's a good idea to try to define Performance Measures that do, in fact, provide a monotonic relationship with Customer Satisfaction Performance.

In Figure 9-4, not only have we modeled the relationships as monotonic, we have modeled them as *linear*. In other words, as the Performance Measure moves in the direction of goodness, Customer Satisfaction Performance continues to improve *at the same rate*. As we have seen with the Kano model, however, the relationships for Delighters and for Dissatisfiers are not linear (see Figure 2-7). Delighters have the potential of disproportionately increasing customer satisfaction as the SQC moves in the direction of goodness,

	Technical Response X	Technical Response Y	Raw Weight
Need A	◎	Δ	15
Need B	○		20

Figure 9-3   Amount of Impact

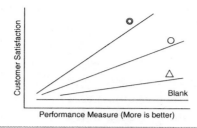

Figure 9-4 Customer Satisfaction Performance as a Function of a Performance Measure

while Dissatisfiers have the potential of producing disproportionately increasing customer dissatisfaction as the SQC moves away from the direction of goodness.

This type of disproportionate, nonlinear behavior cannot be easily modeled in QFD. The way to handle it is to treat such SQCs as if they were linear, but to note their special characteristics when the time comes to set target values. We'll be discussing target value setting in Chapter 12.

9.1.2 Judging Impacts of Other SQCs

It is generally more difficult for development teams to judge impacts when the Substitute Quality Characteristics are not measurable. This is because a simple diagram such as Figure 9-4 cannot readily be used as a conceptual model of the relationships. There is no clear and simple "dial" to turn when the SQC is a product function or process step.

Typically, development teams think of non-measurable SQCs as either *present* or *absent.* When determining an impact in this context, they try to judge whether Customer Satisfaction Performance will be high if the SQC is present, and low if the SQC is absent. If the effect on Customer Satisfaction Performance level is predicted to be large, the team will assign ◎ as the impact.

In fact, it is generally unrealistic and counterproductive to think of product features or service elements in this binary fashion. Let's take an example from a software development application. Consider the product function "File Open command" and its linkage to the customer need:

Can mix material from many documents

A software developer would find the question, "What is the linkage between 'Can mix material from many documents' and 'File Open command'?" meaningless, because the software must include the "File Open command" in order to be functional at all. In other words, at first glance, since there must be a "File Open command" in any case, there is no need to ask the question. Lurking behind the developer's response is the assumption that the only choice is between "provide an Open command" and "don't provide an Open command."

In fact, there is a continuum of choices. For example, the Open command function could be designed to allow the user to open many documents at once, thereby strengthening the linkage to customer satisfaction. Or it could be designed to open only one document, forcing the user to invoke the command several times in order to access several documents and thereby lowering Customer Satisfaction Performance for this need.

Another variation in the Open command design might be to provide the user with a convenient list of files that have been associated with any currently opened or recently opened files.

Many other variations on the possible capabilities of the Open command could be imagined by the creative software engineer. These creative options do not occur on a simple linear continuum. However, the developers would do well to create a model in their minds of a simple continuum of design possibilities ranging from "stripped-down Open command" to "deluxe Open command." Such a model provides the team with a realistic range of possible technical responses for the SQC. If several high linkages show up for a product function, then a "deluxe" version of that function may be called for.

9.2 IMPACT VALUES

In Figure 9-2 we saw the most common numerical values that QFD teams assign to the linkage symbols. Some early QFD applications in the U.S. used 5, 3, 1, and 0 to represent levels of impact from "strong" to "none." Over time, QFD facilitators felt the need to create a stronger contrast between "strong" and the other impacts, so that strong impacts would have more influence on the ultimate prioritizations. The value 9, which Don Clausing had seen in Japanese applications in the early 1980s, was rapidly adopted for this purpose. The value "9" is a comfortable multiple of "3" and serves the purpose of making the strong impacts dominate the matrix. The choice of numerical system should be based upon the goals of those doing the QFD work. More often than not, QFD work is aimed at setting product-development priorities *based on customer importance or value.* That being the case, it is better to use a 0, 1, 3, 9 scale because it differentiates product functions better than a linear scale like 1, 3, 5, 7, 9 would provide.

Some facilitators favor "7," because it is a compromise between "5" and "9." Others favor "10" because it serves much the same purpose as "9" but provides for easier manual calculations than the other choices.

The greater the difference between the values assigned to "strong" and "moderate," the less likely it is that an SQC with only "moderates" assigned to it will be assigned a technical importance greater than an SQC with at least one "high." Most QFD practitioners in the U.S. feel that is the way things should be.

There is no universal standard for any of the choices, but mathematically you want there to be clear differentiation; hence a non-linear scale like 0, 1, 3, and 9 works rather well. The impact values simply provide a way for the development team to express its judgment on the relative impacts of SQCs on customer needs. The SQCs can then be differentiated in terms of their overall contribution to Customer Satisfaction Performance.

9.3 PRIORITIES OF SUBSTITUTE QUALITY CHARACTERISTICS

Once the development team has determined all the impacts, or linkages, some simple arithmetic provides one of the key results of QFD: the relative contributions of the Substitute Quality Characteristics to overall customer satisfaction.

These represent the priorities of the SQCs, and are placed near the bottom of the HOQ, as shown in Figure 9-5. Figure 9-6 illustrates how the figures are calculated.

The impact of Technical Response X upon Need A is high. We multiply the numerical value for "high" (9) by the Normalized Raw Weight for Need A (.43). The result of 3.9 has been entered above the diagonal of the cell. This value is called the **relationship** of Technical Response X to Customer Satisfaction Performance on Need A. After computing

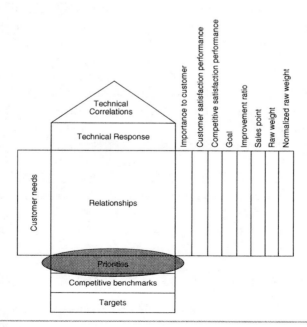

Figure 9-5 Priorities

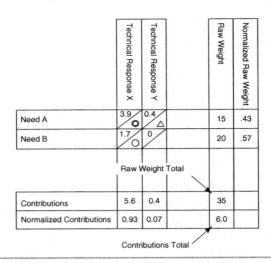

	Technical Response X	Technical Response Y		Raw Weight	Normalized Raw Weight
Need A	3.9 ◯	0.4 △		15	.43
Need B	1.7 ◯	0		20	.57
	Raw Weight Total				
Contributions	5.6	0.4		35	
Normalized Contributions	0.93	0.07		6.0	
	Contributions Total				

Figure 9-6 Contribution Calculations

all relationships, we add all the relationships for a Substitute Quality Characteristic and put the result in a Totals row at the bottom of the matrix. These Totals are called the **contributions** of the SQCs to overall customer satisfaction. The larger the contribution, the more influence the SQC has on Customer Satisfaction Performance, and therefore the more important it is for the product or service to do well in the implementation of that SQC.

If the SQCs and their contributions are to be transferred to the left side of another matrix (as in Figure 4-24, Chapter 4), it is useful to convert the contributions to normalized figures, using a normalization method analogous to the one described in Section 7.7, and to use the normalized contributions in the next matrix. QFD practitioners must take care to gain insight from the numbers, but not treat them as the final decision makers, lest they follow numbers blindly at the expense of their own intuition and expert design guidance. The point of all the VOC work preceding QFD is to understand and focus on meeting those needs the customer deems most important, even if the numbers somehow show one key need slightly lower than another. If the VOC says Need A is the most important, we dare not ignore or compromise it much.

9.4 NEGATIVE IMPACTS

It happens occasionally that a Substitute Quality Characteristic is found to have a *negative* impact on Customer Satisfaction Performance for a certain attribute. In other words, as the SQC "dial" is conceptually rotated in the direction of goodness, Customer Satisfaction Performance for a certain attribute is judged by the development team to go down.

Such negative relationships can happen when an SQC has been selected for its positive impact on one or more attributes, but the attribute with which it is negatively linked has not yet been considered. Some examples:

- In computers, faster internal clock speed may have a positive impact on the customer's need to get work done faster, but it may also have a negative impact on the customer's need for system reliability. Faster clock speed implies higher internal operating temperatures, which generally cause parts to deteriorate faster.

- In service, a wider range of services may have a positive impact on the customer's need for getting all transactions processed from a single source, but it may have a negative impact on the customer's need for a simple method to use the entire range of services (since more choice implies more difficulty in making a selection).

- In automobiles, thicker steel sheets on the doors may have a positive impact on the customer's need for safety in case of collision, but may have a negative impact on the customer's need for good fuel efficiency (since the weight of the car would go up with thicker steel sheets).

Such negative impacts complicate QFD computations. They also require some care on the part of the development team when analyzing the QFD results. Here is a good way to handle negative impacts:

1. Define special symbols to represent the negative impacts—for example, ◎, ○, and Δ.
2. Define corresponding numerical values for the negative impacts (-9, -3, and -1 for strong negative impact, moderate negative impact, and possible negative impact are most common). Multiply the appropriate impact by the Raw Weight to calculate the relationship (as described in Section 9.3). Along with the positive relationships, some of the resulting relationships will now be negative.
3. For each Relationships section column containing negative impacts, compute *two* sums: the algebraic (signed) sum of the relationships, and the sum of the absolute values of the relationships.
4. If the difference between the algebraic total and the absolute-value total is small (as for Technical Response X in Figure 9-7), then the effect of the negative impact is small and can probably be disregarded.
5. If the difference between the algebraic total and the absolute-value total is large (as

	Technical Response X	Technical Response Y		Raw Weight	Normalized Raw Weight
Need A	2.3 ⊙	0.8 ○		15	0.25
Need B	1.0 ○	-3.0 -⊙		20	0.33
Need C	-0.4 △	0.4 △		25	0.42
Raw Weight Total ⟶				60	
Algebraic Totals	2.8	-1.8			
Absolute Value Totals	3.7	4.2			

Figure 9-7   Negative Impact Calculations

for Technical Response Y in Figure 9-7), then the effect of the negative impact cannot be ignored. The team is being confronted with a **breakthrough opportunity.** That is, the team is being challenged to define one or more Technical Responses that provide positive impacts to replace the one that contains negative impacts. One way to try to solve this problem is to define Technical Responses that have narrower ranges of impact.

An alternative method for dealing with negative impacts that is favored by some QFD facilitators is

1. Express all impacts on Customer Satisfaction Performance as positive
2. Study the negative impacts as they are reflected in the Technical Relationships (the roof) section of the HOQ (see Chapter 10)

Regardless of the approach, it's better to get rid of negative impacts than to find ways of handling them in QFD. If negative impacts show up in the QFD process, try to find new SQCs that have a positive impact across all customer needs.

9.5 MANY-TO-MANY RELATIONSHIPS

As a final point, we should note the process we use in QFD to finally arrive at the contributions of the SQCs. We normally start with customer needs. From each need we generate one or a few SQCs (see Figure 8-4, Chapter 8). We might expect that when we determine the impacts, we will see high impact for SQCs that were generated to relate strongly to the need from which they originated, and low impacts elsewhere—basically a one-to-one relationship between SQCs and customer needs, as in Figure 9-8.

In practice, the SQCs tend to take on a life of their own once they are placed in the matrix. Their impact on *all* customer needs must be evaluated, and many of the surprises of QFD spring from these evaluations. The SQCs invariably are found to relate to many customer needs, sometimes even more strongly to customer needs that did not suggest the SQCs in the first place (see Figure 9-9).

Unexpected strong and weak relationships come up. SQCs thus emerge as important because of relationships with customer satisfaction that no one had previously understood.

The matrix structure of QFD helps the team systematically evaluate all possible relationships, even those that "could not possibly matter." Thus, we see how the QFD process helps a team to ask questions it might not otherwise ask, and therefore arrive at answers it might otherwise overlook.

After all the relationships have been evaluated, the design team that is following a DFSS approach would identify the top-ranked SQCs as the key inputs to be followed up with Critical Parameter Management or CPM.

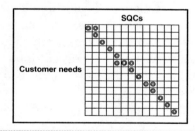

Figure 9-8   Pattern of Impacts after SQCs Have Been Deployed from Customer Needs (One-to-One)

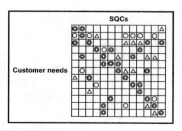

Figure 9-9   Pattern of Impacts after All Relationships Have Been Evaluated (Many-to-Many)

9.6 SUMMARY

We have now seen how the elements of the Technical Response (the Substitute Quality Characteristics) can be prioritized: by estimating the impact of each SQC on the Customer Satisfaction Performance of each Customer Need. The Relationships section of the HOQ provides the mechanism for displaying all the relationships, and for computing the priorities of each SQC.

The Relationships section lies at the heart of QFD. It provides the development team with a way of mapping the relevance of proposed technical responses to customer needs.

The development team makes a determination, for each SQC-customer need pair, of the impact that SQC has on Customer Satisfaction Performance on that customer need. The impacts are usually represented by symbols. The numbers related to these symbols are a measure of the amount of impact.

The product of the impact and the importance of the customer need (the need's Raw Weight) is called the Relationship. The Relationship is a measure of the importance of the impact to overall customer satisfaction.

The sum of Relationships for a single SQC is called the Contribution of that SQC to overall satisfaction. The relative values of Contributions for all SQCs provides a rank ordering of SQCs that can guide trade-offs and resource allocation for the remainder of the development process.

Judging impacts is never easy, but it is especially difficult when the SQC has not been expressed as a continuous variable. Nevertheless, there are ways of thinking of non-numerical SQCs as ranging from "basic" to "deluxe," which helps to define a continuum of effort or attention. This continuum makes it easier for the team to judge impact, and also to develop creative solutions to meeting customer needs.

Now that we understand how the SQCs relate to customer needs, we have to ask how the SQCs relate to each other. Is it possible that performing well on one SQC might cause us to perform poorly on another? Or perhaps performing well on one SQC makes it easier to perform well on another. The Technical Correlations chapter (Chapter 10) shows us how to do this analysis. This is the beginning of CPM work for those QFD efforts that reside within a DFSS framework.

9.7 DISCUSSION QUESTIONS

- Take a subset of your known customer needs. Create an abbreviated House of Quality matrix and place these needs on the left. Put a few of your Substitute Quality Characteristics on top of the matrix. Fill in the impacts according to your best judgment. Any surprises?

- Show your matrix to a colleague. Do you both agree on the way you have set up the matrix, and on its contents? If you don't agree, what can you learn from each other?

Technical Correlations 10

We have mentioned in earlier chapters that QFD is a key to concurrent engineering because it facilitates team members communicating with each other. This chapter presents an explicit mechanism for mapping out exactly what kind of communication must occur on the development project. The Technical Correlations section (Figure 10-1) provides the mechanism. It shows for which technical areas close communication and collaboration are important, and for which they are not. It also shows us where design bottlenecks may occur, and therefore where design breakthroughs are needed.

The Technical Correlations section is probably the most underexploited part of the House of Quality. Few QFD applications use it, yet its potential benefits are great. Perhaps after reading this chapter, you will be encouraged to use it yourself for the competitive advantage it offers.

10.1 MEANING OF TECHNICAL CORRELATIONS

The Correlations section of the House of Quality is sometimes called the Technical Correlations section. More often, it is referred to as the "roof" of the House of Quality. It maps interrelationships and interdependencies between Substitute Quality Characteristics. This section of the HOQ is probably the least used in today's practice of QFD. However, as this chapter indicates, the analysis of the roof can lead to important insights in the development process.

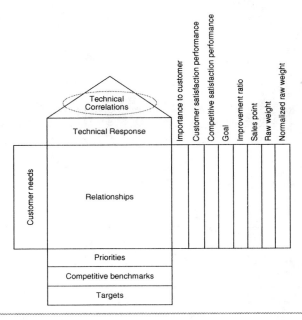

Figure 10-1   Technical Correlations

The Correlations section consists of the half of a matrix that lies above the matrix's diagonal. The SQCs are arrayed along the top and side as shown in Figure 10-2. The matrix is then rotated 45 degrees, and since the SQCs are already available along the top of the HOQ, they double as the labels for both the rows and columns of the roof, making the row and column labels unnecessary (see Figure 10-5).

Very often, especially after a technical concept has been decided upon and is somewhat understood, the developers will be able to see that as SQC_x is moved in the direction of its target, SQC_y will be influenced, either towards or away from its target. The degree and direction of influence can have a serious impact on the development effort. Indications of negative impact of one SQC upon another represent bottlenecks in the design. They call for special planning or breakthrough attempts.

For example, in a case where a producer of integrated circuits was developing an Application-Specific Integrated Circuit (ASIC) for a customer, analysis of the Technical Correlations disclosed that certain customer requirements were technically incompatible. Had this incompatibility not been discovered during the product-planning phase, the ASIC producer believes it would have wasted several million dollars in preliminary development work before discovering the need to redesign.

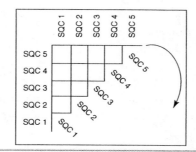

Figure 10-2   Matrix Above the Diagonal, To Be Rotated

Other examples of conflicting technical correlations are the following:

- For an automobile, increased BTU rating of an automobile air conditioner (SQC$_x$, the More the Better) may have a negative impact on automobile weight (SQC$_Y$, the Less the Better). Notice how the SQCs are somewhat solution-dependent, because it is assumed that an air conditioner will be used to cool the auto interior, and that higher BTU ratings imply heavier air-conditioning units.

- In computer software, an increased number of print-command options (SQC$_x$, the More the Better) may have a negative impact on the number of keystrokes and mouse clicks required to invoke a print command (SQC$_Y$, the Less the Better).

- For service, reducing the number of minutes a customer speaks to a service associate (SQC$_x$, the Less the Better) may have a negative impact on the number of calls a customer must make to get a problem solved (SQC$_Y$, the Less the Better).

In QFD, we usually identify five degrees of technical impact, as in Figure 10-3.

These symbols carry no directional connotation. Some QFD practitioners believe that the technical correlations should be treated as bidirectional in impact. Then, if either SQC varies, the other will vary according to the type of correlation identified between them.

I feel it's more constructive to indicate a direction of impact, since a developer can often make a strong argument for impact of SQC$_x$ upon SQC$_Y$, but no impact of SQC$_Y$ upon SQC$_x$.

√√	Strong positive impact
√	Moderate positive impact
<blank>	No impact
x	Moderate negative impact
xx	Strong negative impact

Figure 10-3   Degrees of Technical Impact

For example, for the automobile, increased BTU rating of an air conditioner could raise automobile weight, but increased automobile weight does not necessarily affect the air conditioner's BTU rating. A counter-argument is that increased automobile weight *does* affect the air conditioner's BTU rating, because the air conditioner will have to be more powerful to achieve the same cooling in a heavier car as compared to a lighter car.

If the QFD team wishes to record direction of impact, the symbols in Figure 10-3 can be coupled with arrows indicating the direction, as in Figure 10-4. Bidirectional impact can be denoted by a two-headed arrow ($\leftrightarrow$).

Once the Technical Correlations matrix has been rotated, the redundant rows and columns have been removed, and correlations have been filled in, it might look like Figure 10-5. The interpretation of the topmost cell, for example, is then

> Moving SQC 1 in the direction goodness has a moderate negative impact on SQC 5's direction of goodness.

10.2 RESPONSIBILITY AND COMMUNICATION

One of the most important benefits of the Technical Correlations is to indicate which teams or individuals must communicate with each other during the development process. If a team with prime responsibility for meeting the target value of SQC 1 runs into difficulties, or changes plans, Figure 10-5 tells this team that the team responsible for SQC 4 will be seriously affected, and the team responsible for SQC 5 will be somewhat affected, by their change of plans.

One method for making this information more explicit is to construct a Responsibility Matrix. This matrix would display the SQCs along the left, and the possible responsible teams along the top. A cell in the matrix would indicate the relationship of the team to the SQC. The possible relationships are shown in Figure 10-6. A good management practice is to assign responsibility for an objective, such as meeting the target value for an SQC, to a single individual or a single organization. Therefore, the Responsibility Matrix would have a single ◎ in each row.

A responsibility table for the roof shown in Figure 10-5 might look like Figure 10-7.

√√→	Strong positive impact, left to right
√←	Moderate positive impact, right to left
<blank>	No impact
x←	Moderate negative impact, right to left
xx→	Strong negative impact, left to right

Figure 10-4 Degrees of Technical Impact with Direction of Impact

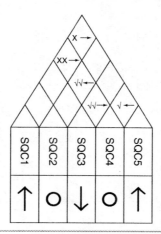

Figure 10-5   Roof of the House of Quality

◎	Primary responsibility
○	Supporting role
Δ	Should be informed

Figure 10-6   Responsibility Symbols

In Figure 10-7, Organization A has primary responsibility for SQC 1. Because changes in SQC 1 strongly affect SQC 4, the organization responsible for SQC 4 (Organization E) must be informed of progress on SQC 1. Organizations C and F have primary and support responsibility, respectively, for SQC 2. SQC 2 has no impact on any other SQC, however, so communication about SQC 2 progress does not have to be communicated to any other organizations.

10.3 CORRELATIONS NETWORK

An alternative but equivalent representation of the correlations in the roof is the Relationship Digraph, or Relationship Network diagram, shown in Figure 10-8. In this diagram, the SQCs are represented by circles, and the SQC affected is shown by arrows connecting the circles. The degree and direction of influence is shown by the √ and X indications alongside the arrows. A more-elaborate formulation of this type of network, called the Interpretive Structural Model, has been developed by John N. Warfield,[1] but is beyond the scope of this book.

---

1. John N. Warfield, *A Science of Generic Design: Managing Complexity Through System Design,* 2nd ed. (Ames, Iowa: Iowa State University Press, 1994).

	Organization A	Organization B	Organization C	Organization D	Organization E	Organization F	Organization H
SQC 1	◎		△		△		
SQC 2			◎			○	
SQC 3		◎		○	△		○
SQC 4			△		◎		
SQC 5			◎		△		

Figure 10-7 Responsibility Matrix

Notice that SQC 1 in Figure 10-8 has arrows emanating from it, and none pointing toward it. This is an indication that SQC 1 is a **driver** in the sense that it influences other SQCs but is not in turn influenced by any SQCs. On the other hand, SQC 2 has only incoming arrows and no outgoing ones. It is called an **indicator**. The status of SQC 1 as a driver indicates that efforts to move SQC 1 in the direction of goodness will affect other SQCs as a byproduct (in this case, negatively). It does not seem worthwhile to invest resources in moving SQC 2 directly, since it is strongly affected by SQC 4 directly and by SQCs 1, 5, and 3 indirectly.

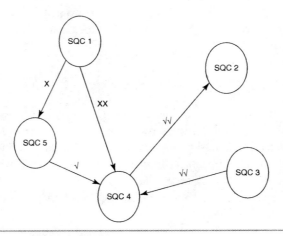

Figure 10-8 Relationship Network Diagram

10.4 OTHER CONSIDERATIONS

There are two possible sequences regarding concept selection and QFD: A concept is chosen either before or after the HOQ. The former sequence is usually associated with a DFSS framework, but not exclusively so. The latter is more traditional QFD, and is sometimes followed by DFSS practitioners as well.

In the first sequence, the QFD work has begun after the design concept has been chosen. Typically, within DFSS the VOC is planned, analyzed, and validated, and then concept exploration and selection are done through Stuart Pugh's Concept Selection matrix and approach (see Section 13.2.4, Chapter 13), prior to starting the HOQ. With a concept already selected, the conflicts in the roof will be explored for possible trade-off studies and/or concept modification. A useful method to help resolve such trade-offs is the TRIZ contradiction matrix, first created by Gerhardt Altshuller and highlighted in QFD practice by Noel Leon-Rovira.[2] TRIZ stands for Teoriya Resheniya Izobretatelskikh Zadatch, Russian for "the theory of solving inventors' problems." One aspect of TRIZ uses a contradiction matrix that explores 40 underlying functional principles across 39 features that can be traded off against each other. These principles were determined from extensive patent study, and the QFD practitioner is encouraged to undertake further reading if working in a trade-off situation on an existing design concept. While there is still some debate as to the order (HOQ first or Concept Selection first), I believe as long as the concepts are iterated to solve or deal with trade-offs, that is all that really matters in the end. The use of Stuart Pugh's Concept Selection Matrix and development process explores the solution space rather nicely, as many DFSS practitioners have observed in practice. It ensures that the best possible hybrid solutions in the solution space framed out by customer needs are explored.

The second sequence is the completion of the HOQ prior to concept selection. In this sequence, when there are strong conflicts in the technical correlations, these conflicts may be influenced by concept choice. They can be reduced or eliminated depending upon which concept is under consideration. Often, what separates concepts is the uniqueness of how functionality in the product design is split or allocated among the product elements. This uniqueness may impact trade-off situations where some increased functional independence exists between one design concept and another. As a result, there may be fewer trade-offs in some concepts. In such a scenario, it is possible to do a what-if analysis with each of the key concepts and develop multiple versions of the HOQ. Consider filling out an HOQ for each of the top two or three concepts. This is not

2. Noel Léon-Rovira and Ing. Humberto Aguayo, "A New Model of the Conceptual Design Process Using QFD/FA/TRIZ," *Proceedings of the 10th Symposium on Quality Function Deployment* (Novi, Mich.: QFD Institute, June 1998), republished at www.Triz-Journal.com/archives/1998/07/d/index.htm.

often advised, but may be useful when the product introduction may have a high-risk impact on the business. In that case, decisions regarding which concepts will *not* be carried forward must be based upon the most detailed information possible. Exploring the top two or three concepts further with their own HOQs, and even product-design matrices, may be warranted before deciding on the final product concept.

10.5 SUMMARY

The "roof" of the House of Quality shows the impact of work on one SQC on the status of other SQCs. The roof can show the existence and nature of design bottlenecks. Where bottlenecks or opposing SQCs exist, the team must plan a concentrated activity to accomplish both.

The correlations in the roof spell out which organizations or individuals must communicate with which others, to allow for smooth progress in meeting target values of SQCs during development. To ignore these interrelationships is to invite chaos into the development process, because changes in plans in one area may go unnoticed in other affected areas until too late in the game. Without the analysis inherent in the roof, some interrelationships will almost certainly be ignored.

Communication about, and responsibilities for, SQCs can be summarized in a Responsibility Matrix. This matrix provides a clear graphical map of ownership for the SQCs.

An alternative graphical representation of the correlations is the Relationship Digraph. This diagram helps to identify the drivers and indicators among the SQCs.

We have now prioritized the SQCs and evaluated their interrelationships. It's time to set technical targets. But technical targets cannot be set in a vacuum. We must see how well the competition is doing and set targets that will ensure that we are competitive. This implies benchmarking the competition. We'll do this in the next chapter.

10.6 DISCUSSION QUESTIONS

- Given the key Substitute Quality Characteristics in your development work, which are the drivers? Which are the indicators? How did you determine your answer?

- Who needs to be informed of design changes? What analysis have you done that shows who should be informed?

- How will your teams deal with conflicts in the HOQ roof? Is there a standard method? Which tools will be used to illustrate the conflicts and resolve them?

Technical Benchmarks

This chapter introduces the concept of competitive technical benchmarking (circled in Figure 11-1).

No organization would invest in the development of a product or service without knowing enough about the competition to be sure that its design is competitive. But if we have defined 30 or 40 SQCs, should we benchmark all of them? Because they have been prioritized using the Relationships section of the House of Quality (HOQ), as we have seen in Chapter 8, we know which SQCs are most important, and we therefore have a strategy for focusing on and benchmarking only the most important of them.

In this chapter, we'll look at the process of benchmarking two types of SQCs: those that have been formulated as performance measures, and those that have been formulated as product functions.

Once the SQCs have been prioritized, the next step is to set targets for them. The most important ones—those with the highest contribution value—require the most care. The QFD analysis indicates that performance of the most-important SQCs will strongly influence customer satisfaction, so the development team will want to set targets for those SQCs thoughtfully. Since these SQCs have been prioritized based on customer needs, it is important that the needs expressed by customers have been appropriately validated. Subsequent management of the development process will then be oriented toward assuring success in meeting those need-based targets. Additionally, as targets are set higher to meet customer satisfaction, longer development times will likely ensue. This will necessitate a trade-off of time-to-market versus customer satisfaction, so that target setting should be reviewed as well. Lastly, it is important when benchmarking that

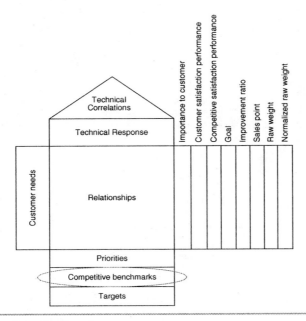

Figure 11-1   Competitive Benchmarks

some estimate be made of where the competitors will likely be when the product is launched, thereby predicting any **future gaps** in performance, feature sets, etc.

A critical set of questions is: How should the targets be set? How aggressive do they need to be? To a great extent, development teams can be guided at this point by the competition's performance as well as their own. If the developers know how well they and the competition are currently performing on the most-important SQCs, they can make crucial strategic decisions regarding whether to match the competition's performance, exceed the competition's performance, or even concede technical superiority to the competition. We'll discuss target setting in more detail in Chapter 12. But before setting targets, the development team would do well to study the competition by benchmarking.

The QFD process provides the basis for strategic competitive benchmarking. The highest-ranking SQCs determine the success of the product or service; therefore, they are the ones to be examined in competitive offerings. The importance of competitive benchmarking cannot be overemphasized; without it, development teams can have many different failure modes. These include **group think**—the self-reinforcement of the team's ideas around issues that need an externally focused examination instead of an internal one. Another failure mode is the ignorance of key ideas or suggestions that do not come from within the group—this is usually called Not Invented Here (NIH). Finally, precious development time and energy can be wasted reinventing solutions to problems that competitors have already solved in standard, non-proprietary ways.

In the process of examining them, the language of the SQCs and the definition of the direction of goodness become important determiners of the competitive benchmarking work.

What do we mean by competitive benchmarking? In general, competitive benchmarking is the process of examining the competition's product or service according to specified standards, and comparing it to one's own product or service, with the objective of deciding how to improve one's own product or service. In QFD terms, the standards are defined as the most important SQCs, as the team identified them earlier in the QFD process.

11.1 BENCHMARKING PERFORMANCE MEASURES

If the SQCs were defined as performance measures, the benchmarking process becomes one of measuring the competition's performance and one's own performance in terms of these measures. To the extent that the performance measures were defined independently of the design of the product or service, the benchmarking process provides ideal "apples-to-apples" comparative data between the competition's and the development team's product or service. The results of measuring the two products or services can be laid down side-by-side (in the HOQ, one above the other) and evaluated at a glance, as in Figure 11-2.

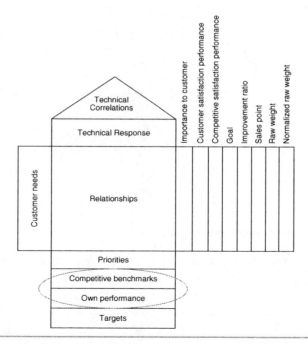

Figure 11-2 Comparison of Competitive Benchmarks and Company's Own Performance

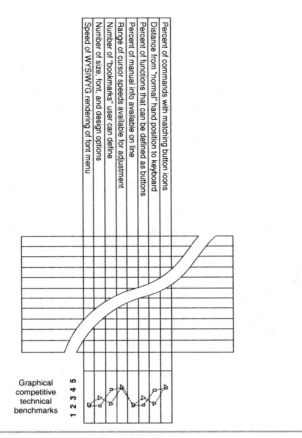

Figure 11-3   Competitive Benchmarks in Graphical Display

Some teams prefer graphical benchmark comparisons, as in Figure 11-3. This is similar to the graphical representation of customer satisfaction performance described and shown in Section 7.3 (Figure 7-16), Chapter 7.

11.2 BENCHMARKING FUNCTIONALITY

If the SQCs have been defined in a more solution-specific manner, with product or service functions explicitly delineated, the comparisons must be much more subjective. This is because the competition's design of a particular SQC is likely to be different, at least in some respects, from the development team's design. One way to deal with these differences in functionality is to decompose the high-ranking SQCs into "sub-SQCs" and compare these to the competition's designs. If the process of developing the SQCs in the first place was by the Function Tree method or the Affinity Diagram method, then the sub-SQCs come from the lower levels of the Function Tree or the Affinity Diagram. In Figure 8-5 (Chapter 8), we see the explosion of an SQC into a long list of subfunctions. Other techniques may be utilized to explore and detail the subfunctions, most notably Critical Parameter Management (CPM), which is described in Section 4.6.1 (Chapter 4). Key subfunctions could be compared to the competition's subfunctions, providing that a design-solution concept has been chosen and its subfunctions can be detailed. Some subfunctions may correspond very closely to the competition's subfunctions; others will not.

The number or percentage of subfunctions that correspond provides valuable numerical information. A listing of the functions that do not correspond provides data for understanding where the competition provides more functionality, or how the competition's design solves the same problem differently from the development team's design.

11.3 SUMMARY

In this chapter, we have seen how QFD provides a method for benchmarking strategically: We use the priorities of the SQCs, determined by completing the Relationships section of House of Quality, to guide us in selecting which SQCs to benchmark. The language of the comparison should be dictated by the language used to define the SQCs.

We can benchmark SQCs described as performance measures, and, with somewhat less precision, we can benchmark SQCs expressed as product or service functions.

Now that we have measured the competition, we are in a position to set target values for our key SQCs. That's the topic of the next chapter.

11.4 DISCUSSION QUESTIONS

- How is competitive benchmarking done in your organization? How do you decide what aspects of the competition's offerings to measure or examine? What is the linkage between what you benchmark and what your customers' needs are? Compare your current method of deciding what to benchmark to the way QFD helps you decide.

- What language is used for making comparisons? Is it numeric? If not, how does it compare to the language used for writing requirements, specifications, and design documents? How effective is it in providing side-by-side comparisons between your products or services and the competition's?

- How much money should be spent on competitive benchmarking? How much does your organization spend? If the process does not exist, how can you get some insight quickly into competitors' products? Who should own the benchmarking process, and own the responsibility of improving it?

12 Targets

This chapter covers the process of setting targets for the key SQCs (Figure 12-1). It takes up the QFD process after the development team has determined the most important SQCs and has benchmarked the competition.

We will look at setting numerical targets for SQCs that have been expressed as performance measures, and at setting function or feature targets for SQCs that have been expressed as features. In the case of numerical targets, we'll look at the possibility of using some simple algebra as a guide. For features, we'll make use of the idea of expressing features in terms of their "sub-features," as we discussed in Section 11.2 (Chapter 11).

Setting targets is of course a matter of greatest interest to product and service developers. Obviously, setting SQC targets will drive all subsequent development activity. Development teams set targets for themselves whether or not they use QFD to plan their projects. Without a process such as QFD for setting targets, the targets tend to be a hodgepodge of customer-oriented and technical goals, with very little linkage or relationship among them. Nor is the prioritization of these targets based on a line of reasoning that all others can follow, let alone agree with.

With QFD, the targets have a context: They are related to customer needs, to the competition's performance, and to the organization's current performance. The rank ordering of the targets is based on the systematic analysis done in the Relationships section (and all the prior QFD analysis). The rank-ordering process is traceable, because all the decisions affecting the rank ordering are recorded in the QFD matrix.

The QFD process itself provides no cookbook approach for setting targets for SQCs. The most vital information not explicitly visible in the House of Quality is the business

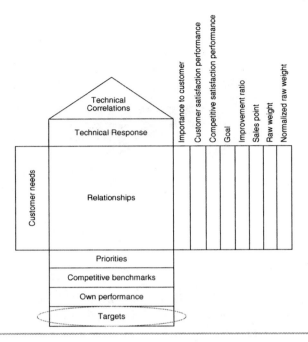

Figure 12-1   Targets

know-how and technical expertise of the development team. However, the HOQ provides much of the strategic information needed, laid out in a compact form.

In Section 9.1.1 (Chapter 9), we referred to the possibility that some of the linkages in the Relationships section might not be linear, because the associated SQCs may be Dissatisfiers or Delighters. The target-setting stage is a good time to deal with these Kano classifications. It is useful at this point, if the team has not already done so, to classify the SQCs according to their categories in the Kano model. For example, with today's increasing fuel costs for automobiles, a competitive Kano classification is overall fuel efficiency. A Delighter in such an environment would be a display of recent fuel efficiency as well as current trip efficiency. Such a feature would allow commuters to track their miles per gallon and learn how to improve it, by varying routes and driving speeds for example.

For those SQCs classified as potential Delighters, the team must decide how aggressive it can afford to be in target setting. There is relatively little downside risk in setting a conservative goal: Customers will not notice the absence of a Delighter. However, the potential gain of setting a goal that beats the competition is high.

For those SQCs classified as Dissatisfiers, the team cannot afford *not* to be aggressive. Customers are expecting perfection in these areas, and anything less than perfection will result in customer dissatisfaction.

For those SQCs classified as Satisfiers, the team can expect that the better it performs on the SQC, the greater the Customer Satisfaction Performance will be for the linked customer needs. The rest of this chapter deals with strategies that can help in setting targets for Satisfiers.

12.1 NUMERICAL TARGETS

12.1.1 COMPARISON WITH COMPETITION

One approach to setting targets is similar to the process of setting Customer Satisfaction Performance goals in the Planning Matrix (see Section 7.4, Chapter 7). The primary inputs to goal setting for Customer Satisfaction Performance in the planning matrix are

- Importance of customer attribute to customer
- Our current satisfaction performance rating
- Competition's satisfaction performance rating

Similarly, the primary inputs to target value setting of SQCs are

- Rank order of SQCs (Priorities)
- Competition's technical performance (Competitive Benchmarks)
- The development team's product technical performance (Own Performance)

The line of reasoning for setting targets is also similar to that used in setting goals in the planning matrix. Starting with the highest-ranking SQC, determine the strength of the development team's position relative to that of the competition. Based on the team's knowledge of the difficulty of performing well on the SQC, the team can decide whether to aim to do better than the competition, to match the competition, or to concede technical leadership to the competition. As a general rule, the goal should be for technical performance that exceeds the best in the world for those SQCs that matter the most to overall customer satisfaction.

12.1.2 MATHEMATICAL MODELING

While QFD is certainly not a precise mathematical model of the relationship between technical performance and Customer Satisfaction Performance, a little bit of simple mathematics could serve as a guide to the development team in setting targets.

In the case of a Substitute Quality Characteristic for which Less Is Better, we may imagine that the relationship

$$\text{Customer Satisfaction Performance}_A = f(\text{Technical Performance Measure}_x)$$

is approximated by a linear function of the form $y = m \times x + b$. The slope m of the line is defined by two known points.

The coordinates of the first point are $(p_{\text{world class}}, s_{\text{world class}})$—that is, the point that corresponds to the best product in the world. Satisfaction performance s is highest, and technical performance p is lowest (best). In Figure 12-2, Customer Satisfaction Performance for the world-class point was measured at 4 (with 5 as the best possible score), and technical performance was determined by competitive benchmarking to be 80.

The coordinates of the second point are (s_0, p_0)—that is, the point that corresponds to the development team's current product. In the example, satisfaction performance was determined by market research to be modest (3.5) and technical performance (based on laboratory measurements) was higher than (not as good as) that of the world-class product (100).

Visual inspection of the line passing through these two points in Figure 12-2 suggests that a good value of p to aim for might be 60 or less, which could result in a Customer Satisfaction Performance level of about 4.5.

The general equation for the line is

$$s_{\text{LTB}}(p) = M \times p + [s_0 - (M \times p_0)]$$

$$\text{where } M = (s_{\text{world class}} - s_0) \div (p_{\text{world class}} - p_0)$$

and:

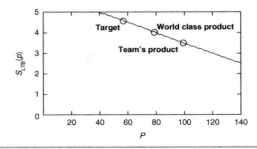

Figure 12-2   Satisfaction Performance $s_{\text{LTB}}(p)$ vs. Technical Performance p (LTB)

$S_{LTB}(p)$	denotes Customer Satisfaction Performance on a customer need as a function of a Substitute Quality Characteristic p of the type Less the Better
$s_{world\ class}$	denotes Customer Satisfaction Performance with the best product in the market
s_0	denotes Customer Satisfaction Performance with the development team's product
$p_{world\ class}$	denotes technical performance of a Substitute Quality Characteristic with the best product in the market
p_0	denotes technical performance of a Substitute Quality Characteristic with the development team's product
p	denotes technical performance of a Substitute Quality Characteristic

A similar but slightly more complex relationship can be modeled in the case of Target Best (TB). Here we might consider that a parabola best describes the relationship between technical performance and Customer Satisfaction Performance, as in Figure 12-3, where the target value for p is 4. (The meaning of the other variables is as before.)

The coordinates of the first point are $(p_{world\ class}, s_{world\ class})$—that is, the point that corresponds to the ideal target value. Since the direction of goodness is Target Is Best, it is not possible to perform better than the target (4). In this example, we have assumed that achieving the target (4) results in the best possible Customer Satisfaction Performance (5).

The coordinates of the second point are (s_0, p_0)—that is, the point that corresponds to the development team's current product. As in the previous example, Customer Satisfaction Performance was determined by market research, and the value of the SQC (4) was determined by laboratory measurement.

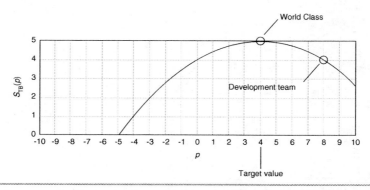

Figure 12-3 Customer Satisfaction Performance s(p) vs. Technical Performance (p), Target Is Best

The general equation for the parabola in Figure 12-3 is

$$s_{\mathrm{TB}}(p) = [|s_{\text{world class}} - s_0| \times (p - \text{target value})^2] \div (-p_0^2 + 2 \times p_0 \times \text{target value} - \text{target value}^2) + s_{\text{world class}}$$

This mathematical model provides some insight for setting the target for a Substitute Quality Characteristic, but care must be exercised in its use. For one thing, there is no guarantee that the relationship between any SQC and the corresponding customer attribute satisfaction performance is precisely linear or precisely quadratic. More importantly, Customer Satisfaction Performance for any attribute is usually the function not of a *single* SQC, but of several SQCs.

Consider the Relationship section matrix in Figure 12-4. The relationship between SQC X and Attribute D has been modeled as one-to-one. The team can consider satisfaction performance on Attribute D to be solely related to SQC X. However, what can the team assume about the relationship between SQC U and Attribute A? Two other SQCs (V and Z) besides SQC U contribute to satisfaction performance on Attribute A. The relationship, if it can be modeled at all, is much more complex than a simple one-to-one function. Even if the development team were to assign one-third of the influence on Customer Satisfaction Performance to each SQC, other complications could muddy the waters. For example, the direction of goodness might not be the same for each SQC.

Target values that are selected must always pass the common-sense tests of practicality and affordability. Customer-importance weightings must be taken into account when deciding which targets will be set more aggressively and which may set to "reasonable." If the degree of difficulty or costs to meet an aggressive target are too high, then the target must be modified to meet practical constraints. If there is no general understanding of the trade-offs in target shift versus customer preferences, a mid-term project survey can be done with key customers to find price sensitivities and feature-set preferences. Such a

	SQC U	SQC V	SQC W	SQC X	SQC Y	SQC Z
Attribute A	◎	◎				◎
Attribute B		○	◎			
Attribute C	△		◎		○	
Attribute D				◎		
Attribute E		△	○	◎	◎	

Figure 12-4 Relationship Section

survey would include a design that allows conjoint analysis to determine customer preference by feature set and pricing sensitivities.

Since there are no specific closed-form methods for setting target values in QFD, one reasonable approach could be the following:

1. Treat each SQC *as if* it were the only SQC contributing to Customer Satisfaction Performance of an attribute.

2. Use the simple models described here to create a first estimate of an appropriate target value.

3. Repeat this analysis for all the customer satisfaction attributes that this SQC is linked to. This creates multiple target values for the same SQC. In Figure 12-4, for instance, we would estimate three target values for SQC W: one each for its relationships to Attributes B, C, and E.

4. Choose the most aggressive of these target values.

12.2 NONNUMERIC TARGETS

Setting targets for Substitute Quality Characteristics defined as features or processes is obviously more difficult than dealing with numbers. A number is one-dimensional, but features and processes are multidimensional and multifaceted.

There are two helpful ways of thinking about targets for nonnumeric SQCs: the continuum model and the sub-feature model.

In the continuum model (Figure 12-5), we may imagine the SQC to be on a continuum as described in Section 9.1 (Chapter 9). This continuum could have as its endpoints "stripped down" and "deluxe." The development team could judge where on the continuum

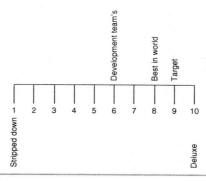

Figure 12-5   Continuum Model

their current offering lies, and where the best in the world lies. To clarify these judgments to themselves and to others, they would do well to make their subjective judgments as objective as possible by documenting:

- The differences between Best In World and *deluxe*
- The differences between Development Team's and Best In World
- The differences between Development Team's and *target*

By using the sub-feature model, the development team can explode each feature to be targeted into its component sub-features, as described in Chapter 8. Each sub-feature could be evaluated according to the continuum model, or could be exploded into lower-level sub-features. Targets could be set by continuum, where Best In World and Development Team's sub-features line up, and by identifying sub-features to be added where the features don't line up. In Figure 12-6, we can see at a glance how the sub-features line up, and how those that are present score via the continuum model. The arguments for setting targets are then easier to justify and explain. The column of continuum targets provides a profile of the target that is at least comparable in detail to the profiles of the Development Team's and Best In World profiles.

In Figure 12-6, the development team's existing product, and the Best in World product, both deliver **Sub-feature a** at about the same level on the continuum model. The team has decided to set a target for this sub-feature to outperform Best in World. In the case of **Sub-feature d**, the development team's level of delivery is not very high (3), but Best in World lacks this Sub-feature. Assuming Best in World will not provide **Sub-feature d** within the product-planning timeframe, the team has decided to continue delivering the sub-feature, but not to improve or expand it. Hence, the Target has been set to the same value at which the sub-feature is already rated. In the case of **Sub-feature f**, Best in World already provides

	Development Team's	Best in World	Target
Sub-feature a	7	7	8
Sub-feature b		5	4
Sub-feature c	6	8	8
Sub-feature d	3		3
Sub-feature f		8	
Sub-feature g	6		6

Figure 12-6   Target Setting by Sub-feature and Continuum Analysis

that sub-feature at a high level. The development team has decided to concede this sub-feature to the competition, and use its resources to compete in other areas.

12.3 SUMMARY

In this chapter, we have seen how to set targets for performance measures and how to set targets for functions or features. Target setting is easier and more objective when Substitute Quality Characteristics are expressed as performance measures. Nevertheless, even with SQCs that are expressed nonnumerically as features or processes, it is possible to set targets in terms of "degree of deluxeness" of implementation, and in terms of the existence of sub-features.

In any case, setting targets for the most important SQCs provides the development team with a systematic method for deciding how to compete against the best in the world.

This completes our description of the elements of QFD. We have explored all the sections of the House of Quality in detail. We have seen how QFD encourages us to listen to the customer and to represent customer needs in a hierarchical format—a particularly useful representation in which we can zoom in or zoom out, and in which we can understand the relative importance of each customer need, as well as see how well we are performing in meeting each need.

We have seen in the Planning Matrix how to use the customer needs and competitive market research to make strategic top-level decisions about what to emphasize in the development project. The primary result of this strategic planning is the Raw Weights that represent the revised relative importances of the customer needs, based on a combination of customer perceptions and the company's business objectives.

We have explored various ways of generating a statement of Technical Response—the Substitute Quality Characteristics (SQCs) that together represent the product or process requirements. The primary approach, but not the only possible approach, has been to "deploy" the customer needs into a set of solution-independent Performance Measures.

We have explored QFD's unique method for establishing the linkage between the Technical Response and the customer needs—the development of impacts in the Relationship section.

Finally, we have examined the QFD approach to prioritizing the elements of the Technical Response by multiplying the impacts by the Normalized Raw Weights of the customer needs and adding them for each SQC. We have also seen how these priorities can be used to shape a benchmarking effort, and to set target values for the SQCs.

We've come a long way in this discussion of QFD, and we've accomplished a lot. What's next?

Many developers have launched into using QFD after studying the parts of the House of Quality as we have just done, only to become mired in details. The key to successful use of QFD is to understand how QFD relates to the organization's overall operations and needs, and to understand how to deal with the practical, minute-to-minute, day-to-day details of QFD that drive a team to despair if not properly managed.

In Part III, we'll look at how the principles of QFD mesh or could be made to mesh with the organization's needs. In Part IV, we'll explore the practical issues of implementing the QFD House of Quality.

Once we've mastered the HOQ, we'll be ready for Part V, a discussion of the different directions a team can go with QFD after completing the HOQ. Join us on this journey toward being the best in the world!

12.4 DISCUSSION QUESTIONS

- In your development process, how are targets set? Can you link these targets to the competition's performance? To customer needs?
- Is your organization more likely to be comfortable with numeric or nonnumeric SQCs? If nonnumeric, does sub-feature analysis exist to provide a foundation for goal setting? If not, how much work would be required to do such analysis?

PART III

QFD from 10,000 Feet

The Larger Picture: QFD and Its Relationship to the Product Development Cycle

This chapter examines QFD from the perspective of organizations that develop products and services. These organizations struggle with development issues all the time, and they have developed strategies for dealing with those issues.

Those strategies include creating phased development models; acquiring various development tools, including CAD/CAM hardware and software; and making use of various budgeting and project-management tools. Despite these attempts at improving their processes, anyone working within one of these organizations knows that more can be done.

In this chapter, we see how QFD can have a positive impact on the development process. First, we look at QFD's role in improving communication between development team members. We even consider the suggestion that QFD and DFSS could be the "glue" that helps the team members to work together more efficiently, even if—*especially if*—they come from different disciplines and understand different jargon. QFD is suited for less-prescriptive organizations, while DFSS is more prescriptive for each project, demoting QFD to a managerial role for flow down, as just one aspect of overall product development.

As a reminder, back in Chapter 2 (Figure 2-3), we drew a holistic view of product development that showed a continuous loop: Planning New Concepts → Designing → Manufacturing & Delivering → Selling → Planning Concepts. In Chapter 3, we showed how DFSS and QFD were related in a DFSS project-tool sequence. Furthermore, we indicated in Chapter 3 (Section 3.2) how QFD played a role in the front end to determine how product concepts as well as service concepts were developed. As other texts[1] do a great job explaining the role of QFD in a DFSS context, that will not be repeated here.

In this chapter, we look at how QFD's original formulation has been extended (as Enhanced QFD) to cover some important aspects of development that have always been difficult to systematize: technology and concept selection.

Finally, we look at reliability engineering, and how QFD can provide guidance in this difficult discipline.

13.1 CROSS-FUNCTIONAL COMMUNICATION

In most organizations, the product- or service-development process was in existence long before QFD came onto the scene. The introduction of QFD is often viewed by developers as an add-on—a tool that must or can be used in addition to the existing development processes. An alternative way of viewing QFD is as an organizer, or as the glue that can bind together the many aspects of development.

In Figure 13-1, the objects arranged in a circle around QFD represent typical organization functions, each of which plays a role in successfully bringing a product to market. In order to execute their functions well, the people undertaking these organizational functions must communicate with each other efficiently, and the functions must all be focused on a common goal.

For example, Purchasing must negotiate for the best materials, parts, and services at the best price. The vendor's goods and services must have exactly the characteristics that allow the developers to deliver the product or service that their customers will prefer over the competition. Purchasing cannot work in a vacuum. The purchasing agents must understand what Manufacturing needs and what Product Design has called for, and must understand how these things relate to customer needs. Small misunderstandings can lead to large errors.

Likewise, the Product Design and Manufacturing people must know what's possible in the way of purchasing from suppliers. Are the specifications realistic? Are the right things being specified? Could better things be specified? Are the costs within budgetary constraints? Only a high level of constructive communication among Product Design, Manufacturing, and Purchasing can guarantee the best answers to these questions, and the smartest decisions at the outset of development.

To a large extent, thinking of QFD as being at the center of the communication process helps each functional group to find out how its work fits in, and to tell all the other groups what it needs from them.

---

1. Most notably, Randy C. Perry and David W. Bacon, *Commercializing Great Products with Design for Six Sigma* (Upper Saddle River, N.J.: Prentice-Hall, 2007).

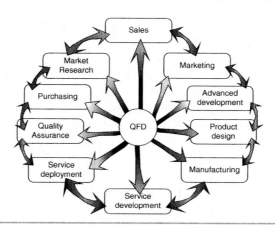

Figure 13-1   QFD as an Organizer of the Development Process

Further, putting QFD in the center can enable concurrent engineering, as described in Chapter 2. Imagine the activities depicted in Figure 2-6 proceeding without a communications tool such as QFD. How could the Manufacturing team and the Design team work toward a common purpose? How could bottlenecks in Delivery be anticipated and prevented? These can be achieved only if the organization has established up-front and ongoing planning that keeps the customer needs in focus and that links actions at every step to these customer needs. If concurrent engineering, or anything resembling it, is to occur in a development organization, QFD will be seen as the principal tool to support it.

Equally importantly, QFD provides criteria for determining the goodness or appropriateness of any decision. These criteria are derived directly from, or can be clearly traced to, customer needs. Hence, the Voice of the Customer becomes the key backdrop against which communication occurs during the development process.

Development cycles are generally rather long for most products and services. Automobile development cycles in the U.S. were about five years during the 1980s, and approached three years during the competitive 1990s. Desktop computers take six to 12 months for development by the most competitive U.S. companies (completely new designs take much longer). The time between software releases is generally one to two years. Brand new services from financial companies often take six to 18 months to launch.

The biggest problem facing developers with these long time frames is dealing with the uncertainties of the future. Who can be sure that decisions made on a particular day will be appropriate in the business climate of one, two, or five years in the future? So many things can change during that time. Customer attitudes could change based on political, social, or scientific developments that could not have been predicted. Even if the events could be predicted, the reactions of customers cannot be predicted with certainty.

Against such a backdrop of uncertainty, how can the developer produce the right product or service at the right time? Two strategies, used in combination, can go a long way toward reducing these risks: reducing cycle time, and staying tuned to the customer. We've discussed the first strategy elsewhere (Section 2.4 of Chapter 2, and many other places in this book).

Staying tuned to the customer means developing a clear and detailed understanding of customer needs during the planning stages of development. It also means checking development decisions against up-to-date assessments of customer needs on a continuing basis.

The longer the development cycle, the more likely it is that the marketplace and associated customer needs will change. The change will most likely be a quantitative change—that is, importance levels or satisfaction performance levels on the customer needs may vary. Less likely, but very significant if it happens, is a qualitative change: A new customer attribute may appear. The QFD matrix or matrices can easily be updated with such changes, and development targets and priorities can be reassessed to determine whether the development work as planned is still on track.

Once again, by placing QFD at the center of the organization's development communication model, any changes to the Voice of the Customer can quickly be assessed against each function's activities.

13.2 ENHANCED QFD AND CONCEPT SELECTION

When QFD was first introduced into the U.S., the QFD model assumed that the selection of appropriate technology for a product, piece part, or service concept was outside the scope of QFD. Don Clausing and Stuart Pugh realized that the process for selecting innovative concepts could and should interact with the translation of customer needs to prioritized technical responses. They embodied their ideas in a process called Enhanced QFD (EQFD).[2,3]

EQFD consists of five broad but interrelated processes: Contextual Analysis and Static/Dynamic Status Analysis, Structuring of Product Design Specification, House of Quality, Concept Selection, and Total System/Subsystem Analysis. In DFSS terms, these interrelated processes are steps in what today is called Concept Development or Concept Engineering.

---

2. *Enhanced Quality Function Deployment* (Cambridge, Mass.: MIT Center for Advanced Engineering Study, n.d.). Series of five instructional videotapes.

3. Don Clausing and Stuart Pugh, "Enhanced Quality Function Deployment," *Proceedings of the Design and Productivity International Conference* (Honolulu, Feb. 6–8, 1991): 15.

13.2.1 CONTEXTUAL ANALYSIS AND STATIC/ DYNAMIC STATUS ANALYSIS

Contextual Analysis in EQFD is a comparison of competitive products for the purpose of deciding at a strategic level how to position the product being planned with respect to both the business and the technical environment.

In contextual analysis, the development team creates a series of pairwise plots of critical functional parameters across several products. The goal is to characterize the current state of product offerings with respect to critical parameters.

Figure 13-2 is an example of a single pairwise plot. Gasoline-powered vehicles of a certain type are plotted using two critical performance parameters: vehicle weight versus fuel efficiency as measured in miles per gallon. The plot reveals which vehicles are most fuel efficient (the ones plotted closest to the upper diagonal bound), and that all vehicles perform within the range shown by the upper and lower diagonal bounds. Clearly, a vehicle that performed lower than the lower bound would be unacceptable in the market (unless it had some major compensating characteristic outside of this analysis), while a vehicle performing above the upper bound would be very competitive.

A wider range between the upper and lower bounds is a possible indicator that the associated technology is **dynamic**, meaning that a relatively wide range of technical solutions exists for achieving the same result. EQFD distinguishes between **dynamic** and **static** concepts.

Static concepts are those that have stopped evolving and have become standardized. The wheel is an example of a static concept—it has been around for thousands of years, and it is almost always used as the locomotor device for vehicles that move over relatively smooth surfaces, such as paved roads or paths. (There are other alternatives, used for specialized situations: runners for ice and snow; treads for uneven surfaces. Most of these are also static for their particular applications.)

The wheel suspension, on the other hand, is considerably more dynamic. Many technical solutions are in use, and for the piece parts of suspensions, a still wider variety of concepts compete with each other.

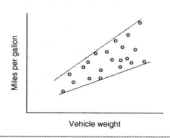

Figure 13-2   Contextual Analysis—Plot of Critical Parameters

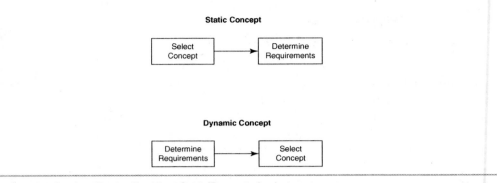

Figure 13-3 Product Design Based on Static/Dynamic Analysis

If a concept is static, the conventional concept is probably the best to adopt. However, a team should make note of any new technologies now available that may offer breakthroughs to replace static concepts. For example, conventional tolls on toll roads have been in place for multiple decades, yet are being moved aside for electronic toll collection. This is an example of a disruptive technology as described by Clayton Christensen.[4] If the concept is dynamic, the team would do well to compare contending concepts and choose the best for their application, or develop an even better one. The process for evolving toward the best solution is called Pugh Concept Selection, and is described in Section 13.2.4.

In EQFD, developers are urged to perform contextual analysis on critical parameters (as determined by QFD), and static/dynamic analysis on technical concepts. Figure 13-3 summarizes the relationship between Static/Dynamic analysis and QFD: If the concept is static, first select the concept, then use QFD to determine the requirements. If the concept is dynamic, first use QFD to determine the requirements (using solution-free SQCs), then use the resulting prioritized SQCs and target values as input into the concept-selection process.

13.2.2 STRUCTURING OF PRODUCT DESIGN SPECIFICATIONS

In EQFD, Clausing and Pugh recommend the use of generic design parameters as a way of beginning the process of developing Substitute Quality Characteristics. Pugh has described these generic parameters in his book, *Total Product Design*.[5]

4. Joseph L. Bower and Clayton M. Christensen, "Disruptive Technologies: Catching the Wave," *Harvard Business Review* (Jan–Feb 1995).

5. Stuart Pugh, *Total Design: Integrated Methods for Successful Product Engineering* (Reading, Mass.: Addison-Wesley, 1991).

The generic parameters include shelf life, packaging, life in service, aesthetics, and more than 25 other factors (more than 30 factors in all). As indicated in Chapter 8, this is one of several methods for developing SQCs.

The advantage of using a generic set of SQCs, rather than generating one by the Affinity Diagram method, is that the generic set is likely to include topics that the development team would otherwise overlook. The team can, of course, use or ignore any of the items in the generic list. But by starting with a very complete set of generic topics, they have the opportunity of building their list from a kind of master list that represents the best thinking on the topic.

13.2.3 HOUSE OF QUALITY

The HOQ in EQFD is the same as in "traditional" QFD.

13.2.4 CONCEPT SELECTION

The Pugh Concept Selection Process is the centerpiece of EQFD. It helps developers select between alternative concepts, or converge upon a concept that's better than the starting alternatives. It helps the developers do this in a way that exploits the best aspects of teamwork, and avoids the worst.

Typically, when developers are confronted with alternative concepts, they align themselves with one or another concept. The process they use is similar to comparing apples to oranges. As the concepts are compared, people tend to focus on one or two strengths, and on one or two weaknesses, of each concept. For example, one automobile design is aerodynamically superior, while another design is less costly to manufacture. This type of unbalanced comparison tends to encourage people to ignore the full complement of criteria that all concepts should be judged by.

When one concept is selected, the other is rejected. The result is that some people "win" and others "lose." The process of Concept Selection becomes contaminated with this "win/lose" attitude, and emotions dominate the decision process.

There are at least two tragedies associated with the win/lose style of selecting from alternative options. The first is that the best aspects of each rejected concept are discarded along with the concept itself. The second is that the losers find themselves forced to support the winning option. The quality of their support is likely to be quite poor, both because they may not understand the winning concept well, and also because of human nature.

The Pugh Concept Selection Process increases the likelihood that the best aspects of all the alternatives will be reflected in the chosen alternative. This usually means that the chosen alternative will be better than any of the starting alternatives. The Pugh Concept

DATUM

Response: quicker is better		+	+	+
User maintenance: fewer tasks is better		−	−	−
Ease of Mfg.: fewer parts is better		S	−	−
Ease of service: clearer messages is better		+	+	−
Risk: older technology is better		−	−	−
Mfg. risk: more off the shelf parts is better		S	−	−
		+ 2 S 2 − 2	+ 2 S 0 − 4	+ 1 S 0 − 5

Figure 13-4   Pugh Concept Selection Process

Selection Process also increases the likelihood that the entire team will participate in evolving toward the chosen alternative, and will then be more willing and able to support its implementation.

The Concept Selection Process was described by Stuart Pugh as early as 1981.[6] Professor Pugh and others have presented the method to hundreds of developers in the U.S., and because of its appeal, it has been widely adopted by QFD and DFSS practitioners alike.

The general principles governing Concept Selection are

- Aim for a world-class concept, settle for nothing less
- Start with the best current concept, even if it's the competition's
- Either beat the best current concept, or use it

Here's a step-by-step procedure for using the Pugh Concept Selection Process (refer to Figure 13-4).

1. Identify the initial alternative concepts. Describe all the concepts at about the same level of detail. Fairly rough descriptions of the concepts will suffice at this stage.

---

6. Stuart Pugh, "Concept Selection—A Method That Works," *Proceedings of the International Conference on Engineering Design,* Rome, 1981 (New York: ASME Press), WDK 5 Paper M3/16: 497.

Represent each concept by a diagram, picture, or cartoon. (Sometimes this may be difficult or impossible—for example, with software concepts. In such a case, identify key words that stand for the concept.) The idea is to make the picture suggest, as much as possible, what is important about the concept at a glance, so that team members can scan the concepts and be rapidly reminded of the essentials of each.

Developers typically start with five to ten concepts—sometimes fewer, rarely more.

2. Develop success criteria. The criteria can be drawn from customer-needs data, from product-specifications documents, and from business requirements. Express each success criterion along a continuum, so that one of these terms will apply: the Larger the Better (LTB), the Smaller the Better (STB), or Nominal the Best (NB).

Developers typically generate long lists of criteria. However, it's best to pare this list down to ten to 20 items, to keep the matrix to a manageable size, and because we want the key criteria, rather than all criteria, to drive the selection process. One way of paring the list is to prioritize, using the Analytical Hierarchy Process or the prioritization-matrix method. After prioritizing, drop the least-important criteria, leaving the most important ten to 20 items. Another method, if higher-level or previous QFDs have already been developed, is to use prioritizations derived from those previous QFDs.

3. Create a large matrix. The physical dimensions of Concept Selection matrices are generally as large as those of QFD matrices: three to six feet high and ten to 20 feet wide.

Place the criteria along the left edge of the matrix. The criteria should be as legible as possible from a distance. In practice, this is hard to accomplish, because the criteria tend to be carefully worded, and are therefore stated in longish sentences.

4. Designate the most promising concept as the *datum*, and place its diagram at the top of the leftmost column. Place the diagrams of the remaining concepts along the top of the matrix, to the right of the datum. The diagrams should be large enough to be seen from the opposite side of the room.

The datum is the concept to which the others will be compared.

5. Compare each concept to the datum. Start at the top row (corresponding to the first criterion) and ask whether the current concept is *better, worse,* or *about the same* in terms of meeting the criterion. If the team's judgment is that the concept is *better,* enter a "+" in the corresponding matrix cell. If it is *worse,* enter a "-." If the concepts are *about the same,* they enter a symbol that indicates equality. Stuart Pugh recommends the use of "S" rather than "=," in order to have a symbol that is unlikely to be confused with "-."

Don't bother to distinguish between *much better* and *a little better*. Occasionally, measurement may be possible, or existing data may be available. For the most part, however, the developers will be using their engineering experience and expertise systematically to make a series of focused consensus judgments. I often remind teams I'm working with that the process is not "rocket science." It's a team method for encouraging creativity tempered by discipline. If the developers can develop a concept that *overall,* across all criteria, is better than any concept they started with, they will have gotten the primary benefit of the Concept Selection Process.

6. After comparing all available concepts to the datum, the team tallies at the bottom of each column the number of "+," "-," and "S" scores. There are no cookbook rules for developing conclusions based on these scores. The score we are aiming for is one that contains only "+" items—that is, we want to develop a concept that is superior to the datum across all criteria. There are essentially two possible paths to discovering such a concept.

   The first path is via the brilliant idea, possibly suggested by the discussions and scorings of the other concepts.

   The other path is via observing the relative strengths and weaknesses of the concepts and aiming for a hybrid concept that may be superior to any yet analyzed. In my own experience working with many teams doing Concept Selection, the "mix-and-match" approach works well, and often, as teams consider swapping features of one concept with another, they get the "aha" they need for a breakthrough. Often we focus on *attacking the negatives* by taking the most promising concept and hybridizing it with features of other concepts that have positives in rows where the emergent concept has negatives.

7. When the team identifies a concept that is superior to the datum, it's a good idea to promote that concept to be the new datum, take a break from the process, and upon reconvening, generate new concepts that will be better still. A view of this process is shown in Figure 13.5. There is probably no end to how good the concepts can get; it's limited by the available time a team has for this process. In my experience, teams often work together in Concept Selection for three to eight long meetings ("long" being a half-day to a day). The additional meetings allow the team to explore fine distinctions between similar concepts, or to branch out to markedly different concepts that may have been suggested by earlier sessions. The meetings may be held on consecutive days. That is sometimes impractical, however, because of other work some team members may be responsible for, or because the team may need to gather information from other sources before the next meeting.

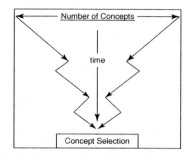

Figure 13-5   Iterative Nature of Pugh Concept Selection Process

A few precautions should be mentioned regarding the use of Pugh Concept Selection. Engaging a skilled facilitator is often a good idea to avoid any team-leader bias in choosing the concepts and hybrids to move forward in each round. Such a person will also have experience to deal with other challenges that may arise from individual team members, and to know when to let discussions continue and when to close them off.

Another precaution: The success criteria listed on the right should all have been validated by customers or at least be derivatives of such, but they may not have equal weights. When concepts are tied or very close to being tied, they need to be re-examined in light of which ones satisfy the more-important customer criteria when choosing one over others. The interested reader or student of decision making may want to peruse the book *Making Robust Decisions* by David Ullman.

13.2.5 Total System/Subsystem Analysis

The HOQ is used in EQFD iteratively, first to develop system-level specifications, then at more-detailed levels of design. In Figure 13-6, we see that the output of the first HOQ is a set of system-level specifications. These become the input to a Pugh Concept Selection matrix that is used to develop the most appropriate technical concept. This concept, along with the system-level specifications, becomes input to a second QFD matrix (called a "design matrix" in EQFD terminology), the output of which is a set of subsystem specifications. Some of these subsystems may involve dynamic concepts, and are therefore suitable for Concept Selection, leading to subsystem concepts. These concepts become the basis for lower-level design matrices for piece parts.

Thus, the EQFD formulation provides a structure that product developers can use to guide their technology choices as well as their designs. Like QFD, EQFD helps developers reduce development time, increase customer satisfaction, lower product cost, create greater cooperation between functions, establish a better corporate memory, and make fewer mistakes. EQFD also provides an explicit road map for the development of complex systems involving multiple levels of design.

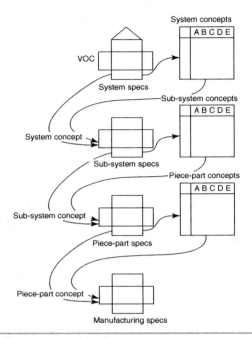

Figure 13-6   Deployment through the Levels

13.3 ROBUSTNESS OF PRODUCT OR SERVICE

No product- or service-development project can escape the necessity of reliability analysis. Some developers plan and design for reliability before launching their products. Others are forced to do so after product launch, by angry customers. Although many developers act as if it is cheaper to skimp on reliability design during development, it is almost always an order of magnitude more expensive, if possible at all, to build in reliability after the product or service has been launched. Proper VOC work at the beginning of design and development efforts as part of either the QFD or the DFSS process should have uncovered the need for a reliable and robust product or service, as well as the magnitude of customer satisfaction or dissatisfaction associated with it. If this work was done up-front, then QFD will make it part of the overall design and development efforts.

The most common and effective reliability-planning techniques are the Process Decision Program Chart (PDPC), Fault Tree Analysis (FTA), Failure Mode and Effects Analysis (FMEA), and Robust Engineering.

PDPC[7] is one of the Seven Management and Planning Tools introduced in Chapter 3. In this method, the team draws a diagram—possibly a flow diagram—that describes how a customer would use or interact with the product or service. At each point in the customer's interaction, the team brainstorms all the possible things that could go wrong. For each brainstormed mishap, the team then develops **countermeasures**. These countermeasures fall into two categories: actions that can be taken *when* the mishap occurs, and—better—actions the development team can take to *prevent* the mishap. PDPC diagrams can get huge; several methods exist to focus the team on the most likely and the most serious possible mishaps.

Fault Tree Analysis[8] refers to the systematic deconstruction of possible product or service failures into their component causes based on the *function* of the system, product or service. Analysis usually begins with a list of the functions, possibly generated from a functional tree. For each functional failure, several causes can be generated by the development team. Each of these causes will in turn likely have several causes. Thus, the development team can build a complete Fault Tree. Having completed this analysis, the team must then decide how to best design the product or service to prevent the failures. Fault Tree Analysis takes into account both qualitative and quantitative analysis, by examining both the branch probabilities for the former and the number of branches for the latter. It is extremely effective in evaluating potential warranty problems or sudden field failures as well as in proactive use before systems, products, or services are launched.

Failure Mode and Effects Analysis[9] refers to the systematic analysis of possible product or service failures based on the *design* of the product or service. See also 4.6.4, in Chapter 4. Analysis begins with the major components of the product or service. For each top-level component or subsystem, the team identifies possible ways that component could fail. The analysis can extend to components within each subsystem, as well as to the subcomponents and parts. Each failure mode is described in terms of the nature of the failure, the possible causes of the failure, and the effect and impact of the failure. In addition, the team estimates for each failure mode the probability of its occurrence and the severity of its effect. Sometimes teams also document safety issues, repair or recovery processes, and plans for preventing or responding to the failure.

---

7. Michael Brassard, *The Memory Jogger II: A Pocket Guide of Tools for Continuous Improvement and Effective Planning* (Salem, N.H.: GOAL/QPC Inc., 1994).

8. Don Clausing, *Total Quality Development* (New York: ASME Press, 1994).

9. J. M. Juran (author) and Frank M. Gryna (photographer), *Quality Planning and Analysis: From Product Development Through Use,* 2nd ed. (New York: McGraw-Hill, 1980).

Robust Engineering[10] is an extension of the statistical field of Experimental Design. This extension, originally developed by Dr. Genichi Taguchi and now practiced increasingly in the U.S., helps engineers identify those design factors that have the dominant impact on the performance of their system. It further helps them select target values for these dominant factors such that the system performance will be robust—i.e., the system will operate consistently—in the face of environmental variation, manufacturing variation, and product deterioration over time.

Detailed discussion of these techniques is beyond the scope of this book.

For each of these methods, the developers are faced with an imposing list of areas where they could devote time and money to improve the reliability or robustness of the product or service. For almost all development projects, it is impossible to perform all the reliability work they have identified. They must select some areas to work on, and leave other areas alone. (Naturally, all life-threatening aspects of the design must be attended to.) Once the reliability analysis is complete and all possible reliability areas have been identified, the team can then turn to QFD to guide them in prioritizing which areas to work on.

13.4 SUMMARY

In this chapter we have examined the impact QFD can have on the overall development process. Its impact spans three major areas: team members' communication; their coordination; and reliability engineering.

Regarding communication and coordination of team members, QFD's power lies in its relationship to the entire product- or process-development cycle. It should be positioned as the primary communication tool, facilitating and even defining communication between functional organizations.

With regard to technology and Concept Selection, Enhanced QFD, or Concept Engineering, is a product concept development schema having QFD at its center, but also incorporating other key tools. It furthers QFD's influence on the development process. In addition to the HOQ, EQFD incorporates Contextual Analysis, Static/Dynamic Analysis, Concept Selection, and Total System/Subsystem deployment through the levels.

---

10. Genichi Taguchi (author) and Don Clausing (technical ed. for the English ed.), *The System of Experimental Design: Engineering Methods to Optimize Quality and Minimize Costs* (New York: UNIPUB/Kraus International Publications, 1987), 2 vols. See also Madhav S. Phadke, *Quality Engineering Using Robust Design* (Englewood Cliffs, N.J.: Prentice Hall, 1989).

Finally, with regard to reliability engineering, a key activity in any development project is designing for reliability. While powerful techniques such as PDPC, Fault Tree Analysis, Failure Mode and Effects Analysis, and Robust Engineering are available, by themselves these methods cannot guide developers as to which aspects of the design most need reliability engineering. QFD as a prioritization tool plays an important role in helping to focus reliability-engineering resources on the critical reliability issues.

Thus, we see how QFD can make the development world a better place to be for developers and their customers. Unfortunately, there are many obstacles along the road to this better place. These obstacles are the same ones that make it so hard for organizations to effectively adopt Total Quality Management or any of its aliases.

Given these obstacles in *our* organization, should we give up on QFD? I recommend no such thing. QFD will have a reduced positive effect in an imperfect environment, but the effect will still be positive. In the next chapter, we'll discuss what it might be like to use QFD under less-than-ideal circumstances.

13.5 DISCUSSION QUESTIONS

- What product-development information gets communicated across functional lines? What mechanisms support this communication in your organization?

- Which functional groups regard themselves as the least influential? How might QFD help them to change their self-assessment? What benefits might accrue if these groups regarded themselves as more influential in product-development decisions?

- How does the Advanced Development function decide which technologies to invest in? What activities does Advanced Development pursue that resemble Contextual Analysis and Static/Dynamic Analysis? Classify the key concepts in your product or service according to the Static/Dynamic dimension. Share your results with your colleagues. Do they agree with your analysis? Given your analysis, how do Concept Selection activities in your organization match up? What are the advantages and disadvantages of your organization's Concept Selection activities versus Pugh Concept Selection?

- How is reliability engineering done in your organization? Compare your organization's tools with PDPC, Fault Tree Analysis, and Robust Engineering. How are priorities set? How might QFD change priority setting for reliability engineering?

QFD in an Imperfect World

This chapter covers the real-world scenarios of corporate life, where the customer is *not* king, and where quality processes are *not* uppermost in every employee's mind.

In this chapter, we'll look at scenarios where some team members resist QFD. How can the visionary manager or QFD facilitator help people to benefit from QFD in the face of opposition? While the topic of organizational change is beyond the scope of this book, anyone attempting to introduce QFD into an organization will do better by anticipating the possibility of resistance.

Much of the resistance to QFD comes from its multifunctional nature. Because it helps team members from different functional groups to communicate, it appears to some members of these groups as if other team members are taking over their jobs. In this chapter, we'll explore these "turf" issues, and look for constructive ways of introducing QFD.

When Lean Six Sigma and Design for Six Sigma initiatives have been deployed, QFD is often included in the methods and tools. In such cases, the acceptance and practice of QFD is increased, compared with companies that have not done any work with Lean Six Sigma or DFSS. However, unless clearly spelled out, ownership of various parts of the QFD process may still become contentious.

The QFD process—especially the planning of QFD, acquiring the VOC, and developing the HOQ—has the effect of broadening each person's concept of what a successful product or service must consist of. Some product developers welcome the implied broadening of their responsibilities; others don't. Welcome or not, however, the broadening effect exists.

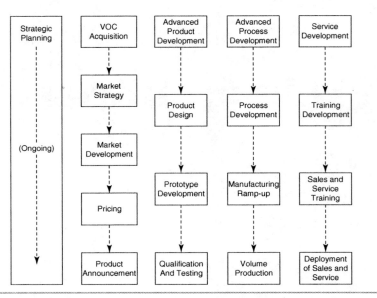

Figure 14-1   Uncoordinated Corporate Product Development Processes

Figure 14-1 shows a simplified view of processes within a corporation that relate to developing products and services. As with many organizational charts, this figure does not contain horizontal arrows or lines to indicate communication and decision-making paths that cross, or that should cross, organizational boundaries. In the ideal organization, the organization's infrastructure would support such cross-functional communication and decision-making, and QFD would be part of that support.

In many situations, however, the organization has not evolved to the point where cross-functional communication is easy and natural. QFD, if it exists, is used in isolated areas, and therefore its full benefits are not gained. For many QFD advocates, this is a fact of life. All is not lost, however: While the full set of possible benefits is not available, some benefits may be. It's a good idea for the realistic manager planning to use QFD to understand the possible partial-benefit scenarios. Let's take a look at some of the most common of these scenarios.

14.1 MARKETING FUNCTIONS IN ENGINEERING-DRIVEN ENVIRONMENTS

Marketing departments react to QFD in various ways, depending on the organizational culture.

In technically driven organizations, product definition emanates from engineers and scientists. Hence, the activities of the marketing department are normally centered on overall sales strategy and management of key accounts, rather than on product definition.

QFD is often viewed with suspicion by people in such marketing roles, because the QFD paradigm explicitly suggests that product definition be shared with other departments, in particular with the technical organizations that actually perform the functions. In other words, QFD may be seen as a threat to turf that the marketing team claims but is insecure about.

The rational marketing professional's response to QFD in this case might be: "QFD provides me a way of having more influence on the product-development process than I currently have." Should a QFD champion encounter resistance to QFD by marketing departments, very likely fears of losing control over coveted turf are at issue.

In a DFSS deployment or DFSS culture, where VOC is jointly owned by marketing and engineering functions, QFD efforts often proceed very smoothly. Since both functions are well aware of the what, how, and when of VOC collection and processing, the HOQ goes especially smoothly. As pointed out earlier, if QFD has been previously utilized before the advent of DFSS, its usage often increases and QFD gains more acceptance due to the deployment of DFSS.

14.2 ENGINEERING FUNCTIONS IN MARKETING-DRIVEN ENVIRONMENTS

In marketing-driven organizations, it may be the engineers or the technical establishment that is suspicious of QFD. There may be technical leaders accustomed to the role of approving or dashing to the rocks project proposals that emanate from the marketing organization. Typically these technical gurus retain control of murky technologies that no one else in the organization understands. They tend to be unfamiliar with customers or with the wide range of customer needs that bear little relationship to the technical aspects of the product.

The rational technical leader's response to QFD might be: "QFD provides me with a way of off-loading the uninteresting nontechnical aspects of product definition on others who can combine their knowledge of the customer with my knowledge of the technology."

Once again, in a DFSS and also in a "Marketing for Six Sigma" (MFSS) environment, the natural joining point between the two functions of marketing and engineering is at the collection of VOC. In more than 30 DFSS full deployments, we have seen that the joint teamwork in the VOC efforts often does wonders for cross-functional teamwork and interdepartmental cooperation.

14.3 QFD IN ENGINEERING-DRIVEN ORGANIZATIONS

Engineering-driven organizations generally develop products that are centered around advanced and dynamic technical concepts. Examples are:

- Computers that are faster and cheaper than previous models, because of advanced semiconductor technology
- Tennis racquets that are more lightweight than previous models, because of advanced materials development
- Medical instruments that permit surgery without cutting open large areas of the patient's skin, based on breakthroughs in the combined areas of medicine, optics, and micromechanical engineering

These products may have mixed success in the marketplace. Sometimes the dominant technology fills customer needs, as with lightweight tennis racquets. In other cases, the dominant technology is irrelevant to market needs, as with mainframe computers in the early 1990s, with demand continually growing for workstations and desktop computers.

The typical problem in engineering-driven organizations is that engineers, for one or several reasons, don't listen very much to the Voice of the Customer. The most common reasons are these:

- Engineers believe they already know what customers want
- Engineers don't regard listening to customers as "part of the job"

Better buy-in is usually obtained by showing engineers the processes of preparing to collect, collecting, and analyzing the VOC. A high-level view of such a VOC front end for QFD, as shown in Figure 1-1 (Chapter 1), may be of help in obtaining buy-in from the engineers.

A common approach to QFD in engineering-driven organizations is for engineers to consider QFD a high-level design tool, the use of which need not affect other functions in the organization. Many engineering groups have developed the Affinity Diagram of customer needs according to their own beliefs of what the customers want. They have gone on to estimate customer importance of needs and customer satisfaction, usually by consensus of the engineering team, and have developed the entire HOQ on their own (see Figure 14-2).

This approach would seem to sidestep the primary purpose of QFD, as a tool that is meant to drive the VOC into product design. Indeed, as we have seen (with U.S. auto manufacturers in the late 1970s and early 1980s, for example), many large companies have suffered greatly by not responding to their customers.

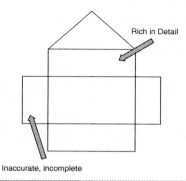

Figure 14-2 Engineering-Driven QFD

Despite this misapplication of QFD, there are still some benefits to building a House of Quality, however incorrect the basic inputs to it are. The main benefits are these:

- The group's attempt to create an Affinity Diagram of all customer needs usually results in a far broader, better-structured description of the customer than any single group member could have developed alone.

- The entire team gains a shared, even if incorrect, vision of the customer's needs (and how the Substitute Quality Characteristics relate to them). Most other methods of top-level product specification begin with an elite few engineers or scientists. These technocrats conceive of a product concept while closeted away in a private area. Once they have worked out their ideas, they announce their concept to the rest of the team, and attempt to win them over. A common alternative to this approach involves some iterated series of presentations of their ideas to management and to the rest of the team, in which they accept comments and objections, and respond to them in subsequent versions of their ideas.

Whether the remainder of the team eventually comes to understand their vision, or agrees with it and is willing to support it, is a function of the leadership and the persuasive skills of the original elite designers. This "selling of the product vision" is a hit-or-miss affair, in which the elite designers win if their idea is accepted, and lose if it is not. Conversely, those who must be sold often resist the new idea in order to avoid conceding a win to the elite engineers, since conceding a win implies they must have lost. Of course, it's the organization that suffers the greatest loss, since this kind of selling process generates imperfectly understood concepts along with winners and losers. No organization needs imperfectly understood concepts, and no organization needs losers.

By contrast, with the QFD process, a much larger team participates in developing the concept from the start. Its members work together on it, and emerge with a common understanding of it. They jointly develop a detailed understanding of the relationship between customer needs (as they see them) and corresponding Substitute Quality Characteristics. The customer needs may be inaccurate, but at least the whole team can support the project. A product developed by a team with a common vision is more likely to succeed than is a product developed by a divided team.

Thus we see that even when some of the basic principles that QFD supports (starting with the correctly obtained and analyzed VOC and using cross-functional teams) are violated, the use of QFD can have some benefits. Furthermore, when teams *do* violate basic principles in the use of QFD, their errors become visible to them, creating motivation to do the job right the next time.

14.4 QFD IN MARKETING-DRIVEN ORGANIZATIONS

Marketing departments in strongly marketing-driven organizations tend to regard downstream activities such as purchasing and manufacturing as irrelevant to product planning.

In these environments, QFD activities often involve the market-research and product-marketing functions with little or no representation from engineering and manufacturing.

The QFD activity may be strongly driven by the VOC, but because of technical under-representation may also poorly reflect technical and implementation aspects of the product or service. The result, as with engineering-driven QFD, will be a product-planning process that is off the mark (Figure 14-3). Nevertheless, some benefits are still available to the participants: common understanding of the customer and of a desired direction, as well as increased ability to communicate with each other.

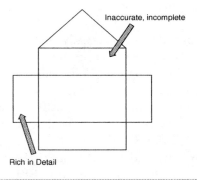

Figure 14-3   Marketing-Driven QFD

14.5 Manufacturing and QFD

The manufacturing organization can benefit from QFD, even when it is used in isolation, in much the same way that engineering-driven QFD can provide benefits to the engineering team.

The manufacturing manager or manufacturing engineer needs to know how the manufacturing function's work affects customer satisfaction. It pays to create a QFD matrix in which the Whats are expressed as either the VOC or a derived language linked to the VOC.

The Hows can be used to describe manufacturing's activities expressed in a suitable language. As with every other part of QFD, it's best if the Hows are expressed as performance measures.

Sometimes when manufacturing works in isolation, a matrix will be constructed that reflects that isolation. Such a matrix may, for example, have a piece-part specification for the Whats and process parameters for the Hows, as in Figure 14-4.

Manufacturing organizations are not alone in having multiple customers and suppliers. Possible partners with the manufacturing organization include those functions shown in Figure 14-5. In some cases the partner is primarily a customer, in others primarily a supplier; in still other cases, the partner is both a supplier and a customer.

For each relationship with another group in which manufacturing is either a customer or a supplier, the What-versus-How analysis will lead to a better understanding of how the relationship should work.

The standard QFD models indicate that the VOC can and should be deployed all the way from high-level system specification down to the factory floor. Back in 1991, survey data indicated that fewer than one QFD application in five used QFD for production.[1] The advent of DFSS and Six Sigma has made great strides in aligning process and product improvement efforts with the VOC, creating demand for the use of QFD.

14.6 Sales and QFD

Besides the potential supplier/customer relationships that every sales organization has with groups internal and external to the corporation, and the benefits possible from What-versus-How analysis, the sales organization may have a particularly interesting opportunity for using QFD if its role involves generating sales proposals of any complexity as a normal part of business.

1. Amitabh Pandey, "Quality Function Deployment: A Study of Implementation and Enhancements" (Master's Thesis, Massachusetts Institute of Technology, 1992).

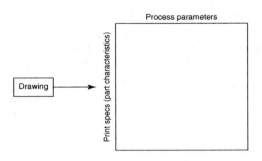

Figure 14-4 Piece-Part QFD When Manufacturing Is Isolated

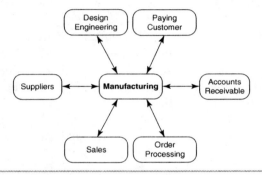

Figure 14-5 Manufacturing, Its Customers, and Its Suppliers

Some sales teams have successfully collaborated with their customers to create a set of customer requirements (the Whats), and have used QFD to demonstrate how their proposals would meet their customers' needs (the Hows).

14.7 SERVICE AND QFD

We've already seen several examples of QFD used for planning service activities. Internal service groups, such as Human Resources and Site Management, have gained insight into their functions by defining their internal customers, interviewing their customers to determine the attributes of customer satisfaction, and then evaluating their existing programs against these internal-customer wants and needs.

Figure 14-5 may also be used to illustrate how QFD can link and capture a wealth of information from various functions. In this way, an archived QFD project and documentation can become a database of knowledge which may be revisited to shorten derivative product design cycle times as well as future product design cycle times. In the last few years, some critical applications of QFD have been implemented in a DFSS environment to handle certain challenges in healthcare, including rapid deployment of new services. One example of this was in addressing customer queuing problems in patient outreach and service centers at Quest Diagnostics. The benefits obtained included improved implementations in over 40 geographically separated service centers.[2]

14.8 EXERTING INFLUENCE BEYOND ORGANIZATIONAL BARRIERS

When product developers embark on a QFD project, their first step is to acquire the structured customer wants and needs, the Voice of the Customer. This process often makes developers, especially design engineers, uncomfortable, because the VOC tends to include elements the developers feel they have no control over. When collecting the VOC, developers often learn that customers care about "easier order processing," "on-time delivery," or "defect-free manufacturing," topics that product engineers typically have little experience with or influence upon. Listening to the customer often has the effect of widening the scope of the problem to cover functions and skills that are not currently involved. What is a product developer, or any other QFD team member, to do with such customer needs?

The answer obviously varies with the organization. In some cases, there is simply no one available to listen to and respond to these needs. The QFD team may not be truly cross-functional; for example, there may be no one present to represent the order-processing function, so the team may feel powerless to improve the order-processing system. This is an example of how important it is to develop products and services with cross-functional teams that truly cover all functions. If each functional group really has an impact on the customer's perception of the product, then each group's inputs into the product from the very beginning will have the greatest positive effect on the end result.

Let's take customer dissatisfaction with order processing as an example. Order processing may be seen by product developers as a fixed, unalterable process of great weight and complexity. Its vagaries may be regarded as difficulties that everyone, product developers and customers alike, must put up with. The attitude that a process owned by another department cannot be altered is of course a symptom of the departmental barriers that

---

2. Michael Scutero, Six Sigma Master Black Belt; Sonja Draganic, Six Sigma Black Belt; and Angela Rylsky, Six Sigma Black Belt, all of Quest Diagnostics Inc., USA, "Challenges in Rapid Deployment of New Services in Healthcare," *Transactions of the 18th Symposium on QFD* (Ann Arbor, Mich.: QFD Institute, 2006).

build up in larger organizations. These barriers lead to inflexibility and paralysis in organizations and ultimately make them unable to compete.

It doesn't have to be this way. These are the very barriers that QFD can help break down.

Product development teams can actually influence such processes as order processing far beyond what they may believe, in at least two ways. The adage "fight 'em or join 'em" provides a useful way of looking at how product developers can deal with the problem.

The *fight 'em* approach aims at forcing or inducing an improvement to the order-processing function. The *join 'em* approach aims at accepting the order-processing function as-is and making the product robust against the process's weaknesses.

The Fight 'Em Approach. This usually means motivating ("It's the right thing to do") or forcing ("Do it or else") the owners of the order-processing function to fix their system.

Motivating often starts with the VOC. The developers can confront the order-processing system's owners with their VOC findings. Assuming the data is credible, reasonable, open-minded order-processing system owners are likely to want to change their processes to better meet the needs of their customers. They can (and should) use QFD and concept selection to develop the desired process changes. If the order-processing system owners are so inclined, they can align their QFD activities with the product-development activities, and begin breaking down barriers between departments.

Forcing changes generally involves going over the heads of the order-processing owners by revealing the VOC data to levels of management that can mandate order-processing changes. Forcing changes in this way is obviously a method of last resort, but I have seen such methods work after other attempts have failed. Forcing changes is most effective when the VOC data is credible, easy to understand, and kindles an emotional response. (See Table 6.2, Chapter 6; all of Chapter 17; and the case studies in Chapter 21.)

The Join 'Em Approach. This means adapting the product design so that it avoids or sidesteps the weaknesses of the order-processing system. The most-common situation relates to products that can be customized to many configurations, such as automobiles, computers, or office panel systems. These products come with many options. Some options are independent of all other options, as for example tinted windows in automobiles: These windows could be made available for all car models, and would have no mechanical dependence or interdependence with other options. Other options won't work with, or are dependent on the existence of, yet other options in the product—for example, airbags, which may be installable in certain steering columns and not in others.

The existence of many options, with a multitude of interdependencies, creates untold difficulties for everyone and every process connected with order processing. Salespeople and order takers must be aware of all the complexities and must be able to explain them to customers, who must be capable of absorbing them and evaluating and selecting the options that are right for them. Everyone involved must possess a vocabulary to formulate each option, and to understand the differences between the many configurations.

Pricing is, of course, a major problem. Besides discount structures, which are normally customer-dependent, the prices for various options and configurations of a product can be difficult to compute and justify.

To design a product to be robust with respect to an order-processing system that cannot easily handle these complexities is to design a product that is simple to understand, has relatively few options, and yet satisfies a wide range of customers; or that has decoupled functions for design functionality. Here is a challenge that QFD can help teams to understand.

Listening to the VOC can be disturbing to design engineers, because they will be exposed to aspects of the product that are beyond their normal view of what they are responsible for. This is exactly why QFD can be valuable to a development team. It forces developers to redefine and broaden their scope, and it helps them to design products that more fully satisfy customers. In the end, these are the products that will keep their company competitive.

14.9 SUMMARY

In this chapter, we have seen that although the ideal environment for successful implementation of QFD may not exist, teams can still benefit from QFD. The principle of linking a team's actions to its customers' needs is a worthwhile one to support under any circumstances. While market-driven organizations differ substantially from engineering- or technically driven organizations, the common element between them is that one function or the other may hold power in product-development planning that it is reluctant to give up to a cross-functional QFD team. This is often ameliorated by joint collection and analysis of the VOC by both engineering and marketing functions.

The discussion in this chapter has shown that even if this obstacle cannot be overcome, teams can use QFD within their own domains, using the best possible data available to them. The difference in the results of using QFD in this limited fashion will stand out in stark relief against the ideal situation. This difference can be exploited to move the organization in the right direction.

In cases where top-down leadership lacks the vision of the customer-driven organization, strong middle managers have introduced QFD. They have used QFD to highlight the lack of good customer information as well as the need for it. QFD can be infectious in such organizations. As it grows in popularity, it can back top management into an uncomfortable corner from which the only way out is to acknowledge the organization's dependence on the customer.

This completes our survey of issues relating to the corporate environment of QFD. We have looked at the need to create efficient horizontal as well as vertical communication in the organization, and at an important enhancement to QFD that provides a framework for managing technology within the context of meeting customer needs. We have also acknowledged that even in less-than-ideal circumstances, QFD can play a vital role.

Now that we've had a look at the trees and the forest, it's time to *walk the talk* by taking a hike along the QFD implementation trail. The next section of the book deals with QFD implementation.

14.10 DISCUSSION QUESTIONS

- Compare your organization to the profiles presented in this chapter. Which profile is closest to, and which is farthest from, your situation? Share your analysis with colleagues. Do they agree? If not, what are the differences between their analyses and yours, and what accounts for the differences?
- How far from the ideal multifunctional team model will you have to diverge in order to implement any kind of QFD project in your organization? What would be a fruitful objective for your organization's first QFD?

PART IV

QFD Handbook

Introduction to the Handbook

Understanding the QFD matrices is not the same as doing QFD. Most people, as they participate in their first QFD activity, get confused. They usually don't know how long the process should take, or what to do next. The process is complex enough so that people get lost among the trees and have difficulty finding their way out of the forest.

Other parts of this book have described what QFD is, and how it fits in with product development and Lean Six Sigma initiatives. The QFD Handbook focuses on how to implement QFD, with special emphasis on the House of Quality. If a development team can gain a level of confidence with the HOQ, the various possible follow-on matrices will be relatively simple to develop.

While QFD has been associated with many successful development activities, there are also many examples of QFD projects that have gotten bogged down and eventually abandoned. Successful QFD projects depend on many factors. Some of these are

- Management support, fueled by faith in the benefits that QFD promises
- Complete, credible customer data, and a process to obtain it correctly
- The right cross-functional development team
- Thorough planning of the QFD project
- A skilled neutral facilitator, or a project leader who can function as one

In the QFD Handbook, we'll expand on these and other factors. Many of these factors are needed to successfully deploy a Lean Six Sigma or DFSS initiative. Those companies that have successfully deployed them will have little trouble engaging teams to follow good QFD practices. The Handbook is intended to show you the range of choices available for structuring a QFD activity. It is also intended to make the QFD process as real as possible for you, so that you can anticipate the many issues and obstacles that may come up along the way.

Before you start your first QFD project, read this Handbook. Then imagine going through the phases described here, one by one. Try to imagine how each team member will react to the work of each phase. Who will support the effort? Who will oppose it? Who will slow things down? Which aspects of the process will be familiar to the team? Which will seem strange and therefore a bit disorienting?

As you review these and other possible obstacles, try to imagine how you will deal with them. Your role could be facilitator, manager of the development group, or member of the QFD team. Whatever role you play, by visualizing what you'd like the process to be, and by anticipating what may go wrong, you will be able to influence it constructively.

Just plunging in to QFD is probably a formula for failure. Careful planning and time management are major predictors of success. The QFD Handbook will help you plan and manage the work.

In the Handbook, QFD will be presented as a *phased activity*. The purpose of describing it this way is twofold. First, the phases refer to distinct types of activities that should be performed more or less in sequence to make the QFD effort as efficient as possible. Second, the phases provide guideposts to help ground the participants as they work their way through QFD.

We've numbered the phases from 0 to 3. Phase 0 is the planning phase. Because the work of QFD hasn't started yet at this stage, it can be thought of as a "pre-phase"; hence the number 0. The other phases are

Phase 1: Gather the Voice of the Customer

Phase 2: Build the House of Quality

Phase 3: Analyze and Interpret the Results

Subsequent phases of QFD—deployment of SQCs into processes, for example— can be thought of as additional phased QFD projects, each with its own plan and implementation.

An important success factor for QFD is the support of management. Besides the support of top management, the value of which cannot be underestimated, the actions and attitudes of the QFD manager are critical. Let's take a look at who the QFD manager is, and what the QFD manager's role should be.

15.1 THE QFD MANAGER'S ROLE

QFD is a complex activity, involving the coordinated efforts of several people. Therefore, QFD should be treated as a project. We'll talk about making time estimates and setting up schedules later, but the key point for now is this: The QFD effort needs a manager. That manager could be the QFD facilitator—the impartial person who runs the QFD meetings. (See Chapter 18 for a discussion of the QFD facilitator's role.) Or it could be one of the team members whose job is closely linked to the success of the development project, such as the marketing manager or the development manager. Whoever the QFD manager is, that person is responsible for the success of the QFD activity, and would do well to manage it with the same professionalism and discipline that would be brought to bear on any other project.

As with any other effort, the outcome will depend a lot on who is on the team and how well they were chosen for their roles. Key people must know what's in it for them before they agree to participate. So some selling to the team members on the part of the QFD Manager is a prerequisite.

Here are some "do's and don'ts" for QFD Managers.

DO:

- Make sure you and your team understand and agree on the benefits you wish to receive from the QFD activity.
- Line up the people who will be needed for QFD. Make sure they know when and for how long they will be needed.
- Establish a schedule for each of the QFD phases.
- Track progress of the QFD activity, and continually seek ways to keep the activity on schedule.
- Create a mechanism for keeping people in your organization who are not on the QFD team up to date. Ensure that their concerns and ideas are represented on the QFD team.
- Ensure that the QFD team members have been brought up to speed in terms of their knowledge of the development project, of QFD, and of the QFD project.
- Make use of time between meetings by assigning data gathering and other research-type functions to the QFD team members.

DON'T:

- Drive the QFD project to a foreordained conclusion. If you cannot keep an open mind as to the outcome of the QFD process, it would be better to simply announce your desired conclusion and skip the QFD.

- Assume that everyone will know what to do. Instead, explain what's going to happen in each meeting well before the meeting day, and at the beginning of each meeting as well.

- Allow the QFD team to make decisions without data. At the very least, point out to team members that they are doing so when you see it happen. Obviously, some development decisions do get made with no data—it can't be avoided completely. The QFD process can be used to make such decisions more visible. This gives the developers the opportunity to consciously decide whether they really want to take such a risk.

- Be so devoted to your functional business role that it biases you against making good choices for the benefit of the whole project and prevents you from participating fully on the team. Don't be a poor ambassador for your function within the company.

15.2 SUMMARY

We've got our project, we've decided to do world-class development (and therefore we're going to use QFD), and we've got our QFD team and its manager. We're ready to launch the QFD activity. Let's start by planning the QFD work.

Phase 0: Planning QFD

This chapter deals with the critical process of planning the QFD activity. In this chapter, we'll break planning down into several key planning activities, and we'll explore each in detail. The key activities are

- Establish organizational support
- Determine objectives
- Decide on the customer
- Decide on the time horizon
- Decide on the product scope
- Decide on the team
- Create the QFD schedule
- Acquire the facilities and materials

This chapter assumes that you know what QFD is and that you have familiarized yourself with the ideas covered earlier in this book. This is a long chapter, because planning—the subject of the chapter—cannot be completely separated from doing. As a result, we'll be discussing a lot of QFD implementation steps along with the planning of those steps.

Planning QFD for a development project is the key to realizing success later on. Developing the House of Quality typically involves considerable expense for gathering the Voice of the Customer (twenty thousand to several hundred thousand dollars for commercial-volume products and services), and then involves a team of eight to 15 people working together in a room for two to ten days (16 to 150 person-days). Why not develop a plan to ensure the success of such a resource-laden activity?

One of the QFD facilitator's jobs is to see to it that the activity is properly planned. He or she should work closely with the design team project leader on the planned activities. This chapter identifies the major areas that must be planned, and the questions that must be answered.

16.1 ESTABLISH ORGANIZATIONAL SUPPORT

Organizational support for QFD is a key determiner of success. The key elements of organizational support are

- Management support
- Functional support
- QFD technical support

Management support refers to a commitment by top management in the organization to provide and allocate the resources needed to complete the QFD activities. The resources may include whatever time and money it takes to gather the VOC if not separately budgeted and conducted, whatever it takes to acquire the services of a skilled QFD facilitator, and whatever it takes to keep the QFD team focused on its task until the desired results have been achieved.

Functional support refers to the commitment of related functional groups to participate in the QFD activities as needed, and during the subsequent development process, to honor the decisions of the development team as a result of the QFD efforts. The functional groups referenced here are those that will be partners with the development team in completing the development of the product or service. For product development, these functional groups could include Purchasing, Manufacturing, Quality Assurance, Marketing, Sales, and Service. For process development, these functional groups could include Purchasing, Training, Marketing, and Finance.

QFD technical support refers to the acquisition of skills necessary to implement the QFD activities. Everyone on the development team will require at least an acquaintance with QFD principles, preferably through a short training seminar, although many team members will be able to learn much of what they need to know as the QFD process unfolds. The QFD facilitator will need to be quite familiar with the various elements and options of QFD, in order to guide the team into using those parts of QFD that will help them achieve their design and development objectives. The QFD facilitator will also need to have good group-facilitation skills in order to help the development team achieve its goals and manage its time well.

16.2 DETERMINE OBJECTIVES

QFD provides an array of possible benefits to the teams that use it. Many development managers imagine that the benefits of the process are self-evident and need not be articulated or selected. Quite possibly, these managers have focused only on those few of the many benefits that matter to them, and have overlooked the others. The facilitator should present the development manager with a choice of the possible benefits and ask him or her to specifically identify the ones that apply to the upcoming project. What follows is a list of possible benefits:

- Understanding customer wants and needs
- Determining quality and business goals for the product/service
- Rank-ordering proposed product capabilities
- Developing a common team vision of the product/service
- Documenting all decisions and assumptions about the project in a single, compact diagram (the HOQ)
- Creating a list of actions that will move the project forward
- Developing clear linkages between technical decisions and customer needs
- Minimizing the risk of mid-project restarts. This benefit accrues because new information made available in the middle of development can be put into perspective by adding it to the existing HOQ and other QFD matrices. Without such a discipline, new customer information or new technical information is often given disproportionate emphasis, leading to overreaction.
- Rapid product planning. Although QFD appears to be time-consuming, most groups find that product planning is quicker, more complete, and more efficient when they use the HOQ structure. This is because QFD provides a structure that can be managed and planned for, in contrast to most product-planning activities nowadays.

All of these benefits have been discussed elsewhere in this book. By knowing in advance which benefits are most important to the development manager, the facilitator will be able to allocate more time and attention to those parts of the QFD process that support those benefits. Clearly stating the objectives to the entire team helps everyone to use the process to best support those objectives.

16.3 DECIDE ON THE CUSTOMER

16.3.1 THE IMPORTANCE OF CLEAR DEFINITIONS

During the QFD process, the team members will be making many judgments. They should be guided by a pre-existing design team project charter and documented business case. They will be estimating the relationships between product or service capabilities and customer needs, for instance. In order to make these judgments meaningfully, the team will need to make clear and consistent definitions if they do not exist in the project-charter and business-case documents.

For example, to the extent that a product capability or performance measurement is vague, the team runs the risk of tacitly defining it one way in the morning, another way in the afternoon. Sometimes during a discussion, team members will discover their disagreement is because some are assuming one definition, while others are assuming another. The resulting values the team puts into a quality table will reflect one view from the morning exercise, another from work in the afternoon. If the team is not clear about its underlying assumptions as it builds the quality table, the results will be inconsistent and will have little value.

The team's most important underlying assumptions will be those about the customer. In my experience, it may be difficult for product-development teams to agree on who their customer is. QFD becomes a magnifying glass for disagreements among team members. The most common disagreements tend to be about customers' needs—first, because team members don't really know their customers well; and second, because different team members tend to focus on very different customer types, without realizing it.

Some people assume the customer is the person who makes the decision to buy the product or service. Others assume it's the person who actually uses the product. These may be very different people. For example, the buyer of a toy is usually the parent, but the user of the toy is usually a child. In a more-complicated situation, the buyer of a fleet of 50 delivery trucks may be a fleet manager, but the users of the trucks may be 50 drivers. In addition, five mechanics may maintain these trucks. Who is the customer? There may be more than one type of customer involved in a value chain or supply chain. The end user usually is the one who creates the demand, and thus her or his needs cannot be ignored. However, the purchasing decisionmaker's needs must also be addressed. It is rare that they are in opposition; rather, forgetting one group's needs is more often a cause of complications.

Another complication lies in market segment-definition. Returning to our toy example, is it intended for boys or girls? What age group? How much education? What type of physical dexterity? And for our trucks, will they be driven by the same person each day, or will they be rented by the day to different drivers? What type of cargo will they be carrying? Will they be used in the city or the country?

Obviously, if the team has not discussed and come to consensus on these questions, team members will be talking at cross purposes as they proceed through the QFD process. If they *have* discussed these issues and come to consensus, they will work more efficiently and harmoniously with each other.

16.3.2 IDENTIFY ALL POSSIBLE CUSTOMERS

The first step in defining the key customer is to make a list of all possible candidates. This is usually done by QFD planners or market-research experts. The Affinity Diagram is a useful tool for managing this list of customers.

Start by brainstorming all possible customers of the product or service you are planning. Put them onto Post-it Notes. Wherever possible, be specific. Use the names of actual customers, even the names of people. When mentioning companies, company types, or other groups, try to replace these abstract entities with people or roles within the organizations. Remember, the quest for customer satisfaction is a quest for satisfying people, not organizations.

Use the Affinity Diagram method to group the brainstormed items. General categories of markets, user types, or product-application types may emerge. From these categories, work toward a list of clearly defined customer groups. One test for clear group definitions is to identify a handful of actual customers and see if all team members place them in the same groups.

16.3.3 IDENTIFY THE KEY CUSTOMERS

Now that we have identified several customer groups, we must focus in on the key customers. The idea is to optimize our product-design decisions around these key customers, and then try to include as many additional customers as possible in our plans. Here are three ways the team might decide on the key customers:

- Everyone agrees quickly
- Prioritization Matrix method
- Analytic Hierarchy Process[1]

---

1. Thomas L. Saaty, *Decision Making for Leaders: The Analytic Hierarchy Process for Decisions in a Complex World* (Pittsburgh, Penn.: RWS Publications, 1988); and Thomas L. Saaty, *The Analytic Hierarchy Process: Planning, Priority Setting, Resource Allocation* (New York: McGraw-Hill, 1980).

Everyone Agrees Quickly

Once the customer groups have been identified, deciding on the key customers is some-times easy: Everyone glances at the list of customer groups with little or no disagreement, and they are able to decide on the key customers.

More often, the number of customer groups may be pretty large. Or, the types of groupings may not be comparable (for example, European banks vs. information-systems managers at banks). Under these circumstances, teams typically have considerable difficulty coming to consensus on the key customer group.

If everyone cannot quickly agree on the key customer group, one of the other methods for selecting it may be useful.

Prioritization Matrix Method

The Prioritization Matrix method uses a technique similar to the weighting and multiplication of columns in the QFD HOQ Planning Matrix.

We identify one, two, or (rarely) more factors or criteria by which to rate all the customer groups. The following are typical factors:

- Revenue potential of this customer group over the next three years
  - ○ High revenue potential
  - ○ Moderate revenue potential
  - ○ Low revenue potential[2]

- Current revenue derived from this customer group over the past three years
  - ○ High revenue derived
  - ○ Moderate revenue derived
  - ○ Low revenue derived

- Sales force familiar with this customer group
  - ○ High: Most of the sales force familiar
  - ○ Moderate: Some of the sales force familiar
  - ○ Low: Almost none of the sales force familiar

---

2. Where information or estimates are available, quantify "high," "medium," and "low" with actual amounts. For example, "high" = $50M or higher, "moderate" = $5M to $49M, "low" = less than $5M.

For each customer group, we then evaluate that customer's importance based on the high/moderate/low judgment in the dimensions we have chosen. We assign numerical values as follows: high = 3, moderate = 2, low = 1.

For each customer group, we multiply the numerical values for all factors we have assigned. The resulting product is an indicator of how important the customer group is to us (Figure 16-1).

If we have evaluated customer groups according to two factors, as is most common, possible products of factors are 9, 6, 4, 3, 2, and 1. The customer groups with factor products of 9 and 6 are our key customers.

Some refinements of this method are occasionally used. To treat one criterion (column) as more important than another, increase the range of numerical impact values. Instead of using 3, 2, and 1, for example, use 5, 3, and 1. Some prefer to add the impact values rather than multiply them. Multiplying creates greater differentiation in the importance column; adding creates less. Some teams use the "spend-a-buck" method in each column. They require that the values in a column add to a fixed total, such as 1, 10, or 100.

If one of the columns represents a Less the Better factor, the team must convert the factor to More the Better. For example, if "degree of difficulty" is the factor, it should be converted to "ease." If the Less the Better factor is cost, and the costs can be estimated, then the factor could be the inverse of the estimated cost.

The Prioritization Matrix, of course, can be used wherever a team needs to rank-order a set of comparable elements, and the team has primarily its own judgments to serve as the basis for rank-ordering.

Analytic Hierarchy Process

The Analytic Hierarchy Process (AHP) is an alternative method for deciding on the key customers, given a list of all possible customers. AHP is a highly developed mathematical

	Revenue potential	Recent revenue	Importance
Segment A	3	2	6
Segment B	1	1	1
Segment C	2	1	2
Segment D	3	3	9

Figure 16-1 Prioritization Matrix

Importance Judgments					
	Customer Group A	Customer Group B	Customer Group C	Customer Group D	
Customer Group A	1	3	5	7	
Customer Group B	1/3	1	3	5	
Customer Group C	1/5	1/3	1	1	
Customer Group D	1/7	1/5	1	1	
Totals	1.68	4.53	10.00	14.00	

Normalized Importance Judgments						
	Customer Group A	Customer Group B	Customer Group C	Customer Group D	Raw Weights	Normalized Raw Weights, Converted to Percentages
Customer Group A	0.60	0.66	0.50	0.50	0.56	56
Customer Group B	0.20	0.22	0.30	0.36	0.27	27
Customer Group C	0.12	0.07	0.10	0.07	0.09	9
Customer Group D	0.09	0.04	0.10	0.07	0.08	8

Figure 16-2   Analytic Hierarchy Process

system for priority setting. It is explained in detail in Thomas L. Saaty's two books on the subject.[3] We will describe only a simple version of AHP here, for dealing with this limited problem of deciding on the key customers.

The team creates a matrix with the list of all customers along the left side and also along the top. (See the example in Figure 16-2.)

The team members then estimate the importance of customers pairwise. They enter the degree of importance of Customer Group A compared to Customer Group B in the cell in row 1, column 2. The value they will put in this cell is one of the following:

9:    Customer Group A is extremely more important than Customer Group B

7:    Customer Group A is very strongly more important than Customer Group B

5:    Customer Group A is strongly more important than Customer Group B

---

3. Saaty, *Decision Making for Leaders* (1988) and *The Analytic Hierarchy Process* (1980).

3:        Customer Group A is moderately more important than Customer
          Group B

1:        Customer Group A is of equal importance with Customer Group B

Use reciprocals 1/3, 1/5, 1/7, and 1/9 when the relationships are reversed.

Having made a judgment about the importance of Customer Group A compared to Customer Group B, the judgment of importance of Customer Group B compared to Customer Group A is automatic; it is the inverse of the first value, and the group places the inverse in the cell in row 2, column 1. Thus, as the team fills in the cells above and to the right of the cells on the diagonal, all the rest of the cells are automatically determined.

The team then normalizes the values in each column. To normalize a value in a column:

1. Compute the total of the values in the column.
2. Replace each value by the result of dividing that value by the column total. The result will be a "normalized" set of values that total to 1.

This method is described in more detail with regard to Normalized Raw Weights in Section 7.7 (Chapter 7). The normalized values bear the same proportions to each other as do the original values.

The team then averages the weights in each row and places the averages in the Raw Weights column. Finally, the team normalizes the Raw Weights by dividing the Raw Weight in each row by the Total Raw Weight, and converting it to a percentage (by multiplying the normalized value by 100). The Normalized Raw Weights give a good measure of the team's judgment of the relative importance of all the customer groups.

Comparison of Prioritization Matrix Method and the Analytic Hierarchy Process

The Prioritization Matrix method and the Analytic Hierarchy Process each have advantages and disadvantages. The Prioritization Matrix method generally requires fewer team judgments than does the AHP. For n choices (company groups) and two factors, the number of team judgments required is $2 \times n$. For n choices (company groups) and three factors, the number of team judgments required is $3 \times n$.

With AHP matrices, for n choices (company groups), the number of team judgments required is $\frac{1}{2} [n \times (n-1)]$. As the number of choices increases, the number of required judgments becomes very large. Figure 16-3 shows these relationships.

Very often, however, Analytic Hierarchy Process judgments are easier and quicker to make than the Prioritization Matrix judgments. This is because AHP judgments (as

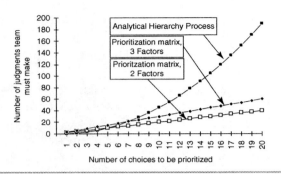

Figure 16-3   Number of Decisions versus Number of Factors

described in this chapter) are more intuitive, less focused on a single criterion (more elaborate applications of AHP provide a method to include any number of criteria in the judgment-making process), and less constrained than Prioritization Matrix judgments. For that very reason, the judgments may also be less consistent than Prioritization Matrix judgments. Some people prefer the types of summary judgments required by the AHP. Others prefer the more-structured approach of the Prioritization Matrix method. There is no best method.

We have found that selection of key customers within one group is often constrained by practical considerations, or **factors**. For example, who is willing to work with us? With whom do we have the most flexible, best working relationship? Which customers have the greatest needs aligned with our development timing? These factors or considerations can be placed into the Prioritization Matrix, perhaps giving that method a more-practical focus in the selection process over the AHP approach.

16.3.4 WHEN THERE IS MORE THAN ONE KEY CUSTOMER GROUP

Often, a team will decide there are several key customer groups, and none of them can be ignored. This might be the case, for instance, if the product is sold to a distributor, who then sells it to retailers, who finally sell it to end users. Many consumer products fall into this category. For example, packaged food, toys, office supplies, consumer electronics, and household hardware are all items that manufacturers normally sell to distributors, which in turn sell to retail stores, which in turn sell to end users.

Customer satisfaction may mean very different things to each of these groups, as indicated in Figure 16-4.

The success of the product depends on customer satisfaction of the distributor, the retailer, and the end user; the product must be designed to meet the different needs represented by these three groups. However, where conflicts arise, the end user's needs must

Customer Group	Needs and Benefits
Distributor	Ease of ordering Just-in-time shipment from manufacturer Easy handling of shipped materials Easy stacking and storing
Retailer	Attractive package Minimum shelf space Product description on package (reduce customer queries) Long shelf life
End User	Easy to learn how to use Long shelf life Long product life Safe to use Environmentally correct packaging

Figure 16-4 Comparison of Different Customer Groups' Needs and Benefits

be given preference, for they are the ones who create demand for products. In addition, price and margin considerations must be evaluated, as they influence the overall value proposition across all parts of the supply chain.

In this kind of situation, some teams have decided to construct an HOQ that acknowledges multiple customer groups. The general form of the HOQ looks like Figure 16-5.

The product features and capabilities are evaluated against the needs of all customer groups. In addition to the obvious increase in the size of the HOQ, there are complications when dealing with multiple customer groups, especially with respect to the relative weighting of the needs and benefits. We'll have a more-complete discussion of these complications, and alternative strategies for dealing with them, later.

Each customer group has its own set of needs or benefits. Within a customer group, it is possible to develop a rank ordering of needs by various survey methods. But it is not possible to determine the relative importance of a need in one customer group compared to a need in another customer group.

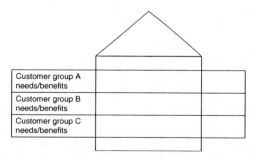

Figure 16-5 House of Quality for Three Customer Groups

The reason it is not possible is because we cannot expect a representative from one group to fairly compare the importance of his or her needs to the needs of a representative from another group—these representatives can speak only for themselves.

Instead, the product-development team must decide how important it is to satisfy each of these groups. Factors that would influence these judgments include:

- Which customer group's satisfaction is directly related to the greatest short-term revenue or profit potential?
- Which customer group's satisfaction is directly related to the greatest long-term revenue or profit potential?
- Which customer group has the greatest influence on sales of the product? For example, in large companies, decisions on the purchase of office furniture are often made by facilities managers and top management.
- How does communication between the customer groups influence purchase decisions? For example, astute purchasing agents may check with end users on the past performance of a product or service before authorizing additional purchases of that item.
- Does long-term satisfaction of a customer group eventually influence purchase decisions, even if not in the short term? For example, many complaints about the quality or flavor of a cafeteria food item might cause a cafeteria manager to stop ordering that food item.

Any method for deciding on the relative importance of the customer groups would start with the development of a set of criteria that would relate to the groups' importance. These criteria would normally be similar to the list of factors just cited, but could include many other business factors, such as business risk, available capital, or ease of accomplishing certain goals.

The goal in deciding relative importance of the customer groups is to develop a weight for each group that expresses the team's judgment of the importance of satisfying that group. These weights can then be multiplied by the importances of the customer needs within each group, to establish relative importances of customer attributes across groups. This is illustrated in Figure 16-6.

Notice that the importances of needs within a group are arranged in descending order of importance *within the customer group,* and that the weights for these needs are normalized; that is, within each customer group, the importances add to 1.

In the case illustrated in Figure 16-6, the team decided to use the ratios 3:2:1 to express the relative importances of the customer groups for its project. These values were determined by one team for a specific application, and they are not necessarily typical of other market segment weightings. By multiplying "Importance of customer group" by

Customer group	Importance of customer group	Need	Importance of need within group	Weighted importance of need
Distributor	0.50	Ease of ordering	0.32	0.16
	0.50	Just-in-time shipment from manufacturer	0.28	0.14
	0.50	Easy handling of shipped materials	0.24	0.12
	0.50	Easy stacking and storing	0.16	0.08
Retailer	0.33	Attractive package	0.30	0.10
	0.33	Minimum shelf space	0.25	0.08
	0.33	Product description on package	0.25	0.08
	0.33	Long shelf life	0.20	0.07
End user	0.17	Easy to learn how to use	0.28	0.05
	0.17	Long shelf life	0.24	0.04
	0.17	Long product life	0.17	0.03
	0.17	Safe to use	0.16	0.03
	0.17	Environmentally correct packaging	0.16	0.03

Figure 16-6 Weighted Customer Importances

"Importance of need within group," the team effectively scaled the importances within each group, so that the resulting weighted importances of need were also normalized. More importantly, the importances could now be compared across groups, and could be re-sorted in descending order of importance (as in Figure 16-7).

The Distributor group was assigned the highest weight, so it is not surprising that most of the needs at the top of the re-sorted list come from that group. There is one important time-saving observation we can make: One need, "Long shelf life," appeared in the lists for two groups (End User group and Retailer group). Assuming that these identically worded needs actually are the same need, we can combine them by adding their weights.

The benefit of combining like needs from different segments is a shorter list of customer needs to deal with throughout the QFD process. Reducing the number of rows by even one is a time-saver, because the Relationships matrix is usually much wider than it is high. Therefore, small reductions in row count reduce the overall matrix by quite a bit.

If we choose not to combine identical needs from different segments, we can still save some time in our QFD process by taking note of these identical needs. Suppose we have

Customer group	Importance of customer group	Need	Importance of need within group	Weighted importance of need
Distributor	0.50	Ease of ordering	0.32	0.16
Distributor	0.50	Just-in-time shipment from manufacturer	0.28	0.14
Distributor	0.50	Easy handling of shipped materials	0.24	0.12
Retailer	0.33	Attractive package	0.30	0.10
Retailer	0.33	Minimize shelf space	0.25	0.08
Retailer	0.33	Product description on package	0.25	0.08
Distributor	0.50	Easy stacking and storing	0.16	0.08
Retailer	0.33	Long shelf life	0.20	0.07
End user	0.17	Easy to learn how to use	0.28	0.05
End user	0.17	Long shelf life	0.24	0.04
End user	0.17	Long product life	0.17	0.03
End user	0.17	Safe to use	0.16	0.03
End user	0.17	Environmentally correct packaging	0.16	0.03

Figure 16-7   Weighted and Regrouped Customer Importances

two identical needs, N_A and N_B—one in segment A, the other in segment B. When we determine the impact of a Substitute Quality Characteristic on N_A, we can automatically use the same impact value with respect to N_B. Because the impact of any SQC on N_A will be the same as its impact on N_B, N_B behaves as a kind of "shadow" customer need, always mimicking the behavior of N_A.

To complete this example, if we combine the two occurrences of "Long shelf life," we get a combined weight of 0.11, and the new overall rank-ordered needs become as in Figure 16-8.

Note that "Safe to use" started with a relatively low importance in the End User group, and after scaling Importance within Group by multiplying with Importance of Customer Group, it had a low importance overall. Should the developers therefore ignore this customer need? Obviously not. This need is an example of an Expected Need in the Klein grid (Chapter 17). Likewise, safety features are Dissatisfiers in the Kano model (Chapter 2). Customers generally take safety for granted until a product or service proves to be unsafe, at which point safety will be ranked as extremely important.

Customer group	Importance of customer group	Need	Importance of need within group	Weighted importance of need
Distributor	0.50	Ease of ordering	0.32	0.16
Distributor	0.50	Just-in-time shipment from manufacturer	0.28	0.14
Distributor	0.50	Easy handling of shipped materials	0.24	0.12
Retailer/ distributor		Long shelf life		0.11
Retailer	0.33	Attractive package	0.30	0.10
Retailer	0.33	Minimize shelf space	0.25	0.08
Retailer	0.33	Product description on package	0.25	0.08
Distributor	0.50	Easy stacking and storing	0.16	0.08
End user	0.17	Easy to learn how to use	0.28	0.05
End user	0.17	Long product life	0.17	0.03
End user	0.17	Safe to use	0.16	0.03
End user	0.17	Environmentally correct packaging	0.16	0.03

Figure 16-8 Weighted and Combined Customer Importances

16.4 DECIDE ON THE TIME HORIZON

A time frame for shipping a product is a useful constraint for planning purposes. Without a time constraint, a team may set impractical goals for itself. It may decide on objectives that take so much time to complete that the delay in shipping the product outweighs any positive impact on Customer Satisfaction Performance and thereby renders the product noncompetitive.

A clearly defined time horizon for the QFD process helps keep the planning realistic. The explicit time horizon contributes to better communication by helping all team members to focus on the same objectives. This should be evident in the project charter and business case. It has been my experience that most design teams know exactly what their time horizons look like. While it's useful to avoid extremely impractical choices when planning a product, team members can sometimes fall into traps by too quickly ruling out seemingly expensive and time-consuming objectives. Some objectives appear at the early stages to be too expensive to be considered, but closer study may reveal an alternative, inexpensive approach.

It may be useful during early stages of product planning to regard desirable-yet-expensive objectives as **breakthrough opportunities**. By labeling these objectives as breakthrough opportunities, the team frees itself up to search for innovative solutions, using methods such as the Pugh Concept Selection Process (see Chapter 12).

For planning purposes, it's always better in QFD to include all objectives, even the somewhat unrealistic ones, as long as possible, so that breakthrough opportunities are not accidentally discarded.

A good rule of thumb for deciding which objectives to include in the QFD is

If the possible objective could be developed by the team by using *all available resources* during the time frame of the project, include the objective.

In any case, the team must decide on a time horizon for the project (and therefore for the QFD activity), and use it consistently during the QFD activity.

16.5 DECIDE ON THE PRODUCT SCOPE

An important QFD principle—and indeed an important principle in most creative design activities—is to postpone detailed design work as long as possible. In this way the focus remains on the objectives, and the range of possible solutions for meeting those objectives stays unconstrained to allow the most creative and efficient solution to emerge. This means the team must keep the technical solution vague and free of design specifics as long as possible.

Nevertheless, there is such a thing as being *too* vague. In any given situation, certain customer needs are likely to be irrelevant. For example, if the product being designed is a camera, the customer's needs—and related Substitute Quality Characteristics—outside of those related to using a camera do not need to be considered. Therefore, as the VOC is being collected, questions on such topics as healthcare needs or financial-management needs would be distractions.

On the other hand, questions on the selection and preferences for preset camera modes, on choices for lighting, and on subjects to photograph all could provide insight to the development team. The customer needs elicited by these questions could help the team develop SQCs that point them to an innovative and competitive camera design.

These remarks indicate a need for *setting the scope* of the QFD activity. The scope defines what's *in* and what's *not in* the QFD discussions to follow. The development team needs to decide for itself, or to determine from management, how much freedom it has for developing its solution. Knowing the scope helps the team to ignore irrelevant data and to include all relevant data and ideas.

16.6 DECIDE ON THE TEAM AND ITS RELATIONSHIP TO THE ORGANIZATION

In some cases, the QFD process is used to make decisions that affect a very small group. For instance, most consulting teams that I have been a member of (since we learned of QFD in 1986) have used QFD to determine the relationship the consulting team itself should have with its clients. In cases like this, the entire team affected by the QFD results—the consultants themselves—developed the QFD matrices.

In most cases, however, the QFD results are intended to affect the activities of an organization much too large to work all together on the House of Quality.

The team that develops the QFD matrix will be making key strategic decisions about how a product or service should look or be. Every good manager knows that the people implementing a decision will be more motivated, better informed, and generally more able to do the job if they have had a part in forming the decision in the first place. The general principles that come into play are **ownership** and **informed decision-making**, both of which are strongly enhanced in people who move in a direction they themselves have set. Thus, when product or service planning employs QFD, it's best for the implementors to be as closely involved in the initial planning as is practicable.

Beyond that, the ideal QFD team should include representatives from all the important functional groups involved in designing, developing, building, delivering, and servicing the product or service. There's a tendency for product developers to exclude certain functional groups from their initial product planning—a relic of the old "throw it over the wall" method of product development—on the grounds that those whose responsibilities are downstream from the initial work lack the knowledge or expertise needed to usefully influence the upstream decisions. In a sense, however, QFD stands for the opposite point of view.

In the QFD paradigm, everything important about a product or service gets decided at the beginning. Whatever customer-satisfaction aspects of the product are ignored upstream are likely to be difficult or impossible to fix downstream. Hence, the QFD team composition is crucial to the overall success of the product or service.

For manufactured products, a typical list of functions that should be represented on the QFD team would look like this:

- Marketing (and in certain settings, representatives from the customer base)
- Sales (if separate from Marketing)
- Product design
- Supplier management/purchasing (and sometimes, representatives from key supplier organizations)

- Manufacturing engineering
- Manufacturing production
- Order processing/fulfillment
- Service

Some of these functions may themselves be multifunctional. For example, for computer design, various subfunctions might include:

- Processor/logic design
- Packaging and power design
- Design-tools (CAD) design
- Product documentation

For many marketing functions, the subfunctions might include

- Domestic marketing
- Overseas marketing
- Market research
- Sales support and liaison

Allowing for such multifunctional subdivisions in other functions as well, the list of possible representatives may be far too large. This situation occurs very often in larger organizations.

A high-level manager would probably appoint representatives from each function to represent all the subfunctions in their group. These representatives would then have the responsibility of communicating the QFD team's work to their constituents. Criteria that would influence the choice of team members would be

- Knowledge of the customer
- Experience in developing similar products or services
- Willingness to participate in the disciplined planning process implied by QFD
- Accountability for the subsequent development work
- Authority to make decisions and commitments on behalf of the organization represented
- Enough knowledge to represent the concerns and issues of the organization represented

An alert QFD facilitator will assess the cultural norms for communication in the client's organization. At the outset, he or she will establish mechanisms for information to flow from the QFD team to the key parts of the organization that will be affected by the QFD results.

The mechanisms will also allow for reactions, opinions, and data to flow from the outside organization back to the QFD team. A healthy two-way flow of information between the QFD team and the rest of the organization assures that the QFD results will not come as a surprise and will not create undue controversy. This, in turn, assures that the QFD team's results will be rapidly accepted and more easily implemented by the organization.

One team I worked with was using QFD to develop a set of top-level product development criteria for several families of new products to be developed in a rather large corporation. The QFD process was scheduled for five one-day and two-day meetings over a period of six weeks.

The QFD team's results were expected to create some radically new ways of thinking about the company's products. The goal was that these results should cause a minimum of controversy, and take no one by surprise.

At the outset, each team member took the responsibility for briefing a key manager or Vice President in the company after each one-day or two-day QFD meeting. The team members were to explain the major issues the QFD team was working on, and what new or surprising ideas or points of view the QFD team had developed. They were also expected to get the manager's reaction and bring that back to the QFD team. In this way, a two-way dialogue occurred between the QFD team and the rest of the corporation, as the QFD team did its work.

As the QFD effort continued, the team did indeed develop many novel ways of expressing product performance (based on new customer data). These novel ways were bound to entail new measurement methods and new approaches to benchmarking the competition. Rather than inciting controversy, however, the communication process set up by the team had the effect of creating *excitement* around the company. People who were unable to work on the team were impatient for the results, and in one case began making their own product plans based on partial results of the QFD team, which had not yet completed its work.

16.7 CREATE A SCHEDULE FOR THE QFD

The first time I ever explained QFD to a product-development team, I was asked how long the QFD activity would take. I hadn't the foggiest notion, having never been through the process myself. I learned the hard way, working with that team and others shortly thereafter, that

- QFD requires time
- QFD can be shortcut
- QFD should be a managed activity, just like any project

16.7.1 QFD REQUIRES TIME

Every one of the steps detailed below could be a part of a QFD process. Every one of them takes time to perform. The most rapid product QFDs I have facilitated have taken about two days. Drastic surgery was done to the QFD process in order to keep the process that short, and it was done only because I judged that within the environment of this team's operations, the QFD team could not allocate more time. It seemed to me that two days of QFD-style product planning was better than no structured planning at all.

The speed of the team's activities depends upon the team members' familiarity with QFD. Some training up front, through e-learning or with the facilitator, will help remove many questions about the QFD process itself and allow the team to be more productive in performing QFD activities.

In the QFD Estimator Chart (Figure 16-9), I've identified the most common steps in a QFD process for developing the House of Quality. Next to each step, I've provided some rules of thumb for estimating the length of the step. The third column in the table shows typical sizes of tables of data for the step, and the fourth column shows the resulting estimated time for the step. The rest of this section provides explanations for the contents of this figure.

The steps are described in much more detail throughout this book; I've added comments here to explain the rules of thumb and how they lead to the estimates I've provided.

QFD Estimator Chart

Some general rules of thumb seem to hold for several of the important steps in the QFD process. Some of these have to do with the amount of data to process, others with the rate at which teams can process the data. A good way to estimate the time needed for QFD can be based on the following rules of thumb:

1. For any market segment, the number of tertiary customer needs will usually be between 75 and 150.

2. For any market segment, the number of secondary customer needs will usually be 20 to 30.

3. Most teams will generate about three Substitute Quality Characteristics for each secondary customer need.

Activity	Rule of thumb	Typical size	Typical time
Decide on the customer	½ day discussion	15 customer categories identified, 3 are key categories	½ day
Gather the qualitative needs	1 hr. per interview, 4 interview locations, 3 hrs. analysis for each hr. of interview	20–30 interviews	15 days
Structure the needs	Team does it: ½ day. 200 customers do it: 3 weeks.	150 unique needs at tertiary level, 25 needs at secondary level	½ day to 3 weeks
Quantify the needs	Team does it: ½ day. Customers do it: 1 to 3 months.	25 needs at secondary level	½ day to 3 months
Set performance goals	25 secondaries, ½ day	25 secondaries	½ day
Generate Substitute Quality Characteristics	3 SQCs for each secondary attribute: 2 days for 25 secondary attributes	25 secondaries, 75 SQCs	2 days
Determine impacts— SQCs to needs	1 minute per SQC/ secondary pair	25 secondaries times 75 SQCs = 1875 cells	All cells done by entire team: 3.9 days
			Reduced matrix assigned to teams: 2 days
Determine technical correlations—SQC	1 minute per SQC/ SQC pair	$(75 \times 74) \div 2 = 2775$ comparisons	All cells done by entire team: 5.8 days
			Reduced to SQC matrix assigned to teams: 2 days
Benchmark	Varies widely, depending on products, services, technology, and market conditions		
Set targets	Varies widely, depending on products, services, technology, and market conditions		

Figure 16-9 QFD Estimator Chart

4. Smaller teams will process data much faster than larger teams. The fastest processing can be done by a single person—no arguments at all are likely to occur! The larger the team, the more likely it is that there will be differing points of view, and therefore the more likely it is that there will be discussion. One of the important benefits of QFD is the common understanding the team develops of the customer and of the problem they are trying to solve. This common understanding is the natural by-product of the disagreements and the resultant discussions. However, the development team will have to balance the benefits of the discussions against the time they

take. Larger groups mean more discussions and deeper common understanding. Smaller groups mean faster QFDs.

5. Teams always work very slowly at the beginning of the QFD process. Factors contributing to the slower pace include lack of familiarity with the process and lack of experience with the thought processes that the QFD discipline promotes. In many cases, the QFD team members may not know each other well. Even if they do, they may be unaccustomed to interacting with each other on the range of topics and types of customer needs that QFD deals with.

   The time estimates provided in Figure 16-9 represent rough averages of time periods that actually vary widely, starting with very long time periods (half a day to make a single decision) when the team is inexperienced, and ending with very short time periods (two or three minutes to make a single decision) as the team gains experience and confidence.

6. Teams almost always work very rapidly as they near the end of the process. Contributing factors are increasing familiarity with the QFD process; increased familiarity, comfort, and trust among the QFD team members; and development of a vocabulary and concept kit that simplifies QFD judgments.

Following are the main steps in the QFD process. Special considerations affecting time schedules are detailed with each step. The QFD Estimator Chart (Figure 16-9) summarizes the major points involved in scheduling.

Customer Needs/Benefits Matrix

- **Decide on the customer.** There is usually quite a bit of disagreement at first in deciding on the key customers. In my experience, teams have spent anywhere from one to eight hours discussing this point. In some cases, the question is not settled by discussion, but rather by sizable market-research studies aimed at defining useful market segments. Such studies can take weeks or months. One alternative is using historical data to save time, especially for derivative or leveraged products. The QFD Estimator Chart (Figure 16-9) assumes the issue can be settled by discussion.

- **Gather the qualitative wants and needs.** There are so many methods for doing this that no single guideline can apply. Assuming the method involves interviewing, I suggest you estimate the length of the interviews, take travel into account, and provide time for analysis of the interview data. The primary output of the analysis would be the customer wants and needs at the verbatim (tertiary) level. There is usually much more information available from these interviews than just the wants and needs; for estimating purposes, however, I've excluded analysis of everything but the wants and needs.

- In the table, I've assumed one hour per interview, 25 interviews, four days of travel (amounting to 48 person-hours), and three person-hours for analysis for each interview hour.

- **Structure the needs.** If the development team creates the structure by using the Affinity Diagram process, 100 to 300 needs can usually be structured in two to six hours. I have worked with groups that started with as many as 750 needs; the process took about eight hours, and it was exhausting. As with all group processes, the smaller the group, the faster the process. You can include customers in the group and still get the task done within these time frames.

  The best method for structuring needs—having many customers do it, and then constructing a composite Affinity Diagram representative of all the customers—is generally much more time-consuming, although more representative of the customer, than any other way to structure the needs.[4] For more on methods for customers to affinitize customer needs, see Section 17.5.1 (Chapter 17). Factors that affect the schedule are

  - Time to recruit customers

  - Number of customers participating in the process

  - Time for sending the lists of needs to be sorted to the customers and receiving them back

  - Method used for analyzing the received sorts

Given the right statistical tools, this process can be done in as few as three weeks, but more commonly takes about ten weeks.

Planning Matrix

- **Quantify the needs.** At the secondary level in the hierarchy, determine the quantitative importance to customers of the wants and needs, the company's performance relative to meeting those needs, and the competition's performance. The easiest way to do this is to "invent" the numbers—that is, to use the judgment of the QFD team to determine the importance and performance numbers.

---

4. Robert Klein, Applied Marketing Sciences, Inc., "New Techniques for Listening to the Voice of the Customer," *Transactions from the Second Symposium on Quality Function Deployment*, June 18–19, 1990 (Novi, Mich.: GOAL/QPC, 1990).

This process usually involves considerable debate within the QFD team. When a team is making judgments, there can be no right answer other than consensus. Therefore, the goal is consensus. Most teams can come to consensus on importance, satisfaction performance, and satisfaction performance of the competition for 20 to 30 secondaries in about half a day. With a strong facilitator, most teams can come to consensus within almost any time frame they choose.

The other approach to quantifying the needs is to collect data from the customers, usually by survey, as discussed in Chapter 7. This process assumes that the qualitative research has already been completed—that is, the customer wants and needs have already been determined. Given a set of wants and needs, customers are asked to rank-order them, or to identify the few most-important needs. They are also asked how well their current vendor meets each need. If a sufficient sample is chosen, enough customers will respond to fairly represent satisfaction performance of the competition.

Parameters affecting the time required to survey customers are

o Number of wants and needs

o Number of customers surveyed

The number of wants and needs customers are asked about must be kept fairly small (conventional wisdom suggests 20 to 30). Larger numbers require more time than survey respondents are willing to give. Larger numbers are also confusing to customers; they often find they cannot easily differentiate among as many as 40 or 50 needs. This is one of the reasons why QFD is usually done using the customer needs at the secondary level: The number of secondaries rarely rises above 25 to 30 for a single market se gment.

The number of customer needs influences the time required to develop the survey, the time customers need to complete the survey (and therefore the return rate), and the time to perform analysis of the survey results.

The number of customers surveyed, often called the sample size, is critical to the confidence people will have in the results. In general, the more customers surveyed, the more survey results will be returned or responded to. However, the QFD team will normally balance the cost and the time required to survey larger numbers of customers against the advantages of higher confidence levels.

The following factors are affected by the sample size:

o Time to recruit survey respondents

o Time to present the survey to customers (time to print and mail survey forms, or to make telephone calls, or to visit customers)

- Time to analyze the survey results
- **Set Customer Satisfaction Performance goals.** This group-consensus process has a far-reaching impact on all the QFD results. As with other consensus processes, there is no right answer. As a consensus process, factors that affect time to complete include the size of the QFD team, the abilities of the group facilitator, and the number of items to come to consensus on. It is rare to spend more than half a day on this process; frequently only a couple of hours are required.

Substitute Quality Characteristics

- **Generate Substitute Quality Characteristics.** This process is quite variable. It depends on many factors, including:
  - The nature of the SQCs: performance measurements or metrics; product requirements; or product features or capabilities
  - How SQCs are generated: from the customer wants and needs; or already available (from prewritten Product Requirements Documents or from a previous QFD process, for example)

In general, if the QFD team must create ideas during the QFD process, it takes longer than if the ideas have already been prepared. Whether the ideas are created *before* QFD or *during* QFD, it still takes time to do the work. However, we're accounting for the QFD time itself, so we've made that distinction. The QFD Estimator Chart assumes the team will generate SQCs during the QFD process. (This is the most common practice.)

When the wants and needs are used as the basis for generating SQCs, the clarity and purity of the SQCs is very important. Assuming they are clear to the QFD team, and each one expresses a single need, not multiple needs, my rule of thumb is that the team generates an average of three performance metrics for each customer want or need. The work often starts extremely slowly (the first performance measure might take one-half day), and then accelerates as the team's confidence grows.

Impacts

- **Determine impacts of Substitute Quality Characteristics upon needs.** Once the team understands what types of judgments they must make for this step, the process generally moves quickly. The requirements for a quick process are, of course, that the customer wants and needs are clear and unambiguous to the team, and that the SQCs are also clear.

  Often the relationship matrix is sparse—that is, the number of high and medium impacts is small com pared to the total number of cells. When the matrix is sparse, the work goes rapidly, because the team can usually agree in a flash when no relationship exists.

Two important time-saving strategies are available to the team:

o dividing into subteams

o filling in a reduced-size matrix

Both strategies are described more fully in Section 16.7.2.

When the team divides into n subteams, the work takes much less than $1/n$th the time of the whole team working on the matrix together. The new element that accounts for this greater-than-expected increase is that smaller teams are able to come to consensus much more rapidly than large teams.

The reduced-size matrix—the matrix consisting of only those rows relating to the most important customer wants and needs—is often one-third the size of the full matrix, and therefore requires proportionately less time to complete.

Technical Correlations

- Assess the strengths of supporting or opposing correlations among SQCs. In terms of time estimates, this process is similar to the process of determining impacts.

  Quite of few of the correlations cells will be empty and will be rapidly processed by the team. Some teams limit their analysis only to the highest-priority SQCs. Some teams assign the work to subteams. Many teams skip this step entirely. Some teams limit the analysis to the most highly ranked SQCs.

Perform Competitive Benchmarks and Set Targets

- Determine—by benchmarking the competition, by laboratory experiment, or by some other method—what targets to set for the key SQCs. The nature of this work varies so widely that no reasonable time estimate can be given.

16.7.2 QFD CAN BE SHORTCUT

In Section 16.7.1, QFD Requires Time, we mentioned two techniques for shortening the process:

- Matrix reduction
- Dividing into subteams

Let's look more closely at these techniques.

Matrix Reduction

Many QFD practitioners advocate the use of small matrices in QFD. They suggest that rather than spend all available time filling out a single large HOQ, the team would gain more benefit by filling out just the critical parts of the HOQ, and then just the critical parts of other matrices for downstream parts of the development process (such as the Design Deployment and Manufacturing Planning Matrices, described in Chapter 19).

To keep the HOQ small, some choices must be made to eliminate some of the detail in the VOC. By working with fewer rows of Whats, the team will generate fewer columns of Hows. Even if the team has already generated a lot of How columns, just reducing the number of What rows will substantially reduce the matrix size.

In addition to selecting a higher level of the VOC hierarchy to start with (as described in Section 4.2, Chapter 4), there are two other ways of reducing the number of VOC rows: judgment and Raw Weight analysis.

Using Judgment to Reduce Matrix Size

The development team may be working on a focused problem that is narrower than designing a total product. In this case, it can decide which customer needs pertain to its problem and drop the rest. The team could develop a Prioritization Matrix (see Section 4.5, Chapter 4), where a significant criterion used for rank-ordering customer needs would be "Applies to our problem."

In one popular method, sometimes called the Nominative Group Technique (a term that applies generically to many group voting processes), each member of the team is given a certain number of votes, perhaps five. The member can cast the votes for as many or as few customer needs as desired, including casting all five votes for a single customer need. The highest-scoring customer needs will then be the ones ranked most important by most of the team members.

A similar method is called the Multipickup Method. It has been described by Professor Shoji Shiba.[5] In this method, a selection round involves each team member marking those customer needs they consider worth retaining. At the end of the round, all unmarked customer needs are removed.

Using Relative Raw Weights to Reduce Matrix Size

In any Prioritization Matrix (of which the Relationship matrix is one example), the more heavily weighted Whats will have a greater influence on the weighting of the Hows than

---

5. Shoji Shiba, Alan Graham, and David Walden, *A New American TQM: Four Practical Revolutions in Management* (New York: Productivity Press/Center For Quality Management, August 1993).

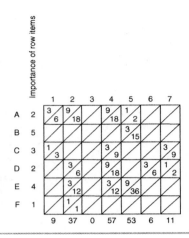

Figure 16-10   Prioritization Matrix

will the less heavily weighted Whats. We can use that simple fact to cut significant work out of the QFD process. Consider the Prioritization Matrix in Figure 16-10.

The rank-ordering of the Whats and Hows is shown in Figure 16-11.

Suppose we ignore the impacts in the matrix row containing the least important What (Row F). How would that change the ordering of the Hows? In setting the importance of the least-important What to zero—which is equivalent to not filling in the impacts in that row—we can see (in Figure 16-12) that the rank ordering and the weights of the Hows change very little. In fact, it may take you a moment or two to see any change at all.

If we ignore the *two* least-important Whats, D and F, the change in rank order and the weights of the Hows is more noticeable (Figure 16-13), but the three most-important Hows are still at the top of the list, while the three least important remain at the bottom of the list.

Taking the process a step further (Figure 16-14), and setting the importances of the *three* least-important Whats to zero, we continue to see relatively little shift in the rank ordering, although the relationships between normalized importances have indeed changed quite a bit.

The more less-important Whats we drop out, the more the resulting importances of the Hows diverge from the original case, where all the Whats are taken into consideration. However, the divergence is gradual. By dropping a single row, sometimes very little of the final result—the rank ordering of Hows—changes. To the extent that our goal is just to prioritize the Hows, we can sacrifice some detail in the rank ordering of the Hows and still get the prioritization that relates strongly to high impact on the Whats.

What	Importance	How	Importance	Normalized importance
B	5	4	57	.33
E	4	5	53	.31
C	3	2	37	.21
A	2	7	11	.06
D	2	1	9	.05
F	1	6	6	.03
		3	0	.00

Figure 16-11   Relationships Using All Whats

What	Importance	How	Importance	Normalized importance
B	5	4	57	.33
E	4	5	53	.31
C	3	2	36	.21
A	2	7	11	.06
D	2	1	9	.05
F	0	6	6	.03
		3	0	.00

Figure 16-12   Relationships Eliminating Least-Important What

What	Importance	How	Importance	Normalized importance
B	5	5	53	.38
E	4	4	39	.28
C	3	2	30	.21
A	2	7	9	.06
D	0	1	9	.06
F	0	6	0	.00
		3	0	.00

Figure 16-13   Relationships Eliminating Two Least-Important Whats

What	Importance	How	Importance	Normalized importance
B	5	5	51	.53
E	4	4	21	.22
C	3	2	12	.13
A	0	7	9	.09
D	0	1	3	.03
F	0	6	0	.00
		3	0	.00

Figure 16-14   Relationships Eliminating Three Least-Important Whats

All this leads us to this time-saving strategy: If there may not be enough time to complete the matrix, be sure to begin evaluating the highest-ranking Whats first. As you get closer to completion, the rank ordering of the Hows will not change very much and you can safely drop some of them.

To get an idea of the most-dominant Whats, arrange them in descending order of importance and compute the normalized importance and cumulative normalized importance.

The Whats with cumulative normalized importance (B and E in Figure 16-15) greater than approximately .50 dominate the entire set of Whats and should always be included in the matrix. Beyond those, the more Whats the better. If you want to estimate the effect of the lower 50 percent of Whats, you can generate a worst-case scenario in which you set impacts for all Whats in the lower 50 percent to the highest value (usually 9). Compare the resulting normalized weights of the Hows to the case where the impacts are

What	Importance	Normalized Importance	Cumulative Normalized importance
B	5	0.29	0.29
E	4	0.24	0.53
C	3	0.18	0.71
A	2	0.12	0.82
D	2	0.12	0.94
F	1	0.06	1.00

Figure 16-15   Normalized Importance of the Whats

set to zero. If there's a big shift in normalized importances, you must complete more of the HOQ. If there's very little shift, you can safely conclude that you will learn very little new by completing more of the HOQ. Be sure to use *normalized* weights when looking at the importances of the Hows. Absolute values are guaranteed to change, but the relative differences between the weights are best represented by normalized values.

Typically, for one customer segment an HOQ will have about 25 Whats, and the normalized importance of about the top one-third of them will accumulate to 50 percent of the total normalized importances. Most HOQs are considerably wider than they are tall, so removing a few Whats from the matrix usually amounts to substantial reduction in matrix size. In Figure 16-16, by eliminating the three least-important Whats, we have reduced the Relationship matrix by 3 × 17 = 51 cells, with very little loss of critical information.

Dividing into Subteams

It is a common adage in organizational dynamics that in order to reach consensus, a group of n people must have n^2 conversations—that is, each person must have a conversation with each other person in the group. While most groups are not too careful to compute and keep track of the number of conversations or interactions they have, it is certainly true that the larger the group, the more time it takes for the group to come to consensus on any issue. (Of course, there is no guarantee that a particular group *will* come to consensus on a particular subject!)

Taken at one extreme, consider the group of size one. Most individuals can make judgments for themselves—at least the focused individual judgments that come up in QFD—very rapidly. No discussion is needed. None of the arguments in favor of a particular opinion need be debated; they can be reviewed within the mind of the single group member almost instantly.

With a group of size two, the situation is quite different. All the private thoughts that a single person did not have to discuss with himself or herself now must be articulated to the other group member. The cost in time to communicate these thoughts is extremely high compared to a single person "talking to himself," and the chances for misunderstanding are much higher. Misunderstanding in turn leads to the need to explain the ideas yet again,

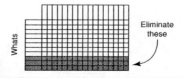

Figure 16-16 Eliminating Least-Important Whats (Whats in Descending Order of Importance)

requiring even more time. If these costs are not already high enough, let us not forget that the communication must go in both directions. Each person must communicate her or his ideas to the other, and both together must converge to a consensus.

With groups of three or more, the number of interactions certainly goes up, even if not as the square of the size of the group. Furthermore, the larger the group, the less intimate the environment, so that the quality of the communication must become more impersonal and more formal. Finally, each person gets less time to state a position, further contributing to misunderstanding.

For all these reasons, the way to get decisions made quickly is to assign them to individuals. If speed were the only requirement when doing QFD, we would ask a single person to fill in the HOQ—and it could probably be done in just a few hours.

The largest matrix (in terms of cell count) in the HOQ is the Relationships section. On my first QFD project, I failed to perform the obvious simple operation of multiplying the 110 customer needs by the 155 SQCs we had generated. Had I done so, I would have been profoundly impressed by the resultant product (17,050). It is almost impossible to get a team to sit together in a room and come to agreement on 17,050 judgments. Even if the team had had the patience and fortitude to come to consensus 17,050 times, the time to complete the task would be overwhelming and unacceptable. Suppose that an average of three minutes would be needed for discussing each cell. The process would take 852 hours, or 106 full working days.

Figure 16-17 illustrates the impact of using subteams on the overall time required to fill in the Relationships section. Suppose that I had asked each of the ten people on the

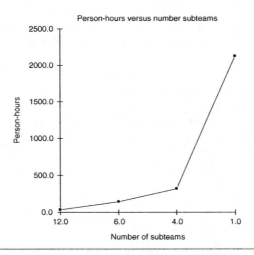

Person-hours versus number subteams

Figure 16-17   Person-Hours to Complete 17,050 Cells with 12 People, Divided Into 12, 6, and 4 Subteams, and Working as a Single Team

team to fill in a portion of the Relationships section. Most people can, on their own, make the required judgments at least as quickly as one per minute. As one becomes familiar with the elements on the left and on top of the matrix, one can make the judgments much faster, usually three or four per minute. Assuming a conservative rate of one per minute, with ten people working in parallel, the team could complete the 17,050 cells in only 3½ days.

Suppose instead that the group had been divided into five groups of two people each. The average time to make each judgment might be twice as long as an individual would need—say two minutes. The overall time to make the 17,050 judgments jumps to fourteen person-days.

A matrix of 17,050 cells is an extreme example. I have never seen a matrix that large be completed. But these simple calculations suggest how much the overall time is affected by dividing the QFD team into subteams.

Of course, completing QFD rapidly is *not* the only requirement. (If it were, the quickest way is to not do QFD at all!) The quality of the QFD decisions is an extremely important consideration, as is group buy-in, or development of a common group vision. Since the surest path to group buy-in is to give the group plenty of opportunity to discuss the problems, the QFD facilitator must continually balance the benefits of group discussions against the cost in terms of time.

If we do divide into subteams, how can we optimize the quality of the decisions? Most groups adopt a hybrid strategy: Perform the task partly as a single large team, and partly as subteams. When I facilitate QFDs and decide it's necessary to divide into subteams, I usually have the entire team work together for a few hours, to help create a standard for how teams will determine impacts.

As facilitator, I can help the group get accustomed to the questions they must ask and answer in order to make the judgments that QFD calls for. If there is a problem in the matrix—for example, unclear Customer Wants and Needs, or imprecise SQCs—there's a good chance the problem can be dealt with while the entire group is assembled. That will make for smoother sailing when the group breaks into subteams.

If possible, I have the subteams work together in different parts of the same room. Then, if one team runs into a problem or discovers the need for clarification or addition of an SQC, it can be easily and rapidly communicated to the other subteams. I have also found it helpful for each subteam to present a few of its judgments to the other subteams at "checkpoints" along the process. This ensures that each subteam is making judgments that the other subteams will have confidence in.

In the discussion above, I have focused on the Relationships matrix as the place to break into subteams. However, it's not the only place where QFD time can be effectively reduced. Almost any part of the HOQ where team judgments must be made is an effective candidate to be sped up by the use of multiple smaller teams working in parallel.

However, it's a good idea to have the entire team work together at the beginning of the process, no matter how slow the progress. The team needs to develop a mutual understanding of the QFD process and of each other before its members will have confidence that breaking into smaller groups will produce results acceptable to the whole team.

It's important to bring the subteams back to work as a single team often enough so that the sense of "teamness" remains. Finally, as the HOQ is completed, the team must analyze the results and determine a course of action. This must be done with all team members together.

16.7.3 QFD SHOULD BE A MANAGED ACTIVITY

After facilitating several QFDs, I came to realize that the amount of work in each of these steps can be estimated in advance. I also discovered that the process can use up the time of many highly paid people—the QFD team members. As a facilitator, I was responsible for helping them to use their time effectively. The best way for me to do that was to lay out a detailed schedule for the entire QFD process, making use of estimates of the size and number of QFD matrices that were likely to be built during the process.

I eventually adopted the practice of establishing a detailed schedule, sometimes predicting task completion to a specific hour. The schedule helped to create a sense of time pressure among the team members, and effectively discouraged the team from wasting time.

I recommend that any facilitator lay out the QFD process just as if it were a complex project. Based on the objectives of the team leader, determine the number of matrices that will be constructed, and other activities that will be part of the process. Make assumptions about the sizes of various matrices (see Figure 16-9, the QFD Estimator Chart), whether the team needs to divide into subteams, and the average time for the team or subteam to make each judgment. Lay out the tasks on a GANTT chart or Task Dependency chart (such as an Arrow Diagram). Estimate start dates and times.

Don't be afraid to micro-estimate the process. It's like establishing the agenda of a long meeting, or a series of meetings. You plan at what time events start and end, in which room, and with which individuals. Things may not go as you plan; some things will take longer, others will go more rapidly. But the detailed schedule gives the facilitator a guide for determining when to intervene and reschedule. The overall schedule helps establish the team's expectations, and thereby helps keep people motivated for the entire process.

Most importantly, the carefully completed time estimate is a key to ensuring that the QFD will be completed within the time frame people expected, and will achieve the desired objectives.

16.8 ACQUIRE THE FACILITIES AND MATERIALS

16.8.1 LOCATION

The QFD process can be spread out over many days or weeks, or it can be concentrated into just a few days. In either case, the team will find the process absorbing and energy-consuming. Unscheduled interruptions should be avoided, as they seriously slow down the process and lower the quality of the results.

To encourage concentrated involvement, many teams choose to locate the QFD activities away from their normal workplaces. They discourage incoming phone calls, and try to make it difficult for team members to make outgoing calls, even during breaks.

Occasionally, however, the team will need to refer to materials located away from the QFD site. These materials might be marketing reports, product test analyses, or material that will help the team answer some question that has come up.

If the team has chosen to locate the QFD activity away from the workplace, it will be helpful to establish some method for accessing materials that normally reside at a distance. Access methods could include facsimile devices, couriers, or a computer connected to some critical project database. Without such an access method, the work of the team can be delayed.

16.8.2 ROOM

The room chosen for QFD must have plenty of bare wall space. QFD matrices are often three to five feet high, and ten to twenty feet long. Entries are made into the matrix while it's hanging on the wall. Architectural columns, corners, molding, and pictures over which the matrix is hung will make the matrix difficult to read and to write on.

The wall surfaces must allow for hanging of the QFD matrices with tape or other adhesive, or with pushpins. Some facilitators like large "white boards" because Post-it Notes stick directly to them. This author prefers Post-it Notes placed on large sheets of paper, which can be removed with the Post-it Notes still in place on them.

There should also be comfortable chairs and adequate working surface space for people to lay out papers or sections of the QFD matrix.

The room should be brightly lit, and the acoustics should allow people to hear each other easily. The author facilitated one QFD process in a room with extremely high ceilings. People's voices echoed enough so that voices tended to "get lost" in the room. Often one person was unaware that another was speaking. People interrupted each other without realizing that they were doing so. The meeting frequently broke down into several smaller, simultaneous meetings.

It's best to use the same room for the entire QFD process. Assuming the room's contents can be kept away from people with no need to know, much time can be saved, and continuity can be preserved, by not having to take down and put up the wall charts every day.

Whenever possible, inspect the QFD room in advance. If it's not ideal, you'll have time to figure out how to make the room work for you, or to find another room.

16.8.3 COMPUTER AIDS

The matrices of QFD are repositories for hundreds or even thousands of bits of information, including qualitative data (such as customer wants and needs), quantitative data, narrative text (such as assumptions that explain the data, decisions that the group made, or issues yet to be resolved), and even graphical data (such as diagrams of concepts in Pugh Concept Selection Matrices).

Today's desktop and laptop computers run plenty of excellent software products that can help store, manage, and display this data.

From very early on, a few enterprising software companies developed specialized software to support the QFD process. This software allowed the calculations to be customized, the matrix dimensions to be varied, and the matrix to be printed in a format that looks like a "textbook" QFD matrix. The various software products are more or less easy to learn and use, and they run on computer platforms that meet most people's needs. For example, QFD Designer by Idea Core[6] makes linking QFD to VOC and TRIZ very convenient.

Beyond these specialized software products, computer spreadsheets can be customized to store the QFD data, compute the formulas, and print the matrix. Many teams prefer to use a series of QFD forms built on Microsoft Excel worksheets that are linked together, and most find this an acceptable choice. Templates can be found for free online.[7]

The following are critical advantages of using computer software for QFD:

- The software serves as a central repository for all decisions, judgments, comments, issues, and notes generated during QFD.

- Computations are performed automatically, and, one may presume, correctly.

- It greatly simplifies, perhaps even makes possible, "what-if" scenarios, such as setting importance of certain wants and needs to zero; setting impacts of certain cells

---

6. See: www.ideacore.com/v1/Products/QFDDesigner/.

7. QFD Excel templates may be downloaded at: www.QFDonline.com/templates/.

to High; or substituting various values for High/Medium/Low impacts (9-3-1, 7-3-1, 5-3-1, or 3-2-1, for example).

- The matrix can be printed in a relatively compact format (several sheets of 8½" by 11" paper, for example).

A useful way of employing QFD software is to assign a note-taker or scribe to operate the computer in the QFD room. This person would be seated out of the way of the group's activities, yet still be able to see the QFD chart hanging on the wall. The scribe would be familiar with the operation of the computer and software, and would quietly record data into the computer as the team makes its decisions. When needed, the scribe can simply read out the results of computations, which the facilitator can copy onto the QFD chart on the wall.

Some people have used LCD displays or LCD projectors that can display the computer screen in the conference room for all to see. I have found this to be partly satisfactory with today's technology. The details of the QFD matrix are hard to read, even on a brightly displayed projector screen. Zoom features on most software can compensate, but not all team members are interested in looking at the same part of the QFD matrix at the same time. Many QFD teams prefer that each team member have a small, personal, up-to-date copy of the QFD chart. The facilitator can print these charts frequently, and photocopies are usually easy to make. This makes the availability of a QFD chart on the wall or projected onto a display slightly less important.

16.8.4 MATERIALS

The QFD process is team-oriented. The team works together, developing the information to be filled into the matrix. To facilitate this team process, the matrix must be visible to everyone on the team, all the time. Ideally, each team member should be able to scan the matrix from top to bottom, side to side, whenever he or she wants to.

The most commonly used materials for QFD do a pretty good job in meeting these needs. Fortunately, these materials are inexpensive, readily available, and familiar to everyone who works in an office.

Flip-chart pads provide large, but still manageable, pieces of paper that can be pieced together to form matrices of any size. Some pads have grid lines preprinted on them. The grids are usually an inch wide. These preprinted grids obviate the need to manually draw rules, something I did often, with makeshift straight-edges, while the QFD team was taking a break or driving home for dinner. I strongly recommend paper with preprinted gridlines.

Masking tape or pushpins are needed for hanging the matrix on a wall, so everyone can easily see it. (Obviously, walls that don't allow for hanging the matrix are to be avoided.) Even if the matrix is held up by pushpins, it's often desirable to tape the sheets together at all the edges, to construct a single large chart. This tactic ensures that the sheets won't get out of order or lost if the matrix has to be hung, brought down, and re-hung several times, as is often the case. If Post-It flip chart paper is available, then it will save time, if the wall surfaces are smooth enough.

Thick-nibbed markers of various colors are used for recording information in the matrix so that they can be seen clearly from a distance. The ability of the team to view the entire matrix, with its patterns and interactions, cannot be overemphasized. To allow a reasonable "bird's-eye" view, the entries must be large and dark.

Some teams favor various color-coding schemes for representing the customer needs segments, or for distinguishing main categories of SQCs, or for recording impacts.

Post-it Notes are invaluable for many purposes. They facilitate team members record-ing data in parallel; they allow for information to be moved from one part of the room to another, or from one part of the matrix to another, without being rewritten; and they allow for information to be organized, grouped, and regrouped as the team's under-standing of the relationships between units of information evolves.

16.9 SUMMARY

In this chapter we have seen what's involved in getting ready for QFD. *Planning* is the single most important action one can take to ensure a useful, timely outcome for a QFD process. A good job of planning involves anticipating the entire process and ensuring that team members' expectations have been appropriately set. It's probably better to err on the side of too detailed a plan rather than too vague a plan.

A good QFD plan involves getting answers to many questions, divided into categories. We've covered the categories of questions by dividing the chapter into these sections:

- Establishment of organizational support
- Setting of objectives for the QFD process
- Determination of the customer
- Determination of the time horizon for the project being planned
- Determination of the product or service concept to an appropriate level of detail
- Determination of the QFD team and how it will communicate to the rest of the organization
- Creating a schedule for the QFD activities
- Acquisition of the facilities and materials needed

As we have looked at these general areas of questions, we've also had to touch on many implementation aspects of QFD. In a sense, planning is like living through the process.

Having established our plan, the next important step is to listen to the customer. We first discussed this in Chapters 6 and 7. Now we're ready to actually meet the customer face-to-face. How shall we proceed? Chapter 17 provides the answers.

16.10 DISCUSSION QUESTIONS

If you have not participated in a QFD activity yourself, identify someone who has, and create with that person a checklist of QFD planning categories. Compare it to the categories provided in this chapter. Reconcile the differences.

Using the QFD Estimator Chart, compare the actual time spent in a previous QFD project with the Chart's projections, and reconcile the differences.

Phase 1: Gathering the Voice of the Customer

This chapter presents guidelines for acquiring the Voice of the Customer (VOC). As we saw in Chapter 6, this is a complex subject. Listening to the customer requires time and good listening skills. As a reminder, we outlined in Figure 1-1 (Chapter 1) the fundamental steps for this key part of the QFD process:

1. Gather the VOC
2. Analyze the VOC
3. Define Customer Prioritized Needs
4. Validate Customer Needs
5. Begin the HOQ Work

No matter how your company or business accomplishes these five process steps—whether formally or informally—they must be done, as QFD will not work well without established customer needs. The better your company does the first four steps, the more value will be added from your QFD efforts. Many companies do the first four steps partially or informally, or across functional groups. It cannot be overemphasized how important they are to developing breakthrough products and services with and for your customers and markets. It is of equal and crucial importance that the VOC be gathered by a team that includes members of the development team. Likewise, all VOC data must come from real customers, not as second-hand information or inferences.

In this chapter, we'll first provide a VOC overview; then we will review the distinction between qualitative and quantitative customer needs data. Next we'll describe a method of classifying the customer needs into categories very similar to the way the Kano model classifies technical characteristics.

We'll then move on to a survey of some of the main methods for capturing the customer's words, and images of customer environments, for adding context to the voices. These methods include focus groups and one-on-one interviews. The one-on-one interviews will be further classified according to whether the interview is conducted at the customer's location (in context), or away from the customer's location. We'll draw enough distinctions between these types of interviews so that a development team could decide which methods to use.

We'll then discuss a completely different, but valuable, method for capturing the Voice of the Customer: analyzing customer complaints.

17.1 VOICE OF THE CUSTOMER OVERVIEW

Figure 17-1 illustrates a preferred, detailed process for acquiring the VOC and for getting it ready for QFD.[1] Sigma Breakthrough Technologies, Inc. (SBTI)—a management consulting firm that employs me—has used this approach to VOC capture for many years. Companies such as Sylvania, TYCO, and American Standard—which are often considered benchmarks of business success—follow variants of this process to a large degree in their customized DFSS programs with SBTI. Many of these product-development practices have been strongly influenced by Professor Shoji Shiba.[2] We are indebted to Joe Kasabula for expanding these thoughts and for helping to develop the process shown in Figure 17-1.

The first VOC stage is to prepare for and visit customers to obtain customer-needs phrases. The second stage is to process the qualitative data so obtained into a set of customer requirements. The third phase is to establish the importance of the requirements. There are three primary methods for obtaining the VOC data, which may be used in various combinations: **focus groups, conference-room interviews,** and **contextual inquiries**.

Using a focus group of customers is generally quick and inexpensive compared to the other methods. However, this method produces the lowest yield of customer-needs information of the three main interviewing methods. Using focus groups is suggested

---

1. Randy C. Perry and David W. Bacon, *Commercializing Great Products with Design for Six Sigma* (Upper Saddle River, N.J.: Prentice-Hall, 2007): 134.

2. Shoji Shiba, Alan Graham, and David Walden, *A New American TQM: Four Practical Revolutions in Management* (New York: Productivity Press/Center for Quality Management, 1993).

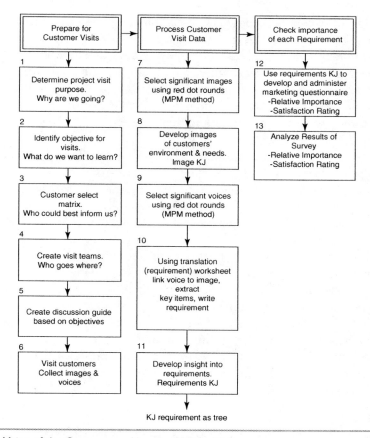

Figure 17-1   Voice of the Customer, as Used by SBTI in 13 Steps

only for small design efforts not usually associated with QFD, perhaps when doing a product fix or redesign. Focus groups may also be utilized to screen multiple product concepts when additional VOC is needed to select a concept. Conference-room interviews—the second approach to VOC data collection—provide an alternative, middle ground between focus groups and contextual inquiries in terms of time, cost, and yield of customer-needs data.

17.1.1 PREFERRED APPROACH—GENERAL DESCRIPTION

Contextual inquiries are extremely valuable, time-consuming, and costly, because members of the development team must travel to the customer's site for each interview. The customer-needs information yield is very high. This is the preferred method when designing complex products or services where significant value will be invested in development, and is illustrated in Figure 17-1 in a fair amount of detail. This is the only method of the three mentioned here where latent, unstated needs may be collected through observation and image capture. This is how breakthroughs in the creation of new products, new product functions, and new services are most often obtained. This approach requires significant investment, in that between 12 and 20 visits are often required to collect detailed customer needs.

When budgetary trade-offs are required, the authors generally recommend a combination of initial contextual inquiries followed by conference-room interviews. It is most crucial that each development team gather the customer's voices and environmental images, which can be partially accomplished by interviews and fully developed by Contextual Inquiry (see Section 17.3.3). Processing the collected customer voices and images to identify needs is accomplished by applying KJ analysis techniques to images and then voices, combined with images that have been translated into product requirements.

Translation to product requirements involves first identifying those statements from the interviews that represent customer needs, and then translating them into solution-free statements of what the product should do. This is similar to the work required in filling out the VOC Table. The team then selects the most important or significant customer needs. This is done by applying team judgment based on the team's understanding of the customer and what the team has heard from a limited number of customer interviews. The preferred approach also involves a survey process, as described in this chapter, and also included in Figure 17-1.

17.1.2 PREFERRED APPROACH—DETAILED DESCRIPTION

Stage 1 of VOC acquisition requires preparation and collection of customer-visit information. Once the interviews have been conducted in Steps 1 through 6, the development team analyzes the transcripts of the interviews and extracts the customer phrases and images, as described in Steps 7 through 11. There are likely to be hundreds of phrases and images. Ideally the team should strive for a minimum of 100 needs phrases and 100 images in order to proceed with KJ analysis. Some transcripts may be duplicates, and some of the phrases as captured will be technical requirements, Substitute Quality Characteristics, and other phrases that are not actually customer needs. The team will have to sort these phrases, using a mechanism like the VOCT (see Chapter 6). All of the phrases are important, but only the customer-needs phrases and needs images are needed at this early stage.

The customer-needs phrases can be transcribed onto cards or Post-it Notes and then analytically structured using the KJ method for complex product or service designs, or the Affinity Diagram process (Chapter 4) for simpler efforts. Ideally, the KJ diagram would be strongly influenced by the customer, by inviting customers to assist in the affinity or KJ diagramming process. Lacking customer involvement, the development team would create the KJ or Affinity Diagram. If no customer involvement is included during the KJ, greater effort must be expended when doing the validation in Steps 12 and13 of Figure 17-1.

Next, the team restructures the Affinity Diagram into a tree diagram. This is partly a clerical process of redrawing the diagram. However, it also involves ensuring that the structure is complete at all levels, because it's possible that a few of the customer needs will have been missed in the interviewing process.

The team next decides which level of the tree diagram will be used in the House of Quality. Generally the secondary level is used, but depending on the size of the tree diagram and the problem the team is trying to solve, other levels, or even subsets of a certain level, may be used. For example, if the project is to improve the packaging of the product, only those customer needs pertaining to packaging would be used in the QFD.

The final phase is to determine the importance and Customer Satisfaction Performance of the needs to be used for QFD in Steps 12 and 13. This is normally done by survey. Whereas the interview stage necessarily involves no more than 50, and often fewer, customers, the survey would reach hundreds or even thousands of customers in order to assess the majority of customers and to understand the way they are segmented.

After the marketing survey, the team develops a rank-ordering of the remaining customer needs. One approach is to classify the needs via the Kano model. The team then generates metrics for the customer requirements. In our language, the team generates performance measures for the SQCs. It then creates a QFD matrix (which Shiba *et al.* call a **quality table**[3]), in which the customer's needs are the Whats and the performance measures are the Hows.

There are many variations on acquiring the VOC using techniques similar to those described here. Many variations on the methods described in this book are possible. The best strategy is to understand the many techniques and methods available, and to choose the ones that best meet the needs of each development team. It is far more important to work through the basic steps outlined at the beginning of this chapter than to debate which methods are the most efficient. Practitioners are encouraged to sidestep these debates in favor of moving forward with obtaining, analyzing, and applying the VOC.

---

3. Shiba et al., *A New American TQM.*

17.2 QUALITATIVE DATA AND QUANTITATIVE DATA

The QFD process requires that customer data be represented as a list of product or service attributes that are important to the customer. The attributes, or needs, are potential benefits that the customer could receive from the product or service. Each attribute in the list is to have some numerical data associated with it: relative importance of the attribute to the customer, and the Customer Satisfaction Performance level of similar products with respect to that attribute.

We call the attributes **qualitative** customer data, and we call the numerical information about each attribute **quantitative** data. Figure 17-2 indicates where in the HOQ these types of data go. The methods for arriving at the qualitative and quantitative data are different. In addition, there are many alternative methods for arriving at each kind of information.

A full discussion of the various methods for discovering the qualitative and quantitative customer data would be a topic for one or more books on market research. In fact, many QFD endeavors begin with complex and expensive market-research projects. However, many product- and process-development teams cannot afford the luxury of

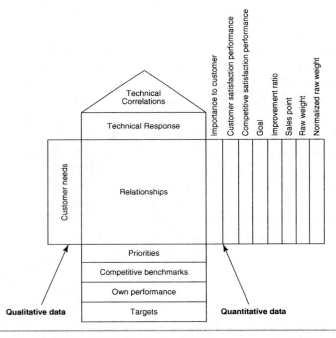

Figure 17-2   Qualitative and Quantitative Data in the HOQ

hiring market-research companies, and must therefore gather their customer data for QFD themselves. Whether they use a market-research department or firm or do the work themselves, it is important for development teams to understand how the data acquisition is being done, and to understand what it means and what it *doesn't* mean.

If you think of the QFD process as a kind of machine that accepts customer data as input and which produces a product or service as output, then it's clear that with faulty input, there's a substantially reduced chance for worthwhile output. Therefore, the VOC-acquisition stage is crucial for the success of the entire endeavor.

In this chapter, we'll describe some important techniques that development teams must be aware of (whether they use the techniques themselves or others use them) in order to make informed decisions about the VOC-acquisition activity.

The general procedure for VOC acquisition is:

- *first* determine the customer attributes (qualitative data);
- *then* measure the attributes (quantitative data).

Qualitative data is generally acquired by talking to and observing customers, while quantitative data is generally acquired by surveys or polls.

17.3 GATHERING QUALITATIVE DATA

17.3.1 CUSTOMER NEEDS AND TECHNICAL SOLUTIONS

One of the most important, yet difficult, distinctions product developers must make is between customers' needs, on the one hand, and technical solutions to meet those needs, on the other. Let's take a few examples to illustrate this.

A person shopping for an automobile may ask for a car with tinted glass. Is this a customer need? In all likelihood, no one actually *needs* tinted glass. In the equatorial parts of the world, tinted glass may be highly desired, however. Really, people need some of the benefits that tinted glass may provide. Tinted glass can reduce glare from bright light, such as direct sunlight or headlights of other cars. Tinted glass can also provide privacy to passengers, by making it difficult or impossible for people outside the car to peer in. Tinted glass can also help to keep the auto interior cool, by reducing infrared radiation from the sun and from warm objects near the car.

For each of these possible needs—reduced glare, privacy, and cooler interior—various technical solutions are available, some of which may be more effective or less expensive than tinted glass. Glare can be reduced by altering the shapes and finishes of surfaces that reflect glare, such as the dashboard and the hood. Privacy can be provided by custom-fitted

Venetian blinds or curtains, or by having no windows at all. A cool auto interior can be achieved by ventilation and air-conditioning methods.

In an example from the world of computers and software, central-processor speed is often expressed in **gigaflops,** billions of floating-point calculations executed per second. Another measure of processor speed is **clock speed,** usually expressed in gigahertz, or billions of clock ticks per second. Some technically oriented computer buyers will specify processors according to their speed ratings in gigaflops or gigahertz.

But it's the rare computer user who actually needs 20 gigaflops or 2-gigahertz clock speed. Instead, users need computers that perform certain tasks within certain time frames. Most of us just want to know, for example, how long the PC will take to start up, which is usually driven by the software installed, among other factors. The speed at which tasks are performed—rapidly recalculating the contents of a spreadsheet of a certain size, or printing or displaying a complex graphic rapidly, for example—is to some extent related to the processor speed, but most computer tasks also depend on other technical factors, such as main-memory capacity (usually referred to as RAM), input-output capacity, floating-point speed, and other factors. Depending on the real need of the computer user, as well as on the design of the software that performs these tasks on the computer, these other factors may actually have a greater influence on performing the user's tasks than does gigaflops or clock speed.

In order to make design decisions that meet customers' needs, product developers must understand the customers' *real* needs, often buried beneath their surface-level requests. Product developers must be able to distinguish between customers' real needs and the technical solutions to those needs. Just as customers don't make these distinctions, many product developers have not seen the necessity to make such distinctions either.

The ability to make clear distinctions between needs and solutions is a prerequisite to generating breakthrough concepts. The original Sony Walkman met a customer need that had hitherto been poorly met at best: the need for portable, high-quality playback of stereo music. It has since evolved into entirely digital MP3 players like the Apple iPod. This is not to say that by simply distinguishing between customer needs and technical solutions, a product developer is assured of a breakthrough concept. However, the habit of not accepting technical solutions as customer needs will certainly help product developers to find creative solutions that delight customers.

The QFD process puts customer needs on the left side of the HOQ matrix, and technical solutions (SQCs) on the top of the matrix. If these get mixed up, filling in the HOQ will be very confusing. In fact, when people get confused during QFD, I often suggest we check whether needs and solutions may be mixed up. Keeping them distinct is essential for a smooth QFD process, as well as for a successful product or service.

A question product developers must answer before they can design their products is: "Why does the customer want the technical solution being asked for (such as tinted glass or 2.5 gigahertz central-processor speed)?" Product developers must also be sensitive to *performance-race* market drivers that tend to influence buying decisions because *better specifications* may look more appealing to customers.

Very often when customers tell us what they want in a product, they will ask for what they believe is the best technical solution to their unstated need. However, customers are not necessarily the most technically knowledgeable people with regard to the design of your product or service. Thus, there is no *a priori* reason why they should possess the best solution. The product developer is usually better qualified to identify effective technical solutions to meet customer needs, but he or she must have a clear idea of those needs in order to do so.

Many product-development engineers will say that customers don't really know what they want. What probably supports that assumption is that many customers tend to ask for inappropriate technical solutions, rather than state what they ultimately want.

In fact, customers are the experts in what they need, but they are rarely given the opportunity to discuss their needs with product designers. One reason for this is that product and process designers often don't have the time to discuss customer needs with customers. When they do, they don't necessarily ask the right questions.

For example, asking the direct question "What do you want?" is an invitation to a customer to provide a technical solution. And yet this is a very natural question for a product designer to ask a customer. The important follow-up question, "Why do you want that?" is not one that product developers have been trained to ask.

A skilled interviewer uses probing questions as a tool to uncover the unstated needs of the customer. Useful variants to "Why do you want that?" are

- If you had that, what would it do for you?
- Have you ever had that? How did it work out?
- What's good about that?

Another key is to listen for words and phrases that should invite a probing follow-up question. Customer terms such as "good," "bad," "easier," "don't like/do like" are invitations to the alert interviewer to probe more deeply to find the real customer wants and needs.

The most important thing the interviewer must do during customer interviews is to persist in pursuing the customer needs by probing beyond the surface responses of the customer. Then, when analyzing the customer's responses, the team can make clear judgments as to which customer responses represent needs and which represent solutions.

17.3.2 FOCUS GROUP INTERVIEWS

The focus-group process involves assembling a group of customers (respondents) together in a room, and facilitating a discussion in which each respondent will state his or her views to the group and can hear and respond to the other group members' comments. Focus group interviews are so named because the group is directed in its discussion to *focus* on certain topics determined by the group facilitator.

The number of respondents in a focus group is generally between five and 15. The larger the group, the more skillful must be the facilitator in order to keep the discussion on the desired topics.

The most commonly stated advantages of focus groups are:

Synergy—one group member's comments may prompt another group member to discuss a topic no one would otherwise have thought of.

Cost and efficiency—several customers are interviewed in the time it would otherwise have taken to interview only one customer.

Some generally acknowledged drawbacks to focus groups are:

Tendency toward inter-subjectivity—that is, an emphasis on those beliefs that the group members share—at the expense of intra-subjectivity, an emphasis on those beliefs specific to a few but not all of the group members.

Lack of adequate "airtime" for any one group member—since the available time must be shared by all group members.

The goal of a focus-group interview is the same as for most other forms of interview: to probe beneath the interviewee's surface-level comments to arrive at the true needs. Many of the same considerations mentioned elsewhere about interviewing style apply to focus groups.

Additional interviewer techniques are also important. The interviewer must act as a facilitator—that is, he or she must control the interactions among the group members, as well as the interactions between each group member and the interviewer. One group member may contradict another, and the one being contradicted may take offense and seek to defend his or her comment. Such a discussion is not likely to generate additional customer wants and needs, and must therefore be controlled by the facilitator.

Some group members may be voluble, others may be reserved, and yet the objective is to capture all of their needs. The interviewer must find ways to encourage the talkative ones to quiet down, and the reticent ones to speak up.

The possibility for two or more people to speak at the same time creates difficulties for the note-taker or transcriber, so the facilitator must exercise a kind of "traffic-control" function, aimed at allowing only one person to speak at a time.

There are special circumstances where the focus-group approach may be the only workable way to collect customer wants and needs. For example, the peer relationships between group members may dictate that the focus-group style should be used. The group members may be more comfortable speaking to each other than speaking to an interviewer in a one-on-one setting. Consider a group of physicians, who may share a common vocabulary and attitude toward their profession that makes them unwilling to discuss their feelings and beliefs with an interviewer whom they may regard as a layman.

Generally, focus-group interviews are conducted in specially constructed rooms. Such a room contains a table around which the interviewees are seated. The facilitator may sit at the head of the table or in some other position that indicates he or she has a special role to play during the interview. At one end of most focus-group rooms, the wall has been fitted with a large window of coated glass, a *one-way mirror*. Behind the mirror, and therefore out of sight of the interviewees, there may be a video camera and an audio recorder, and also several people who wish to observe the interview but who want to influence it as little as possible. Most interviewees—or respondents, as marketing specialists like to call them—become accustomed to the fact that they are being watched and recorded, and seem not to be intimidated by that fact.

17.3.3 ONE-ON-ONE INTERVIEWS AND CONTEXTUAL INQUIRY

Conference Room Interviews

The most common form of one-on-one interview takes place in a conference room or office, sometimes in a neutral area (neither the interviewer's nor the interviewee's normal location). Depending on the availability and ease of scheduling interviewees, either the interviewer travels to the interviewee's site, or the interviewee travels to the interviewer, who stays fixed in one location. The latter is cheaper and more efficient for the interviewer, but is not always possible.

The interviewer will have prepared some sort of *interview guide* that acts as a reminder about what topics to discuss during the interview. The typical interview guide indicates the following:

- Important points for the interviewer to make when beginning the interview, such as explaining the reason for the interview, explaining the style of interview to be conducted, and asking permission to audiotape or videotape the interview
- A suggested interview opening question (usually a general, open-ended question)

- A list of special topics that should be addressed during the interview
- Closing questions and comments

The interviewee will have been selected according to screening criteria that ensure the type of experience, knowledge, attitude, or other characteristics that are needed to discuss the topics that the interviewer wants to cover. Many market-research groups have developed interviewee recruiting and screening techniques to ensure that the interviewees can provide the information the development team needs.

Generally, the interviews are recorded, on either audiotape or videotape (with the interviewee's permission). However, because of the cost of transcribing or even of reviewing the recording, some interviewers prefer to skip the recording process and simply take notes. In my experience, it is well worth the money and time to record, transcribe, and analyze the transcript. There are two reasons for this:

1. If you know you will be analyzing a transcript, you don't need to take notes. In that case, you can concentrate fully on what the interviewee is saying, and you will be more able to detect important points that the interviewee is making as they are being made, so you can follow them up immediately.

2. Very often, the interviewee will tell the interviewer something very important at the beginning of an interview, but will use vocabulary or jargon that the interviewer will not yet have learned. In such a case, the interviewer may not understand the importance of what was just said. An interviewer taking notes will probably miss this important point completely, because the jargon will not have made any sense. On the other hand, the point will appear in a transcript, because an audio or video recording captures everything that occurs, without regard to whether it makes sense.

   If the interview is recorded, the transcript can be analyzed after the entire interview has been conducted, at which time the interviewer has had a chance to learn the vocabulary or jargon of the interviewee. There will be a much-better chance that the interviewer will understand the significance of the comment.

Open-Ended Questions. **Open-ended questioning** is a critical technique during the interview. The term open-ended questioning refers to questions for which there is no simple *yes* or *no* answer, and for which there is no right answer, nor could the interviewee guess at a right answer. Such questions encourage the interviewee to provide much more information than a simple "yes" or "no" could provide.

So-called **leading questions** are the exact opposite of open-ended questions. They invite the interviewee to provide short answers that add little or no information to the discussion. Leading questions often suggest their answers, or give the interviewee very little choice in answering. Leading questions are often designed to help the interviewer prove a point rather than to gain new information. The interviewer is typically interested not in gaining new information, but in confirming an already-held belief. The following are examples of leading questions:

"Are you satisfied with this product?" (The interviewee may answer "Yes" out of politeness or because he or she is satisfied. Even if the answer is "No," very little information has been provided. In any case, the interviewee has very little room to provide a more helpful answer.)

"Do you use a seatbelt when you drive?" (Most people are aware that seatbelts *should* be worn. Hence, they are likely to respond with the "correct" answer that they use seatbelts even if they don't.)

In Figure 17-3, we turn the same questions into open-ended questions.

Both of these open-ended questions encourage the interviewee to provide much more information than with the leading questions. Also, they avoid suggesting answers.

The best interviews are those in which the interviewer strikes a balance between restricting the interview to just those topics covered by the interview guide and letting the interviewee discuss any topic under the sun. With the first type of question, the interviewer is sure to get responses to the topics identified in advance as important, but quite possibly no others. With the second type of question, the interviewer will probably hear about many topics that could not have been anticipated, but may not hear about some topics identified in advance as important.

It's usually a good idea to give the interviewee as much freedom as possible during the early part of the interview. During this period, the interviewee will have an opportunity to touch on unanticipated topics, and also—without explicit prompting—on some of the anticipated topics. In the latter part of the interview, the interviewer can exercise more control. He or she can lead the interview to those topics from the interview guide that have not already been covered, or choose to focus on the unanticipated topics that came up in the first part of the interview.

Leading Questions	Open-Ended Questions
Are you satisfied with this product?	What have your experiences been with this product?
Do you use a seatbelt when you drive?	What are the steps you take in starting your car?

Figure 17-3 Leading Questions and Open-Ended Questions

An important rule to bear in mind in interviews: If you've learned about a customer want or need just once, from a single customer, you don't need to hear about it again. The purpose of interviews is to uncover all possible wants and needs, without regard to their relative importance to the market as a whole. The quantitative part of the VOC-acquisition process will distinguish between the more-important and less-important wants and needs.

Focus.   What we look for determines what we see. The term for this phenomenon is focus. Interviewing experts understand that we always have a focus, whether we are conscious of it or not. In order to learn what we need to learn from any interview, we must notice what our focus is, and if necessary, redirect it to what best serves our interest.

What focus does.   Try this experiment with two colleagues in your office. Ask one colleague to wander around the office and make a note of all forms of work that he sees. Ask another to wander around the same area, and write down all kinds of activities that she notices.

Now compare the two lists. Not surprisingly, they will be quite different, even though each person was looking at the same area at the same time. The difference is due mainly to what they were focusing on. Other sources of difference will be their different personalities and how each person understood the task. But the prime difference is focus.

We have choices about setting focus. We can choose to look at a wide range of things, or to concentrate on just one or two aspects. We can choose to focus on people to the exclusion of machines, or on processes to the exclusion of results. Whatever we choose, and whatever we focus on, we will see what we focus on, and we will not see what's outside of our focus (Figure 17-4). Since our choice of focus both reveals and conceals, we must manage it.

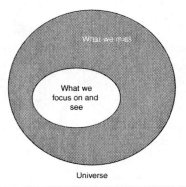

Figure 17-4   Focus

Focus helps us decide which questions to ask and which information to seek. It determines what we will understand out of all the data we are bombarded with.

What makes focus vary. While it's important for us to be aware of and to control our focus, unexpected events often take over. New data received during an interview may cause us to reevaluate some basic assumption.

For example, say we want to understand our customer's perceptions about the reliability of our product. We may have chosen to focus on the implications of failure rate for our customer's daily routine. After discussion with our customer, we may discover that the failure *rate* is not the customer's main concern. Instead, it's our service technician's response when a failure occurs that is much more important to the customer. Obviously, our original focus would not have been on the mark.

This type of surprise occurs frequently during interviews. The fact that open-ended interviewing uncovers unexpected areas of focus is one of its strengths. In order to take advantage of such surprises, we must be prepared to alter our focus.

Sometimes it will not be possible to alter the focus on the spot—we may suddenly find ourselves interviewing the wrong person, or conducting the interview in the wrong location. In such a case, our best approach is to re-plan our objectives for later interviews.

Another way to create the possibility for focus to vary is to share your interpretations of what you are observing with the customer. Your understanding may be just slightly different than what the customer was intending to explain. When you offer your interpretation, the customer has the chance to correct any misunderstandings. Should your misunderstanding be significant, you will be able to adjust your view, and you may have to change your focus.

As you continue your inquiry with the customer, you may develop some design ideas. It's a good idea to share these with the customer. The customer may be able to confirm the validity of your ideas, or show you in what way you're off the mark. Again, this can lead to change of focus.

In summary, recognize that your focus both reveals and conceals, affecting what you will understand during an interview. Be prepared to interrupt the interview to ask for concrete evidence of what is not visible, to share your interpretations with the customer, and to share your design ideas with the customer. Don't wait too long to create these opportunities for refocusing. The longer you wait, the more danger there is that you will have focused on the wrong things and missed important data.

Planning the Interviews. Interviews, and especially their analysis, take time. It's a good idea to plan your work so the time is well spent. Here are some recommended steps.

1. **Identify the questions you want answered.** A useful method for identifying these questions is to have the development team brainstorm all questions and then affinitize them. If the number of questions is very large, you may wish to prioritize them, using either the Prioritization Matrix method or the Analytic Hierarchy Process. These questions won't be asked outright of the customers, unless by the end of the normal interview some of them have not been addressed.

2. **Identify your target customers.** Once the questions are known, you'll have a better idea of how you want to segment your market, and what types of companies or individual users you wish to interview.

3. **Select participants.** In many cases, as we indicated earlier in our discussion of key customer groups, the end user of your product may not be the only person who must be satisfied. For example, a parent makes buying decisions for children, and an office manager makes buying decisions for secretaries. In this case, it's a good idea to conduct interviews with people representing all the important roles in the selection, installation, use, and servicing of your product.

   Sometimes a matrix may be useful for determining how to get the coverage you need in the interviews.[4] Try putting the market segments along the left and the roles along the top. In Figure 17-5 we have identified five market segments and three roles. If we don't want to spend the resources for all 15 interviews, we can distribute our interviews across the columns and rows judiciously, and design the type of coverage that will be the best compromise. In this example, we've minimized the number of interviews in Europe, while also ensuring pretty good coverage of Users over three of our five market segments. Also, at least one of each type of User, and at least one of each market, will receive an interview. The total number of interviews is six, versus the potential of 15.

4. **Establish the right environment for the interview.** Make sure that whoever recruits and screens the interviewees explains to them what the general topic is, how long the interview will take, and when and where it will occur. Get permission in advance to record the interview, to avoid having the interviewee balk at the last moment.

5. **Set the focus.** A final step in planning the interviews is to explicitly decide on the focus. The developers will normally plan several interviews at once. The planned focus may differ for each interview, and that in turn may help the developers decide who should conduct each interview—the reliability expert might wish to speak to the maintainer, for example.

---

4. For more about such matrices, see Perry and Bacon, *Commercializing Great Product:* 133–156.

Market Segments	Purchaser	User	Maintainer
Northeast U.S.	Interview		
South U.S.		Interview	
West U.S.		Interview	Interview
Latin America		Interview	
Europe	Interview		

Figure 17-5   Covering the Segments/Users Grid[5]

6. **Re-plan after every interview.** Regardless of all previous planning, experience has shown that the interviewers will want to re-plan the remaining interviews after each interview has been completed. This is because each interview will bring a number of discoveries, and these in turn will modify the goals for subsequent interviews. Be sure to review the results of each interview as soon as it is over, and be ready to modify your plans based on the results.

Telephone Interviews.   Telephone interviews are inexpensive compared to other styles of interviews. The major cost-saving element is the absence of travel expenses. The interview itself is also often shorter than face-to-face interviews, although there is no physical or technical reason—only a psychological reason—why this should be so. The recruiting and screening process for telephone interviewees is the same as recruiting for face-to-face interviews.

Other important similarities with face-to-face interviews:

- The interviewer can (and should) prepare and work from an interview guide.
- The interviewer can, after receiving the interviewee's permission, record the interview.
- The general strategy of asking open-ended questions, and of providing the interviewee freedom to discuss any topics, is still the best strategy.

The main differences between telephone interviews and face-to-face interviews are these:

- The technical quality of a telephone-interview recording is rarely as clear as that of a face-to-face interview, because the dynamic range and frequency range of voice on

---

5. The segment/users grid in Figure 17-5 resembles a three-factor "Designed Experiment." Taguchi's orthogonal arrays, or other fractional matrices, could be used to plan out groups to be interviewed.

a telephone is limited. Nevertheless, inexpensive equipment is readily available for recording audio.

- The visual clues of facial expressions and body language are not available to clarify to the interviewer what the interviewee is attempting to express.

- The same visual clues are not available to guide the interviewer as to when to switch to a new topic, when to probe, or when to terminate the interview.

- It is not possible over the phone to show the interviewee product samples or simulations, nor to observe the interviewee using or reacting to such samples.

Contextual Inquiry

Contextual Inquiry[6,7] is the name of style of interview that emerged in the early 1990s as a technique aimed at helping developers to uncover hard-to-identify customer needs, and also to generate innovative solutions to meet those needs.

In one sense, Contextual Inquiry is a variation on the one-on-one interview method described above. Almost all of the principles that apply to one-on-one interviews apply to Contextual Inquiry. The difference lies in the additional emphasis on observing the customer *in the context of* the customer's activities, and on the interactive nature of the relationship between interviewer and interviewee—so interactive that it is called *inquiry* as opposed to *interviewing*.

Summary Data Versus Concrete Data.   Suppose you go to a party. The room is crowded; everyone is talking. Music is playing in the background. In fact, you might wish it to be further in the background than it is, because you can barely hear yourself speak. A stranger strikes up a conversation with you.

"What do you do?" the stranger asks.

Each of us no doubt has a different reaction to this common question. Some people don't like to talk about their careers at parties. Others are happy to discuss their work without limit. You have decided to provide a polite answer. You want to be friendly and

---

6. M. Good, "The Iterative Design of a New Text Editor," *Proceedings of the Human Factors Society, 29th Annual Meeting* (Baltimore, 1985): 571.

7. D. Wixon, K. Holtzblatt, and S. Knox, "Contextual Design: An Emergent View of System Design," Proceedings of the SIGCHI Conference on Human Factors in Computing Systems: Empowering People, Seattle, April, 1990 (New York: Association for Computing Machines, 1990).

informative, but because you are at a party, you don't want to limit your participation in the party to a discussion of what you do.

You could answer the question in many ways. After a moment's thought, you say:

"I'm an engineer."

Summary data. You have just provided summary data to your new acquaintance. We call it summary data because it summarizes a vast amount of information (in this case, about what you do) into a very few words that you have tailored for this situation.

What does your acquaintance now know, based on your answer?

First, what you don't do. For example, you don't remove appendixes in a hospital operating room. You don't drive a truck.

Second, that you do some sort of work—you're probably not unemployed or independently wealthy.

Next, that your work is somehow technical in nature.

There are obviously many ambiguities in your answer. The first relates to which engineering discipline you are involved with. Perhaps your work involves machines, electronics, roads, or aerospace, since there are many types of engineers. Even within these disciplines, the possibilities for erroneous assumptions are enormous. Some mechanical engineers are involved with auto engines, others with hand staplers. Some electronics engineers work with microchips, others with telephone receivers.

The second ambiguity relates to the kind of work you do within your engineering discipline. Some engineers design products. Others develop manufacturing processes. Others manage engineers and engineering operations, and don't actually do technical work themselves.

There are endless possible interpretations your acquaintance may apply to your answer. To what extent your acquaintance will truly understand "what you do" will depend on how much additional communication occurs between you.

Concrete data. Given the context of the party, your simple summary answer may have been appropriate. But suppose you had wanted your acquaintance to know as much as possible about what you do. How else might you have responded?

A completely different, albeit extreme, approach would have been to invite your acquaintance to gather concrete data. You might ask him to meet you at your home the following Monday morning and "shadow" you at your work every day thereafter for the next two months. He would observe your commuting experience, your office; he would learn your secretary's name, what type of coffee you drink, and who your colleagues are. He would see how meetings are conducted, what types of phone calls you receive, what mail you read, what mail you throw out, how you communicate to your customers, how much time you spend at your desk, and what you do while you are there. He would learn whether you conduct work while commuting—for example, by cellular phone; whether you bring work home; how much time you spend away from the office on your work.

Because your acquaintance would have followed you for two months, he would understand some of the cycles in your work. Your acquaintance would have observed periodic staff meetings and their occasional cancellations, as well as follow-up and lack of follow-up from these meetings. He would have seen priorities shifted, emergencies dealt with, deadlines met and missed. He would have observed in passing the effects of people missing from the office, on vacation, on business trips, or ill. He might have seen you absent yourself from the office—to attend a meeting with customers, to attend a training class, to take a vacation.

The ambiguities associated with the summary data received at the party would now be much reduced. By observing the building you work in, the conversations you have, and the mail you receive and send, your acquaintance would know precisely which engineering discipline you are involved with. This knowledge would be bolstered by pictures, models, or actual machines, so that many specifics, such as your current project and your current schedule, would be revealed. It would include information on whether you manage others; and if you do, what your management style is. If you are an individual contributor doing technical work, he would see examples of how you do your work: at a computer, at a drawing table, or in a laboratory. He would know what part of your work you do alone, what part in communication with others, and by what types of communication (by phone, in person, by electronic mail, by fax, or at meetings large or small, short or long).

Not only would this in-depth observation clear up ambiguities, but it would reveal the *affect* of your work as well. In other words, your mood and the moods of the people with whom you work would become evident. Your acquaintance would see moods shift, by the hour and by the week. He would see arguments, celebrations, smiles, and frowns. The *pace* of the work, its rhythm and its urgency would become visible. He would understand which parts of the work are repetitive, which parts are unique, and which parts contribute most to the lengths of the cycles.

All these observations would add to a far richer understanding of "what you do." Your acquaintance would experience your work as deeply as possible, short of actually taking over your job and doing it for you.

This very long description of the observations your acquaintance would make is an illustration of the rich knowledge we can get from direct observation. This type of detailed, fully immersed observation characterizes one of the differences between conference-room interviews and market research in its most common form, and Contextual Inquiry.

One fundamental difference between data gathered at a party or in a conference room and data gathered by direct observation is the *concreteness* of the directly observed data.

By experiencing things concretely, we learn *more* than can possibly be written down. By learning of things in summary form, we learn *only* what can be written down.

Shared Experience: Another Kind of Reality.

One of the main principles behind Contextual Inquiry is that the fullest understanding we can possibly get of our customer's needs is obtained by not just observing, but actually *sharing* in the customer's experience as fully as possible. This shared experience helps us to appreciate the customer's needs almost as deeply as if the needs were our own.

With Contextual Inquiry, this means observing the customer living his life as it relates to our product or service. If our customer uses our product at work, then we must share that work experience with the customer. If our customer uses our product at home, then we must share the home experience with the customer.

What the engineer can share with the customer. The notion of "sharing" experiences with strangers might seem to imply more intimacy than many of us are willing to contemplate.

In practical terms, the sharing is no more than a professional relationship aimed at providing the product or service developer with design data. It includes observation of the customer's activity, at the site of that activity. It also includes discussion with the customer to fully appreciate how the customer experiences that activity. If the product is a leisure product, it involves observation of the customer's leisure as it relates to that product, at the site of that leisure.

What Is Design Data?

By **design data**, we mean information that provides the developer with information that can be used to make the product or service clearly support the customer's goals. Therefore, the requirement "easy to use" would not be design data, since we have no information on what aspects of use are important to the customer.

What might be closer to design data would be the requirement "easy to correct my mistakes." If we had seen our customer make some mistakes and then attempt to correct them, we would understand what the obstacles are, and what about our product might remove those obstacles.

When we seek design data, we have three objectives in mind for our product or service. For whatever are the customer's applicable activities, we want to *support* the customer's needs in pursuing those activities, we want to *extend* his or her pursuit of those activities, and we want to *transform* the customer's pursuit to some level or domain that not even the customer thought was possible.

An example illustrating these three objectives will help clarify the distinctions. Before 1975 or so, the primary device for creating legible documents in the office was the typewriter. The typist's activity was aimed at creating a neat, legible document quickly. The most-frequent disruption of this activity was probably the typing error. When a typist made a typing mistake, he or she had to stop typing, roll the paper back so the typing error was within reach, erase the mistyped material (for example, with a special abrasive eraser), roll the paper back into place, and retype the material correctly.

With the advent of word processors, the typist's work has become *supported* in this area by the delete key. Now, when a typing error occurs, the process of correction is much simpler—just hit the delete key one or more times until all the incorrect typing is removed, and then continue. The issues of overzealous erasure (rubbing a hole in the paper!) and misaligning the paper when rolling it back have been obviated by the word processor. Thus, correcting a typing mistake with a word processor is far less disruptive than with a typewriter.

Word processors have done more than support error correction. They provide pagination capabilities, so that page numbering, top and bottom margins, and headers and footers are automatic. They provide easily adjustable tabulation. Some word processors have made it easy to type special characters, such as mathematical symbols or foreign-language characters. All of these capabilities have *extended* the typist's work far beyond what a typewriter could readily support.

In meeting both of these objectives, *supporting* and *extending,* the underlying paradigm remains the same. First, we compose the document—we generally write it in long-hand, for instance. Then, we take our material to the typewriter or word processor and transcribe it.

Word processors since the late 1980s have gone much further and opened the possibility for a completely new paradigm. By providing rapid "cut-and-paste" functionality, integrated outlining capability, spell checking, and grammar checking, they have *transformed* the process of creating a legible document. Many people now *simultaneously* compose and transcribe their documents. This is a process that was all but impossible in the days of the typewriter. Granted, many people still prefer to write documents in long-hand and then type them. But many others, and a growing number of them, now prefer the transformed process of simultaneous composition and transcription.

To summarize, in Contextual Inquiry, we attempt to get design data that shows us how to *support* the customer's current activity, how to *extend* the activity, and finally how to *transform* the activity into something considerably more useful or attractive.

Accepting the Impossible. To be open to paradigm shifts that can lead to transforming the customer's activity, product and service developers must temporarily accept that anything is possible and nothing is impossible. In order to shift customers' paradigms, we have to shift our own paradigms. Paradigms are determined by our implicit beliefs about what is and what is not, and about what can be and what cannot be.

When listening to the Voice of the Customer, it is the customer's unreasonable, impossible demands that have the potential to help us break out of our current paradigms. Our job is to hear the demands as expressions of the customer's needs. Our job as product or service developers is to use our special technical knowledge to help meet the customer's needs. The more open we are to accepting all customer needs as *possible to meet,* the more likely it is that we will conceive of a product or service capability that will meet a customer need that the competition hasn't understood or acknowledged.

Contextual Inquiry: Context. As indicated in the scenario of shadowing an engineer to understand his or her work, there are many aspects to the context of the customer's use of our product. Some of these aspects of context play major roles in how the customer will use our product. Our job as a developer is to identify these key contextual aspects.

No one can know in advance what all of these contextual aspects will be, or which to pay attention to. We can only be aware of the many possible aspects and be ready and willing to observe them when we visit the customer at the site of product use. Probably the most important aspects of the user's context are the ones that differ greatly from the developer's. Those will be less familiar to the developer, and also most difficult to simulate in a laboratory.

Some of the more important aspects of the user's context are location, people, culture, and values.

Location can be divided into immediate vicinity and more-remote areas. The immediate location could be indoors or outdoors.

If indoors, is it in a cramped or spacious enclosure, is it hot or cold, noisy or quiet, humid or dry? Is air quality important? Are safety, comfort, or privacy important issues? Do things happen quickly or slowly? When using the product or service, is the customer lying down, sitting, or standing up? Is the area brightly lit or very dark? Over what period of time does the use occur—for five seconds at a time, ten times per day, or 24 hours per day, every day?

If outdoors, then the immediate location includes whether the use is underwater, at ground level, or above ground. What type of weather and climate prevail? Is the product used only during certain seasons, or year round? Is the product used at night, or during the day? Is the product in a fixed position, or is it expected to move? Is air quality important? What type of clothing is the customer wearing—gloves, sunglasses, safety goggles, sneakers, steel-toed shoes?

The number of possibilities is limitless. This is why it's so important for the product or service developer to actually experience the places where the product is used, rather than make assumptions.

People include those who are in the immediate vicinity of the customer, and those who are not nearby but have important relationships to the customer. Either at a distance or nearby may be co-workers, colleagues, friends, students, teachers, patients, clients, teammates, or opponents. Other relationships of people to the customer that may also be important are boss, subordinate, parent, child, sibling, customer, supplier, coach, enemy, or lover.

Culture covers the issue of how people relate to each other and to the customer. It includes people's beliefs, customs, methods and styles of communication, decision-making processes, and organizational structures. For each of these aspects of culture, we may need to know what happens between people, when things happen, how they happen, who is involved, and why. All these aspects of culture may have important implications for how our product or service will be used by the customer.

One of the most important aspects of culture that product and process developers often overlook is the customer's language. All too often, developers describe their creations in terms that have technical meanings that the customer does not understand. Likewise, customers will refer to their activities using words that have no meaning to the developer. Under these circumstances, it's unlikely that the developer's product or service will be right on the mark. The developer has a responsibility to learn the customer's vocabulary as it pertains to the product or service under design.

Values relate to what is important to the customer. Examples of values that may be important are what the customer respects, disrespects, enjoys, hates, tolerates, notices, and ignores. Most important to the developer is what represents *success* and *failure* to the customer. In a sense, the developer's job is to enhance the customer's chances for success in whatever the customer is doing that involves the developer's product or process.

Contextual Inquiry: Inquiry. When Contextual Inquiry is described to most people, they quickly grasp the significance of experiencing the customer's context. But they tend to ignore the other key word—inquiry. *Inquiry* is very different from *interview;* we'll try to draw the important distinctions here.

Key concepts of inquiry. The important image of inquiry in this discussion is one of *partnership.* As developers, we have special knowledge of our products and of the technology we use to develop our products. The customer has special knowledge of his or her job or activity and how our product fits in.

In a very real sense, we are both experts, but in very different subjects. Our job as partners is to create a large-enough overlap in our areas of expertise so that our product or service will strongly support, extend, and enhance our customer's activity (Figure 17-6).

Inquiry versus interview. When we approach a customer for an *interview,* both we and the customer conspire to take on roles that work against achieving the full level of understanding we seek. As interviewer, *I* ask the questions. As interviewee, *you* provide the answers. *I* am in charge of the direction of our discussion, and the range of topics we pursue. *You* are inclined to take a more-passive role and simply respond to my questions.

In contrast, the *inquiry* can be seen as a joint search for information or for the truth.

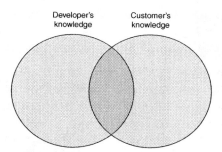

Developer's knowledge   Customer's knowledge

Figure 17-6   Developer-Customer Partnership through Creating Overlapping Areas of Expertise

In the inquiry model, there is little or no difference between our roles. We are inquiring together into the problem of how our product or service can help the customer achieve his goals. We each need to learn about the other's expertise, and we both must *try out* as many design possibilities as we can in our inquiry. Just as the classical market-research interview intentionally separates customer needs from technical solutions, the Contextual Inquiry tries to create technical solutions from the customer needs.

In this relationship, we are striving for a partnership of developer and customer, in which the two together create the product or service concept.

A few simple tools can help us structure our discussions as inquiries rather than as interviews: stage setting, and open-ended questioning.

Stage setting is the process of enrolling the customer in the partnership. This involves explaining to the customer beforehand the main ideas of Contextual Inquiry, with special emphasis on the notion of creating together the ideas that will make the product serve the customer's needs. Most customers are very willing to team up with developers in Contextual Inquiry. Some customers may hesitate because they feel they lack technical competence in product design. Here the developer should reassure the customer that the customer's expertise is an essential ingredient of the inquiry; it is unique and irreplaceable.

Open-ended questions create the best environment for Contextual Inquiry, and for all other forms of interviewing we have described in this book.

Planning the Interviews.   Contextual Inquiry interviews, even more than other forms of interview, take time. The planning outlined for one-on-one interviews applies to Contextual Inquiry as well.

1. **Identify the questions you want answered.** Use the same techniques as described for one-on-one interviews.

2. **Identify your target customers.** Use the techniques described for one-on-one interviews.

3. **Select participants.** Use the techniques described for one-on-one interviews.

4. **Establish the right environment for Contextual Inquiry.** Very often, Contextual Interviews will be set up by account representatives, salespeople, or others who may be in direct contact with our interviewees, but who may not be familiar with the principles of Contextual Inquiry. These representatives may be very successful in getting our users to agree to be interviewed, but the users may still not know what to expect in the actual interview. Importantly, they may not fully appreciate that the interviewer will want to conduct the interview at the site of the activity while the interviewee is performing the activity.

   I've found it best that the developers contact these users directly after they have agreed to be interviewed. During a preliminary phone conversation, the developer can explain the purpose of the interview, and can describe how it will be conducted, how long it might take, and where it will happen. The developer should also get permission to record the interview at this time. Finally, it's a good idea to emphasize how important the interviewee's cooperation will be, and to thank him or her for agreeing to participate. These preliminary phone conversations have proven to be very effective in getting interviewees into the right frame of mind.

5. **Set the focus.** Use the same techniques as described for one-on-one interviews.

6. **Re-plan after every interview.** Use the techniques described for one-on-one interviews.

Interview Steps. It's convenient to think of each Contextual Inquiry interview as consisting of three main stages: introduction, observation, and summary.

Introduction. At the beginning of the interview, it's a good idea to introduce yourself and restate the purpose of your visit. Also restate the need for the interview to take place at the site of the activity.

Next, ask the interviewee to describe his or her work activity to you. Here is where you should start your audio or video recorder.

You should find out the cycle time of the activity—is it repeated every few minutes, or is it repeated every few years (for example, the product-design cycle)? Is it very much the same each time, or is it very different each time?

Ask for information about the environment or the context, including the location, the people, the culture, and the values.

This will be summary data, but it will provide a background for conducting the rest of the interview.

Observe the customer's work or use of the product. Next, ask the interviewee to begin showing you the activity you want to study.

For *short-cycle activities,* you may be able to observe the entire activity from beginning to end, possibly even a few times.

Validating our customer's assertions can be a source for surprises that may cause us to alter our focus. When customers make assertions, *ask them to show you.* Ask to see reports, examples, demonstrations, or even simulations of what the customer has told you.

This is a powerful possibility in Contextual Inquiry. There are two important reasons why we should try to see evidence of our customer's assertions.

The first reason is that we may have misunderstood the significance of what we were told. The second is that the customer may have unintentionally misled us, either by exaggeration or by use of vocabulary we are unfamiliar with. (In rare cases, the customer may even intentionally mislead us.)

For *longer-cycle activities,* such as adjudication of a law case or the design of a product, you obviously won't have time to see the entire cycle. In this case, make sure you understand the entire cycle in abstract terms (as described to you during the introduction phase of the interview). Then, focus on that part of the cycle that is current. Try to observe as much as is practical. Starting with the specifics of what is current, it is often possible to extend one's understanding of cycles that are not currently visible.

For other parts of long cycles, ask to see examples of tangible items that result from those parts. Ask for documents, models, work in progress, demonstrations, diagrams—in short, anything you can actually view first-hand. Ask how the items were produced. Ask for an explanation of each item. Ask for examples of comparable items that may explain what stays the same and what varies from one such item to the next.

Summarize. At the end, summarize the main points of what you learned. Show the interviewee your flow diagram of the activity. At each point in your summary, ask for clarification or correction to your understanding.

Sometimes, during Contextual Inquiry interviews, the developer may see the interviewee experience difficulties that the developer can remedy. It would be disruptive of the interview process to show the interviewee your remedies during the observation phase of the interview, but it would be quite appropriate to help solve the problems during the summary phase.

Simple Prototyping. Just as there is great value for the developer in experiencing the customer's environment and activities first-hand, so also is it valuable for the customer to experience the proposed product or service first-hand. A prototype provides an opportunity for such a "sneak preview."

In many cases, developing prototypes of products or services is prohibitively expensive or time-consuming. Even when prototypes are possible, they may not be available early enough in the design cycle to have a useful impact on the design.

However, inexpensive simulations of products or services can often provide enough realism to help a customer provide useful feedback. The notion of the **paper prototype** was developed along with Contextual Inquiry, and it can be quite useful.

The idea is to simulate on paper, or with inexpensive materials, some essential aspects of the product or service being planned. If such a simulation is possible, then as a customer experiences and responds to the simulation, the developer can redesign the product or service instantly—at the simulated level—and test the redesign on the customer immediately.

Common examples of paper prototypes are

- Graphical User Interface mockups
- Hardware mockups
- Service scripts

Graphical User Interface (GUI) mockups, or *paper mockups,* are made by drawing a large rectangle representing a computer screen on a large piece of paper, such as a desk pad. There are rapid prototyping tools available that can speed this activity, such as MS Visio and Macromedia Flash MX Pro. GUIs consist of many graphic objects including menu bars, windows, icons, and buttons, each with a characteristic appearance. Each object can be simulated by a Post-it Note of appropriate size and shape, with graphics or text drawn on it to resemble its actual appearance on a computer screen.

In Figure 17-7, the desktop, consisting of the menu bar (containing File, Edit, MIDI, Compose, and Setup menus), is drawn on a piece of paper with dimensions 11" × 17". The menus are written in pencil or are each on a small Post-it Note, so they can easily be changed or rearranged. The File menu commands Open, Close, etc. are written on Post-it Notes in pencil. Likewise for the window labeled Rhythms and the button palette containing the triangle and circle icons. The developer will have prepared several additional Post-it Notes, each depicting an object that could appear on the screen sometime during a user's session.

The customer is presented the configuration, and asked to "use" it to perform some task defined by the customer. For example:

- "I would start by selecting the File menu." (The developer would then reveal the File menu options by laying down the File menu commands on the paper prototype, beneath the File menu item, as shown in Figure 17-7.)
- "I want to convert a file produced by Program X, but I don't see a command here that would allow me to do that." (The developer might ask, "What command did you expect to see?"—thus initiating an inquiry into why the customer wants to

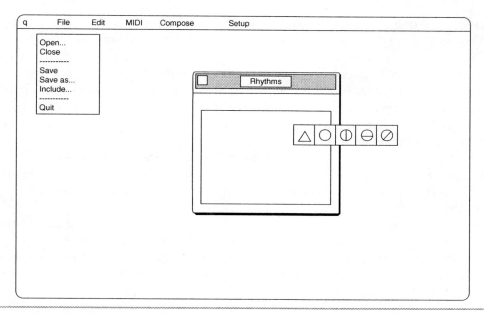

Figure 17-7 GUI Paper Prototype

perform the action at this point, what controls or commands the customer expects to see, and what the customer expects the software will do next.)

If the developer and user decide that a new command is needed in the File menu, the command can be penciled in on the spot, and the simulation can be continued.

This process could continue in as much detail and for as long as both people want it to. The entire interview would be recorded so the developer can capture all new customer needs and product concepts for analysis later.

Hardware mockups are used similarly to GUI paper mockups, but might involve inexpensive cardboard or Styrofoam mockups instead of the two-dimensional mockups used for software simulations. The advent of Stereo Lithography or Rapid Prototyping allows accurate full-sized, three-dimensional mockups to be made quickly. I have seen potential customers' reactions to full-sized resin concept models and concluded that investment in these 3-D mockups results in better reception of the concept by the customer.

Service scripts are flow diagrams that indicate all the steps of the service *from the customer's point of view.* Once again, the developer would ask the customer to invoke the service and simulate it by faithfully following the steps of the flow diagram. At each point, the developer would ask the typical Contextual Inquiry questions, such as: "Is that what you expected?" and "What are you trying to achieve at this point?"

Analyzing Data from Contextual Inquiry. During the interview, the interviewer must be very alert to the customer, trying to understand the "meaning behind the meaning" of what the customer says. It's a good idea to take notes as the interview proceeds, however, to capture important points that you may want to come back to later. Also, experience has shown that creating a flow diagram of the customer's activity can be a useful way of arriving at a joint understanding of that activity. It provides an excellent way for the interviewer to quickly summarize the activity, and for the interviewee to confirm or modify that summary.

Contextual Inquiry interviews should also be captured on video or audio recordings. The recording is the single most important item that you will bring back from the interview. Most of the data derived from Contextual Inquiry comes from analysis of the recordings.

Initially, I recommend that you transcribe the recording word-for-word into a hard-copy document, and then capture the customer data from the transcription (while simultaneously viewing the recording if you have used video). Later on, after developing some experience by transcribing, most teams will skip the transcription step and work directly from the recordings.

Teams of two or three developers scan the transcriptions or the tapes for concrete data, and write each piece of customer data onto a Post-it Note or card. The cards can be affinitized and grouped, just like all other customer data. For more details on classifying the data, see "Analyzing Customer Data" in Section 17.5.

17.3.4 HOW MANY INTERVIEWS?

The purposes of interviews of any type are first to identify all customer needs, and second to identify whatever innovative solutions to meeting those needs customers may offer.

The number of interviews required is based on the yield of the interviews. The first interview will produce a great deal of new information that the interviewers will not have seen before. The next interview will usually contain some of the same information yielded by the first interview, and some new information. With each subsequent interview, we can expect to see an increasing proportion of information that we have already learned, and a decreasing proportion of new information. At some point, the amount of new information from interviews will not be worth the effort to obtain it.

Since the number of interviews is going to be very small compared to the customer base (for most markets), there are no valid statistical inferences that can be made from what is learned from the interviews. After we have heard a customer need once in any interview, we don't need to hear it again. Later, by surveying a sufficient sample of customers we can determine how important the need is relative to all other needs.

John Hauser and Abbie Griffin modeled the rate of new customer needs yielded by successive interviews.[8] They concluded that 20 to 30 interviews were required to get 90 percent to 95 percent of customer needs. This yield rate is based on two important assumptions:

- Customer interviews are one-on-one
- Each customer interview is considered to be independent of the others

The second assumption is very important, because in practice the customer interviews *are not* independent of each other. Any interviewing team will compare notes after every few interviews, and each interviewer will remember the information gathered in that interviewer's previous interviews. By steering each interview into new territory, the interviewer can push the yield rate of new needs much higher than in the Hauser-Griffin model.

Thus, interviewers may regard 20 to 30 interviews as the upper bound of interviews needed to get 90 percent to 95 percent of the needs. The curve associated with this research also shows that after 12 to 20 interviews, the rate of new knowledge acquisition declines substantially.

The Hauser-Griffin research was confirmed by one-on-one conference room style interviews. No research has been done on other styles of interview, such as the Contextual Inquiry. Contextual Inquiry interviewers generally feel, however, that after about ten Contextual Inquiry interviews, the number of new needs identified is not worth the additional effort. Considering the much higher cost of Contextual Inquiry interviews, this informal finding is somewhat comforting.

17.3.5 FOCUS GROUP OR ONE-ON-ONE INTERVIEW?

In many cases, whether to use focus groups or one-on-one interviews is predetermined by external factors, such as the availability of interviewees, or the skills and preferences of the interviewers. But when all else is equal, which is the better choice? The goal is to determine as many unique customer wants and needs as possible (within a budget), so the interviewing style that is likely to deliver the most wants and needs for a given cost is the one to choose.

Hauser and Griffin have studied the relative effectiveness of focus groups versus one-on-one interviews.[9] The effectiveness of the interviews was determined by counting the

---

8. Abbie Griffin and John R. Hauser, "The Voice of the Customer," *Marketing Science* 12:1 (Winter 1993).

9. Griffin and Hauser, "The Voice of the Customer."

number of unique customer needs generated by each interview style, and comparing them to the total number of needs generated by both interview styles.

The study found that four one-hour, one-on-one interviews were about as effective as two two-hour focus-group interviews with six to eight interviewees present at each focus-group interview.

On this topic, Griffin and Hauser conclude: "If it is less expensive to interview two consumers for an hour each than to interview 6–8 customers in a central facility for two hours, then [our data suggests that] one-on-one interviews are more cost efficient."

17.4 REACTIVE VERSUS PROACTIVE MODES

Qualitative customer data is acquired in two contrasting modes: proactively and reactively.

The *proactive* mode involves intentional steps by the development team to seek information. In this mode, the development team takes control over the amount and type of information it will receive, as well as the timing. Surveys, focus groups, and one-on-one interviews are examples of proactive methods of listening to the customer.

The *reactive* mode involves customer data arriving at the doorstep of the development team. In this mode, the customer takes control over the amount, type, and timing of information transmitted. Customer complaints, litigation, and letters of appreciation are example sources of reactive customer data.

Reactive data is generally more plentiful and more readily available to developers than is proactive data. Customer-complaint databases are generally developed to help organizations defend themselves against complaints that arrive faster than the organization can respond to them. Backlogs of complaints often build up. They must be organized to enable specially formed teams to handle them.

Most companies do a good job of organizing and keeping track of incoming negative customer data. Positive data may find its way into sales organizations in order to help with the selling effort, and it may be used to congratulate employees who have contributed somehow to a success story. However, because positive data rarely needs follow-up, it usually doesn't get the same kind of management attention that is given to negative data.

As a result, the available reactive data is generally most useful for discovering what makes customers unhappy. In other words, it provides clues to customer needs that would classified as Expected in the Klein model. Because customers don't separate needs from technical solutions, these complaint databases usually also contain rich information about Dissatisfiers (from the Kano model).

Customer-complaint databases are difficult to analyze from a QFD perspective because of the way they are typically organized. Most complaint databases classify complaints by urgency or priority of response needed. As a rule, the urgency of the complaint is assigned not by the customer, but by the complaint-handling team.

High-priority complaints are those that directly or indirectly will cause a lot of trouble to the complaint-management team if not responded to quickly. High-priority complaints are usually associated with failures of a product or service that render it completely useless to the customer, or that create a safety hazard or other catastrophic impact on the customer.

Lower-priority complaints are generally related to the other categories of customer need (Hidden, High Impact, and Low Impact). Because they are lower priority, however, they sometimes receive little attention, and particularly, little description.

The potential for learning a lot about Expected needs and Dissatisfiers from customer-complaint databases is high, because there is usually a lot of data. Because the databases are not organized to support QFD, however, they require much resource-intensive analysis before they can provide their potential benefit.

Development teams should examine complaint databases and other forms of reactive customer data to determine the cost effectiveness of gaining useful information for QFD. Random sampling of these databases could be a way of harvesting useful information at a reasonable cost.

More importantly, development teams would do well to enhance the structure of these databases for future incoming complaints, so that the incoming data will be easy to access in the future.

17.5 ANALYZING CUSTOMER DATA

After transcripts of interviews or focus-group sessions of any type have been prepared, the development team is faced with the task of making sense of the data. In particular, they must create an Affinity Diagram of customer needs in order to begin constructing the House of Quality.

The overall process to produce the Affinity Diagram is this:

1. Identify phrases that represent customer needs and copy them to cards or Post-it Notes. Wherever possible, use the actual words of the customer. Try to use statements about concrete experiences, rather than statements that summarize feelings. Later in the Affinity Diagram process, a collection of concrete statements will be generalized by the development team into a higher-level customer attribute.

2. Sort the phrases into true customer needs and other types of information, using a construct such as the Voice of the Customer Table (Section 6.2, Chapter 6). During the sorting process, development team members will undoubtedly

develop questions, issues to be resolved, and product-concept ideas. These should be sorted into appropriate categories, to be used or otherwise dealt with as the QFD process moves along.

3. For the remaining customer needs, create an Affinity Diagram.

4. Select the secondary or tertiary level to represent the customer wants and needs in the HOQ.

5. Document all results, especially any unexpected results. Be prepared to inform the organization about these results.

17.5.1 WHO CREATES THE AFFINITY DIAGRAM?

The expert on how customer phrases go together, and what they mean when they are grouped, is the customer. With the exception of the VOCALYST process, described in Section 17.5.2, the Affinity Diagram process can usually be influenced by a few customers at most.

If VOCALYST is not available to the team, a reasonable substitute is to invite a few key customers to take part in the Affinity Diagram process, and ask them either to build the diagram or to participate with the developers in building the diagram.

If customers are not available, the developers will have to build the Affinity Diagram by themselves. The team must be careful to distinguish between customer needs and technical solutions, and to think like a customer as much as possible. There is an obvious risk involved when customers do not participate in this part of the process. Without customer participation, there is inherent risk of inaccuracy until the team's efforts to affinitize the voices are later validated by customers.

17.5.2 VOCALYST

VOCALYST is a proprietary VOC process. It is unique in that it uses statistical methods to create a tree diagram of customer needs representative of how customers have actually affinitized those needs. While VOCALYST is a high-end solution, there are other methods which are less costly. I have had great success using the KJ method with

---

10. Robert Klein, Applied Marketing Sciences, Inc., "New Techniques for Listening to the Voice of the Customer." In *Transactions from the Second Symposium on Quality Function Deployment*, June 18–19, 1990 (Novi, Mich.: GOAL/QPC, 1990): 197.

11. See the VOCALYST product description at: www.ams-inc.com/npd/vocalyst.asp.

team members who have been trained in VOC, following the process depicted in Figure 17-1.

VOCALYST begins with a series of one-on-one interviews of customers or users of competitive products or services. These interviews are conducted in the *probing* style that we have referred to earlier.

Each interview is transcribed word-for-word. Each transcript is analyzed by at least two experienced readers, who independently highlight every phrase that represents a customer want or need. All these phrases are entered into a computer database. Typically, 20 transcripts will yield about 1,000 customer phrases. Using proprietary methods, these phrases, which include many duplicate or similarly worded needs, are reduced down to a set of 50 to 150 unique wants and needs.

Each need is noted on a card, and a full set of cards (usually between 50 and 150) is sent to approximately 100 customers. They are requested to group the cards into piles that seem to them to be similar. In essence, each customer is asked to create a single-level Affinity Diagram. This single level will eventually become the tertiary level of the final customer-need hierarchy.

The customers then pick the need (card) in each pile that seems to them to best represent that pile, and put it on top of the pile. The cards placed on top of the piles are called "exemplars" and are used in later processing to help the VOCALYST analysts determine the names or titles of the secondary and primary levels of the customer-need hierarchy.

Next, the customer picks the pile that is most important to him or her, and marks it 100. All of the other piles are then marked with a number that represents its importance relative to the most important pile (the one marked 100). This step captures weighted importance of the groups of customer needs.

The customer is also requested to evaluate how well the current product meets the need represented by each pile. This is the Customer Satisfaction Performance value for the customer-defined secondary need. The customer then returns the cards to be analyzed. Typically, the return rate is about 70 percent.

Each respondent will sort the deck somewhat uniquely. The VOCALYST process uses proprietary statistical techniques to develop a composite view of the secondary attributes across all respondents, as well as the importance and Customer Satisfaction Performance levels for each secondary attribute. This is exactly what is needed to begin constructing the HOQ.

17.6 QUANTIFYING THE DATA

Once the team has its Affinity Diagram, it is ready to quantify the data. The data elements required for QFD are

- Relative importance of needs
- Customer Satisfaction Performance level for each need
- Competition's Customer Satisfaction Performance level for each need

Most commonly, this data is collected by survey. We've discussed these quantitative methods in Chapter 7.

The key pitfalls in constructing and utilizing surveys are

- Selecting inappropriate samples or sample sizes
- Failing to ensure adequate response
- Wording the survey questions ambiguously
- Analyzing the results incorrectly

Although surveys are most efficiently conducted by professional market-research organizations, many development teams conduct their own surveys. Most "home-grown" surveys take much longer than planned, and thereby engender hidden costs.

Sometimes development teams judge that they cannot afford the time or the money to conduct surveys. This is generally a short-sighted judgment, driven by poor budgeting or cost accounting. Nevertheless, such teams can and do assign the quantitative data by team consensus during the QFD process.

17.7 CLASSIFYING CUSTOMER NEEDS

It is possible to classify customer needs into categories that help development teams make decisions. While the Kano model is useful conceptually for understanding that customers have a range of responses to product *characteristics* based on their needs, we still need a model for classifying customer *needs* themselves. A model for this classification was developed by Robert Klein of Applied Marketing Science, Inc., to whom we are deeply indebted for assistance in explaining his model.[12]

Klein points out that there are two different ways to measure importance of customer wants and needs (attributes): by *directly asking* customers, or by *inferring* importance based on other data.

Directly asking involves one or another of the techniques described earlier in this chapter. All of these techniques have one thing in common: The customer is asked how

---

12. Perry and Bacon, *Commercializing Great Products.*

important a specific attribute is, either with or without reference to other attributes. Attribute importance measured by such direct methods is called *stated importance*.

A method for *inferring* importance is to measure how strongly satisfaction perform-ance with an attribute is linked to overall product satisfaction. We might observe statisti-cally (as in Figure 17-9) that high levels of attribute satisfaction performance correlate to high levels of overall product satisfaction, while low levels of attribute satisfaction per-formance correlate with low levels of overall product satisfaction. We can then infer that the attribute is important, regardless of its stated importance. Using this same reasoning, an attribute whose satisfaction-performance levels are not statistically linked to overall satisfaction can be inferred to be less important, unimportant, or possibly linked to importance indirectly. (See Figure 17-8.) Attribute importance measured by this indirect method is called *revealed importance*.

In Figure 17-8, we see considerable variation in Customer Satisfaction Performance on a particular attribute, but very little variation with overall satisfaction. We can conclude that the satisfaction performance on this attribute is not a predictor or a determiner of

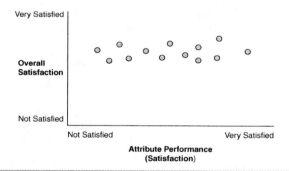

Figure 17-8   Low Revealed Importance

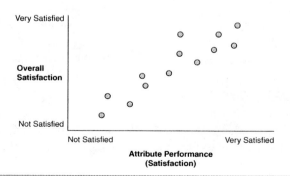

Figure 17-9   High Revealed Importance

overall satisfaction (or at least, is not *directly* a predictor). In other words, the revealed importance of this attribute is low.

In Figure 17-9, the linkage between individual attribute satisfaction performance and overall satisfaction is quite clear. Customers for whom this need was well met were also satisfied with the overall product; customers for whom this need was not well met were also dissatisfied with the overall product. The conclusion is that satisfaction performance with this need is a strong predictor or determiner of overall satisfaction.

Depending on how attribute satisfaction performance information is collected, the data can show a variety of possible relationship patterns. Interpretation of this data must take into account whether the customers surveyed were using one product or several, and whether the sample consisted of customers all of the same general type or from several market segments.

The main idea is to distinguish between customer attributes depending on how strongly they are linked to overall satisfaction. To do this, we must measure both satisfaction performance on each attribute and overall satisfaction with the product. We must then test for statistical correlation between these variables.

The Klein model uses both the revealed importance and the stated importance of each attribute to classify customer needs into four useful categories:

Expected needs: High stated importance, low revealed importance

Low-Impact needs: Low stated importance, low revealed importance

High-Impact needs: High stated importance, high revealed importance

Hidden needs: Low stated importance, high revealed importance

17.7.1 EXPECTED NEEDS

Expected needs are those basic needs that customers insist must be met. If the needs are not met, customers will be strongly dissatisfied. If the needs are met, customers will be only moderately satisfied with the product. Expected needs are very similar to Expected Quality in Kano's model.

An example of an Expected customer need for breakfast cereals might be

Comes in the size package I need

For this need, customers may be very dissatisfied with the product if they cannot get it in the package size they need, but will probably not be highly satisfied with the breakfast cereal just because they are able to buy it in the package size they need.

17.7.2 LOW-IMPACT NEEDS

Low-impact needs are needs that customers may articulate, but which in practice have no relationship, or at least no direct relationship, to the customer's overall satisfaction with the product or service.

An example of a low-impact need would be for breakfast cereals with

Recipes for innovative new uses of this product

People might buy the product regardless of whether recipes are made available. (In the case of breakfast cereals, buying patterns are almost synonymous with satisfaction.)

There is no construct in the Kano model that serves as a counterpart to low-impact needs in Klein's model.

The reader may ask: How could a low-impact need ever emerge as a need at all? This could happen if interviewed customers are aware of the need being addressed by some product or service providers, especially if the way the need is addressed is different with different providers.

17.7.3 HIGH-IMPACT NEEDS

High-impact needs are those that cause high customer satisfaction when they are met, and low satisfaction when they are not met. These are similar to the customer needs for which Kano's Satisfiers or "straight-line quality" are the solutions.

A high-impact customer need for breakfast cereal might be

Cereal is nutritious

People may prefer nutritious cereals, and may exhibit that preference by buying such cereals. If so, the need is high-impact, because cereals with lower nutritional value would not be purchased as often—that is, customers would be less satisfied with such cereals.

17.7.4 HIDDEN NEEDS

Hidden needs are those that customers say are not important to them, or that customers do not mention, but that, if met, strongly affect customer satisfaction. To identify these would obviously provide a product or service developer with a competitive advantage. These are the needs for which Kano's Delighters are solutions. In general, examples of Delighters given in explanations of the Kano model are technical characteristics that meet unspecified needs, rather than needs themselves.

A well-known case from market research relates to breakfast cereals containing sweeteners. Most people *say* that sweeteners are not important to them. This could be represented by the customer (lack of) need

Don't care about sweeteners in my cereal

Yet, their buying habits, and therefore their satisfaction with such breakfast cereals, contradicts that statement—sweetened breakfast cereals generally outsell unsweetened ones. Thus, we have a hidden customer need: stated importance is low, yet revealed importance is high.

17.7.5 KLEIN GRID OF CUSTOMER NEEDS

A simple grid shows the relationships between stated importance and revealed importance, and how they relate to the four needs categories (Figure 17-10).

Knowing whether customer needs fall into one or another of the four Klein categories can help development teams to set Customer Satisfaction Performance goals and determine sales points. Generally, goals would be set high for attributes classified as High Impact and Expected. Attributes classified as Low Impact generally don't warrant high satisfaction performance goals (unless there is some indirect linkage to attributes in other categories.) Attributes classified as Hidden must be looked at carefully to determine the strategic impact of goal setting.

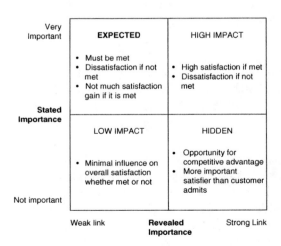

Figure 17-10   Klein Grid of Customer Needs

17.7.6 SEGMENTING & VOC

A final note must be made here before we conclude regarding VOC collection, processing and analysis vis-à-vis differing customer segments. If the final product must cover many different market segments, it is advisable to consider how similar and dissimilar these market segments may be with respect to each other. If they are fairly different, it may be prudent to collect and process each segment's VOC data independently. This will likely be more of an investment, but the payoff will be much greater. Consider that a product that is developed as a family of products may cost-effectively address multiple, differing segments with far less effort if the differing needs of the segments have been identified accurately during the VOC collection. Even if each segment may not be addressed in the first product launch, a series of product extensions can be planned to address the market segments in the prioritized order the marketing and development teams highlighted in the initial business plan.

17.8 SUMMARY

Gathering the Voice of the Customer for QFD consists of two distinct activities: first, gathering qualitative data; second, quantifying the data.

Useful models exist for classifying customer needs. One such, the Klein model, provides a structure for customer needs that mirrors the Kano model for technical characteristics.

There are many methods available for gathering qualitative data: reactive and proactive methods, and interviews of many types. Developers will understand their customers' needs best if they are in direct contact with them. Recording and transcribing interviews with customers frees the interviewer to concentrate on the job of interviewing rather than on taking extensive notes during the interview. It also provides a large volume of unfiltered data for subsequent analysis.

Analysis of qualitative data involves extracting the VOC directly from the transcripts of interviews, and using these phrases to construct an Affinity Diagram. Customer input into the Affinity Diagram process provides the best assurance that the resulting structure truly reflects the way customers think.

Once the Affinity Diagram has been constructed, the data is ready for quantification. The most reliable method for quantifying is a well-designed survey responded to by an adequate number of customers. If a survey is not possible, many development teams estimate these numbers themselves.

Once the VOC has been gathered, sorted into a hierarchy, and quantified, the development team is ready to build the House of Quality. We've already had a good look at the structure of the HOQ in Part II. The next chapter concentrates on QFD implementation topics not yet discussed.

17.9 DISCUSSION QUESTIONS

- Compare the methods your organization uses for gathering and analyzing the Voice of the Customer to the methods outlined in this chapter. What are the advantages and disadvantages of these methods for your organization?
- Taking all factors into consideration, if you were to launch a VOC project right now, which method would you use? What would it cost? How long would it take?

Phase 2 and Phase 3: Building the House and Analysis

This chapter discusses the procedure for constructing the House of Quality. It focuses on the group processes, sequencing of events, and logistics of the team working together. It addresses the practical considerations that come into play. It deals with consensus processes and practical issues such as how long a meeting should last, and how it should be run and organized.

Building the House of Quality is often thought of as "the QFD process." The HOQ is a mandatory tool during Design for Six Sigma efforts designing a new product, service, or process. It encompasses much more than assembling a team in a room and constructing a large matrix. A successful QFD process depends on good planning, good customer data, the right team, and good Substitute Quality Characteristics, all of which should be collected or prepared for in advance. Making good use of these elements during the HOQ-construction meetings is also critical to the success of the QFD process.

18.1 SEQUENCING OF EVENTS

The most useful sequence for building the House of Quality (Figure 18-1) is this:

1. Construct a list of customer needs/benefits
2. Construct a planning matrix and analyze results thus far
3. Generate Substitute Quality Characteristics and analyze results thus far

4. Determine Relationships and analyze results thus far
5. Determine Technical Correlations and analyze results thus far
6. Acquire competitive benchmarks and analyze results thus far
7. Set Targets and analyze results thus far
8. Plan the development project, based on the results of the preceding steps

Notice that analyzing results at every step is woven into the process of building the HOQ (QFD Phase 2). In this book, we have identified analysis as a separate phase (Phase 3) of QFD in order to highlight the activity. In practice, explicit time slots for analysis should be allocated, but throughout the QFD process (at the end of each of the sequenced steps), rather than only once at the end of the entire process.

1. **Construct a list of customer needs/benefits.** These are normally acquired via market research. Hence, they are normally known by the time construction of the HOQ begins.

   If customer needs have been acquired through any method or combination of methods described in Chapter 17, then the team must be familiar with these needs before HOQ construction begins. Otherwise, the first step in HOQ construction must be a familiarization process. The more time the team spends on getting familiar with customer needs, the better.

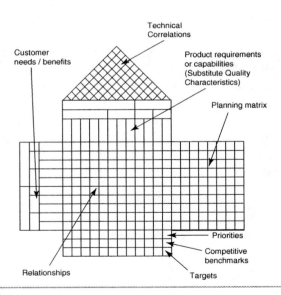

Figure 18-1   House of Quality

At a minimum, the QFD team members must be familiar with the structure of the customer needs—their primary, secondary, and tertiary attributes. They must also understand the relative importance of the attributes at the level being used for the QFD; and they must understand Customer Satisfaction Performance levels. This basic data will be referred to continually throughout the QFD process.

2. **Construct a planning matrix and analyze results thus far.** Customer-importance and satisfaction-performance data normally comes from market research and can simply be transcribed into the matrix. Development of the remaining parts of the planning matrix, setting goals and determining sales points, is a strategic planning activity and should occur before the more-detailed aspects of planning proceed. The end result of the planning matrix is the calculation of raw weights of the customer needs. Very often, especially when time is limited, analysis of the rest of the HOQ proceeds with the most important of these customer needs and may exclude the least important. This selectivity cannot occur unless the planning matrix is complete.

After the raw weights have been calculated, most teams devote some time to analysis of the results thus far. The raw weights (or normalized raw weights) of the customer attributes are often displayed in a bar chart, as in Figure 18-2.

The bar chart in Figure 18-2 bears some resemblance to a Pareto diagram, because it graphically displays numeric information in descending sequence, and because it helps the team to focus on the most-important attributes.

It is not, however, a Pareto diagram. Pareto diagrams are typically created by counting defects that have been classified by category. In contrast, the attribute bar chart of raw weights displays the results of market data and team judgments converted to numbers and multiplied together. Another distinction is that Pareto diagrams typically reveal that the one or two most important items far outweigh all the others (an evocation of the "80-20 rule"). In QFD, it is rare for the most-important items to vastly outweigh the others: the decline is generally more gradual.

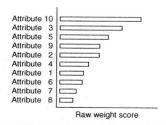

Figure 18-2 Bar Chart of Raw Weights

Analysis usually begins with an assessment of expected and unexpected results. The team will trace the steps it took to arrive at the resulting raw weights, and provide explanations for why the raw weights turned out as they did. It may go back and make adjustments, based on learning what has taken place since the start of the QFD process. It may do some "what-if" analysis to see whether some of the decisions may have had far-reaching implications upon the results (in which case the team may revisit these decisions to be sure it has come to the right conclusion).

The analysis process is very important for team building, as well as for the team to develop confidence in the QFD process. The team members must be comfortable with the rank-ordering of the customer attributes before they can proceed.

3. **Generate Substitute Quality Characteristics and analyze results thus far.** The team uses any of the methods described earlier to generate the SQCs. As has been described in Chapter 8, this is generally the most challenging and time-consuming part of the HOQ process.

Once the SQCs have been generated, the team should allocate one to four hours to consider the process it just went through. In most cases, generating the first SQC takes a long time, as much as half a day. By the end of the process, most teams can generate SQCs almost as quickly as the facilitator can announce the customer attributes that initiate the discussion. Obviously, some sort of learning process has occurred. The learning inevitably has resulted in some new shared perspectives on the product-planning process. This is the right time to make this learning process explicit by conducting a formal analysis session. As with all of the Phase 3 sessions, analysis should consist of the following activities:

- Retrace the steps taken so far
- Consider adjustments to earlier parts of the process, based on what the team has learned since those parts were completed
- Review and document all assumptions made
- Review and document unresolved issues
- Develop action items to resolve all unresolved issues
- Develop a list of next steps

List the benefits the team has derived from the process thus far. The facilitator's job at this point is to focus the team members on what they must do after the QFD process has finished, in order to move the development project along. Some actions will involve preparing for the remaining QFD steps, such as completing the roof. Other actions will directly relate to the development work. Still other actions will relate to informing the rest of the organization about their decisions and findings.

4. **Determine Relationships and analyze results thus far.** This is the natural next step after generating the SQCs. No other part of the HOQ can be completed sensibly until the Relationships and resulting priorities have been determined. Larger teams (more than ten people) generally break into subteams to complete this step.

   When the Relationships have been completed, the team then calculates the priorities of the SQCs (generally by means of QFD software) and performs an analysis similar to the ones previously described.

   Once again, analysis at this point is a critical element that allows the team to understand the implications, benefits, and shortcomings of the process it has just gone through. What-if analysis on controversial decisions will help team members feel comfortable with those decisions, or will persuade them to do further fact-finding or other research in order to come to a better decision.

5. **Determine Technical Correlations and analyze results thus far.** This QFD step is often omitted, although I believe that if the SQCs are chosen properly, analysis of the technical correlations can yield substantial benefits. When it is performed, teams usually start with the highest priority SQCs. They may limit their analysis to the top ten or the top one-third of the SQCs. If they analyze the entire roof, they may break into subteams, although it is difficult to break this matrix into submatrices.

   An analysis at the end of this step is extremely beneficial. This step will likely result in many important action items that affect communication among organizations during the development process.

6. **Acquire competitive benchmarks and analyze results thus far.** Measuring the competition's critical SQCs cannot be done in a conference room. This activity is usually planned after the SQC priorities have been computed, and is executed as a distinct project "off-line."

   Depending on how quickly results come in, one or several analysis sessions will be useful. If the results are coming in slowly, say over several weeks, interim analysis sessions can help the team see how to speed up the process, or how to perform dependent steps sooner.

7. **Set Targets and analyze results thus far.** After all competitive benchmarking has occurred, the team normally convenes to set Targets. An alternate approach is for a subcommittee to propose Targets and present them to the rest of the team for approval.

   The team must not omit a final analysis session to consolidate all gains, to ensure that the development process builds on the QFD results, and make recommendations for more effective QFD processes in the future.

   At this point, the HOQ is complete.

18.2 GROUP PROCESSES/CONSENSUS PROCESSES

The QFD manager's role was discussed in Chapter 15. It's helpful to define other formal roles as well. These are the QFD facilitator, the recorder, and the HOQ scribe. Sometimes more than one of these roles can be taken by a single person. It's worthwhile, however, to define them separately in order to best understand what they are.

18.2.1 THE QFD FACILITATOR'S ROLE

Drive the planning process. Make sure that the elements of planning, as described in Chapter 16, get done. Many facilitators develop a checklist of planning questions and use it to ensure the planning process is completed.

Allocate time. Set a detailed schedule, using the guidelines in the QFD Estimator Chart (Chapter 16, Figure 16-9). Develop an agenda for each group meeting. In the agenda, show the objectives for each meeting and show how these objectives relate to the overall QFD activity. Managing time will be much easier as a result of this type of preparation.

Explain each QFD step. Even though the team may have received some training in QFD, it's a good idea to explain what will happen in the step that's about to begin. In my experience, many teams don't really understand the process until they have experienced it; hence, they may flounder at the beginning of a step. Often it's helpful to practice the step on something simple and unrelated to the development project—although many task-oriented engineers prefer practicing on the project itself.

Get consensus. The QFD process may be thought of as an almost endless succession of consensus events. At every small step of the way, the team must agree on something. Each cell of the HOQ matrix represents a question, and the team must agree on each answer. In the interests of time management, the facilitator must have a supply of tools aimed at helping the team reach consensus literally thousands of times, quickly.

Here are some general principles for getting rapid consensus:

- **Ensure that the group knows exactly what the consensus topic is.** The simplest technique for the facilitator is to state the question the team must agree on, and provide a format for the answer. For example, "If we could move SQC_x, how much would be the change in satisfaction ssperformance on customer $need_A$: big change, medium change, little-to-no change, or definitely no change?"

- **Avoid discussion that doesn't contribute to the team's understanding of the topic.** Often, team members will appear to be arguing, but a careful facilitator may notice that they are actually in agreement (sometimes called **violent agreement**). They may not realize it, because they may be using different language to say the same thing. This is common when the team members come from different disciplines, as is the case in cross-functional teams. The alert facilitator will be on the watch for violent agreement and help the team to see it when it occurs.

A good way of preventing violent or time-consuming agreement is to use silent techniques to quickly sense what everyone is thinking. The silent vote works well:

First, ask the question: "If we could move SQC$_x$, how much...?" Then, ask each team member to silently consider the answer, and silently raise a number of fingers indicating his or her estimation (as in Figure 18-3).

The facilitator can quickly assess, by a show of silent votes, whether everyone is in agreement (no discussion necessary); just one or two people are voting slightly differently than the majority (perhaps they could live with the majority decision); or there is significant polarization.

If there is polarization, it is usually most fruitful to ask a person voting in the minority to present an explanation for his or her vote. Very often, the minority view turns out to have been overlooked by the others. Once they hear the minority view, many people may quickly switch. If the group discusses only the majority view, almost no forward progress will be made, since the majority of the group already holds that view—hence the emphasis on the minority view.

As soon as the minority has had a chance to explain its position, it's a good idea to get the group to vote silently again. Some people may have changed their minds, and others may be able to live with the new majority.

- **Create a sense of urgency.** The enemy of any meeting is time. A good facilitator will set interim goals for the team, such as, "Let's complete 50 cells in the next two hours," and then regularly remind the team of how much time is left. This frequent reminder will help each team member to use the team's time with care.

- **Clarify for the team any differences of opinion.** When the minority *can't* live with the majority, the team needs help to move ahead. One very useful service the facilitator can perform is to understand what the legitimate difference of opinion is, and

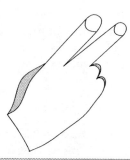

Figure 18-3   Silent Vote

feed it back to the team. Very often, this will help someone on the team to see a way of reconciling the two sides. If not, the conflict can be documented as an unresolved issue. Since the facilitator will see to it that all unresolved issues get assigned actions to resolve the issues, the team will have confidence that the problem will eventually be straightened out, and can move on.

- **Check frequently for shifts of opinion.** Individuals often change their minds silently. Such changes must be explicitly revealed. See "Clarify for the team any differences of opinion," above.

- **Seek realistic consensus.** "Realistic" means that most team members agree, and the remaining members can live with the decision. Realistic consensus, in which team members are generally aligned if not precisely so, is usually good enough for development efforts. Only in a disciplined process such as QFD can such detailed agreements and disagreements be documented and worked through. Before the advent of QFD, development teams did their work in the context of much more severe, and much more hidden, disagreements.

  Whenever a team is concerned about a decision, it can make a note of the concern, and later perform a "what-if" analysis on the completed matrix. The elements of what-if analysis are: Set the value of the cell in question to one extreme value; compute the resulting priorities (or raw weights); then set the value to the opposite extreme and recompute the priorities. If the priorities or raw weights shift only a little, then that decision can be seen to influence the project only by a little. If the priorities shift a lot, then the team knows it must revisit its decision to be sure the members can believe in and support its implications.

18.2.2 THE RECORDER'S ROLE

Record all assumptions, decisions, and action items that arise from the meeting. The recorder's notes should be distributed to the team as quickly as possible after each meeting, since many of the actions may be due for completion before the next meeting starts. Experience shows that the facilitator usually cannot concentrate on the recorder's duties, and therefore needs another person for this task.

Most commonly, development teams rotate the responsibility through the group.

18.2.3 THE HOQ TRANSCRIBER'S ROLE

Determine the mechanism for creating the physical charts. The task of creating a QFD chart that covers the wall, complete with neat rows, columns, and labels large enough for team members to read at a distance, is mundane, yet critical. Because it is so mundane, it can easily be overlooked. Nothing is as embarrassing to a facilitator, or as boring to a team member, as creating the chart at the last minute, while the team watches with nothing useful to do.

Likewise, the calculations are simple enough, but time-consuming and error-prone. No team that has sat with hand calculators performing hundreds or thousands of simple calculations will be motivated to participate in QFD in the future.

These mundane tasks must be planned for, and they must proceed effortlessly during the QFD meetings. In the mid- and late 1980s, when little or no QFD software existed and laptop computers were rare, many teams were forced to manually create the QFD charts and manually calculate raw weights and technical priorities. There is simply no reason any longer to do so. Several QFD software packages exist, and they run on the popular operating system platforms. Most of these packages will print large, empty QFD charts on oversized paper. Even empty charts printed on multiple 8½" × 11" sheets can be enlarged and pieced together into a large chart. These charts can then be taped to the wall and filled in by the facilitator as the QFD process advances. Laptops and LCD projectors together with large projection screens may also be employed; however, some people prefer to use paper, as it is more tangible and better at engaging the team members.

A word to the wise: Don't let this seemingly simple task ruin a QFD activity. Work out the details in advance.

Know the QFD software. Being familiar with any tool helps the tool wielder to do a better job. QFD software does not require software experts to operate it, but the options are generally rich enough to require some preparation and learning before using it in a QFD meeting. Once again, be prepared. One software package that will help make this easier is QFD Designer by IdeaCore.[1] Its suite of solution packages also supports many of the DFSS tools and VOC-gathering methods.

Acquire all necessary physical support. The physical support for QFD has been identified and discussed in Section 16.8 (Chapter 16). As with the QFD charts and software, be prepared.

Publish intermediate versions of the QFD charts. When the QFD process is going smoothly, the team is adding information to the QFD chart continually. Most team members appreciate having in-hand copies of the chart that are as up-to-date as possible. Even with large charts on the wall, many team members prefer to look at a chart up-close, right in front of them. (The chart on the wall still serves the purpose of focusing the team on a single objective.) Keep the team up-to-date with frequent printouts of the matrix or matrices.

---

1. See: www.ideacore.com/v1/Products/QFDDesigner.

18.3 SUMMARY

In this chapter we have seen how to sequence the building of the House of Quality.

We have discussed the benefits of reflecting on the QFD process as it progresses. These benefits include understanding explicitly the implicit decisions and learnings of the team, gaining confidence in QFD and in the team's ability to use QFD for its development work, and helping the team focus on the work it must do to capitalize on the QFD results. This reflection process is important enough to call out as a separate phase of QFD, even though it occurs at the end of each QFD step as well as at the end of the entire process.

We have seen that the sequencing of QFD steps is crucial. The facilitator should carefully map out the sequence in advance, using the sequence recommended in this chapter as a guide.

Many mundane aspects of running the QFD process can create havoc if not worked out in advance. These aspects include preparation of the matrix as a wall chart, the use of QFD software, and the acquisition of the proper room and materials.

We have pointed out that chances for success in QFD can be enhanced by ensuring that certain roles are filled—especially the QFD manager, the QFD facilitator, the recorder, and the HOQ transcriber.

If you have come this far, congratulations! Completing a House of Quality is no easy task, but the effort is well worth it. Now that the HOQ is finished, it's time to consider how QFD can continue to help the development team. In Part V, we'll look at QFD beyond the House of Quality. We can think of Part V as describing *advanced* QFD, a topic for organizations that have achieved substantial maturity in the quality domain. Let's see what advanced QFD looks like.

18.4 DISCUSSION QUESTIONS

Based on your own experience, list the principles of good meeting management. Map these to the techniques for planning and running QFD.

- How closely do they match up?
- Where are they different? What accounts for the differences?

PART V

Beyond the
House of Quality

Beyond the House of Quality

What happens beyond the House of Quality depends upon many things, and in today's Lean Six Sigma and Design for Six Sigma (DFSS) environments, there are even more options than there were when QFD originally gained popularity.

Most organizations that use QFD stop after developing their customized versions of the HOQ. In some cases, groups extend their analysis to an additional matrix, in which performance measures from the HOQ matrix are deployed against features of a product or service. In a very few cases, organizations have gone further, even as far as constructing matrices or tables that describe shop-floor processes and machine settings. However, even in Japan, where QFD originated, the majority of QFD applications stop with the HOQ.

There are numerous reasons why QFD teams don't use the full possibilities of QFD.[1] One reason is lack of specificity in the literature as to how to use downstream QFD matrices. The lack of specificity is an inherent problem in explaining QFD. Real-world case studies are hard to find: Companies are reluctant to share them with the public. Artificial case studies are not very convincing; and in any case, the more specific the example, the less likely it is to resemble any particular reader's experience.

If you are working in a company that supports and is actively deploying DFSS, then the HOQ is just the beginning of deployment of the Voice of the Customer into design activi-

---

1. Andreas Krinninger and Don P. Clausing, "Quality Function Deployment for Production," working paper, Laboratory for Manufacturing and Productivity, Massachusetts Institute of Technology, Cambridge, Mass., August 1991.

ties. DFSS projects and roadmaps will generally require the use of QFD, and in particular necessitate work beyond the HOQ. The reader is referred to Chapter 3, Figure 3-2, where the Critical to Quality (CTQ) items that flow out of the HOQ for further deployment are usually filtered as being either New, Important, or Difficult. While the HOQ and subsequent matrices are the management tools for deployment of customer voices, Critical Parameter Management (CPM) is the technical tool for accounting for the complex $Y=f(x)$ relationships from the HOQ to process and component specifications. Illustration of these complex parameters, one at a time, is done via Product Mapping. QFD, CPM, and Product Mapping form a suite of tools that together place even greater emphasis than QFD alone on developing and detailing the product functions from the customer voices. This is part of the value of DFSS—it creates pull for QFD use—and it goes beyond the HOQ.

This chapter describes the prevailing QFD models. It presents some general principles that should help the QFD practitioner feel comfortable using these models as-is or modifying them for particular applications.

There are two dominant QFD models in the U.S.: the Four-Phase Model and the Matrix of Matrices. While at first glance one may feel required to choose one or the other, in fact they are not in conflict.

There are several important reasons why they are not in conflict.

The first reason is that neither model was ever intended to be used as presented. Any presentation of QFD or any other complex idea has to be specific enough for the audience to understand it. Unfortunately, the specificity carries with it the suggestion that the audience must accept it all or reject it all. In fact, the intent behind both models is to present a basic structure to be customized for each application. So what we really have are two models, each of which can be modified by adding or subtracting matrices, or by redefining matrices presented in the model. It may be stretching things only a bit to say that a team could start with one model, add and subtract some matrices, and find itself implementing the other model.

The second reason why the two models do not conflict is that the smaller, Four-Phase Model is contained within the Matrix of Matrices. Therefore, if you implement the Four-Phase Model, you have implemented a subset of the Matrix of Matrices; if you implement the Matrix of Matrices, you have implemented all of the Four-Phase model.

This second reason leads us to an understanding of the differences between the two models.

The Four-Phase Model is a blueprint for product development in a mature, efficient, disciplined organization (Figure 19-1). The Matrix of Matrices is a blueprint for product development that may be used in such an organization, but also works within the environment of Total Quality Management. The Four-Phase Model covers basic product-development steps. The larger Matrix of Matrices explicitly covers a host of activities that

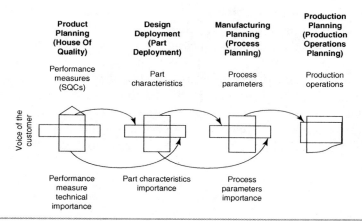

Figure 19-1 Four-Phase QFD Model (Matrix Titles in Parentheses Are Clausing's)

are not explicit in the Four-Phase Model, although at least some of them are probably performed during product development anyway. These additional activities cover such topics as reliability planning, manufacturing quality control, value engineering, and cost analysis. A standard text on product development such as Clausing's[2] covers all of these activities, but they are positioned as adjuncts to QFD rather than as a part of it.

Thus, the differences between the two models can be seen as matters more of style than of content. Serious students of QFD will be familiar with both. They will draw from the simpler or the more elaborate model depending on the needs and receptivity of their audience.

Now let's look at these two models in more detail.

19.1 THE CLAUSING FOUR-PHASE MODEL

Probably the most widely described and used model in the United States is a four-level model known as the Clausing model,[3] or the ASI model. ASI is the American Supplier Institute, an organization that has done much to popularize QFD. Figure 19-1 depicts the model.

---

2. Dr. Don Clausing, *Total Quality Development* (New York: ASME Press, 1994).

3. Clausing, *Total Quality Development.*

Product Planning (House of Quality). The HOQ in the Clausing model is similar to the one described in this book. The SQCs chosen for the HOQ are system-performance measures. The prioritized performance measures are transferred to the left of the second matrix. It is important to understand that design development work is occurring in-between the matrices. A design concept is usually chosen after the HOQ matrix is completed, but not always. Some teams may already be constrained to a general-concept area, and then do more-detailed concept development after the HOQ. In either case, the concept should be more fully developed before beginning the second-phase matrix. Stuart Pugh's Concept Selection Matrix described in Section 4.6.7 (Chapter 4) should be utilized.

The process for arriving at the part characteristics in the design-deployment phase is not obvious from the figure. Part characteristics are obtained through the process depicted in Figure 19-2.

Design Deployment (Part Deployment). The first step in part-characteristics deployment is to develop a function tree as described in Chapter 8 (Section 8.2.1). In Figure 19-2, we see the total product first broken into subsystems, then the subsystems broken into parts. At this point, the important characteristics of each part (the part characteristics) are enumerated. These will be descriptions of elements that are critical to the parts' design. They may include measurements with directions of goodness, which in effect are specification parameters for the parts.

In Figure 19-3, a word processing software package is the total system. Its subsystems include the contents of the carton that the software is to be shipped in as well as the carton itself. In this case, the development team has decided that the carton should serve as permanent storage for the manuals. Thus, "Provides Permanent Storage for Manuals" becomes a part characteristic. Other phrases, such as "Fits on Retail Store Shelf" and "Protects Contents During Shipping," describe other measurable part characteristics.

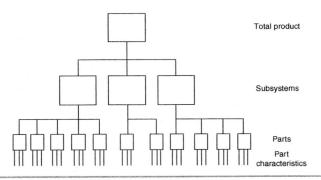

Figure 19-2   Part Characteristics Deployment

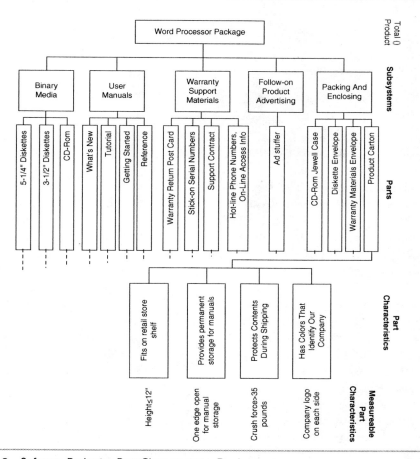

Figure 19-3   Software Packaging Part Characteristics Deployment

The function-tree approach is one way of arriving at part characteristics. Depending on the dominant technology and design culture in various organizations, other methods of parts deployment and part-characteristics generation may be used. In software development, for example, the function-tree approach is feasible with two changes. First, **parts** would be replaced by **modules** or **objects,** which are terms that have specific technical meaning in software-development environments. Second, part characteristics, which are measurable, would be replaced either by module functions, which are verbal descriptions of work that the modules would perform, or by descriptions of data structures and data transformations.

The part characteristics or equivalent elements are placed on the top of the Design Deployment matrix. The team then estimates the impact of each part characteristic on

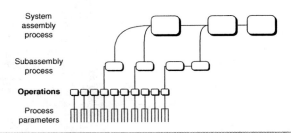

Figure 19-4   Process Parameter Development

the performance measures. The priorities of the performance measures are multiplied by their impacts to compute the Relationships, just as raw weights of customer attributes are multiplied by impacts of SQCs in the House of Quality. The relationships are summed, and the resulting importance values prioritize the part characteristics. This information tells the developers which part characteristics and which parts will be the drivers of customer satisfaction.

For complex products, this process is repeated through as many levels as needed to fully specify each subsystem and part.

Manufacturing Planning (Process Planning). The procedure for manufacturing planning is also not explicit in Figure 19-1. A recommended method for manufacturing planning that would result in process parameters is depicted in Figure 19-4.

The team lays out the main process flow, or system assembly process, and then decides on the subassembly processes needed to feed into the main flow. The operations required to produce each subassembly are next added to the matrix.

This has been described as a top-down process, much like the tree-diagram process. In fact, any combination of top-down or bottom-up processes, including simply using existing process layouts, may be appropriate.

Once the operation steps have been identified, the development team uses its expert knowledge, along with experimentation, to identify the key operations process parameters related to the subassemblies. These parameters are specific to the operations, not the products produced by the operations, so they will likely relate to such measurements as machine adjustments.

The process parameters now become the Hows at the top of the Process Planning matrix. They are prioritized based on their impacts on the part characteristics.

Production Planning (Production Operations Planning). This chart is not a matrix, but rather a table or list that constitutes a checklist of topics or issues that should be considered in planning production steps. Steps suggested by Krinninger and Clausing [4] include

4. Krinninger and Clausing, "Quality Function Deployment for Production."

- Machine settings
- Control methods
- Sampling size and frequency
- Control documents
- Operator training
- Preventive-maintenance tasks

These or similar topics are arranged along the top of the table, and the most important process parameters are arranged along the side. The production planner fills in the table with comments, target values, or any other appropriate language. In this way, production planning can be linked all the way back to the VOC, three QFD levels distant.

19.2 THE AKAO MATRIX OF MATRICES

The Akao QFD model is gigantic and far-reaching. As indicated earlier, the Matrix of Matrices explicitly refers to product-development steps that are not shown in the Four-Phase Model.

In the most accessible English-language source,[5] the QFD structure is presented as a system of 30 matrices, charts, tables, or other figures. Some are the familiar QFD-style prioritization matrices. Others are tables similar to the Voice of the Customer Table or the Production Operations Table. The 30 matrices are generally presented as a grid of matrices. The first four rows are numbered 1 through 4, and the columns are labeled A through F. A fifth column, labeled G, contains six matrices labeled G1 through G6. Thus, any matrix can be referred to by its coordinates—for example, "B3." The layout of the matrices is detailed in Figure 19-5.

Matrix A1 is the familiar HOQ (except for the roof, which is Matrix A3). Some other matrices correspond to the tables in the Four-Phase QFD Model. Depending on the selection of Whats and Hows in each matrix, additional matrices are used for such topics as

- Competitive analysis versus cost
- Parts versus failure modes
- VOC versus failure modes

---

5. Bob King, *Better Designs in Half the Time: Implementing QFD Quality Function Deployment in America* (Methuen, Massachusetts: GOAL/QPC, 1987).

Matrix	What	How	Activity
A1	Voice of Customer	SQCs	Construct Matrix
A2	Functions	SQCs	Construct Matrix
A3	SQCs	SQCs	Construct Matrix
A4	Second level of design	SQCs	Construct Matrix
B1	Voice of Customer	Functions	Construct Matrix
B2	Competitive analysis	Cost	Construct Matrix
B3	Detailed SQCs	Breakthrough targets	Construct Matrix
B4	Critical parts	SQCs	Construct Matrix
C1	New technology	First level of design	Construct Matrix
C2	Functions	First level of design	Construct Matrix
C3	SQCs	First level of design	Construct Matrix
C4	Second level of design	First level of design	Construct Matrix
D1	Voice of Customer	Product failure modes	Construct Matrix
D2	Functions	Product failure modes	Construct Matrix
D3	SQCs	Product failure modes	Construct Matrix
D4	Second level of design	Product failure modes	Construct Matrix
E1	Customer needs	New concepts	Construct Matrix
E2	Functions	New concepts	Construct Matrix
E3	SQCs	New concepts	Construct Matrix
E4	Criteria	New concepts	Construct Matrix
F1			Value engineering
F2			Reliability analysis
F3			Breakthrough planning
F4			Design improvement planning
G1			Quality assurance planning
G2			Equipment deployment
G3			Process planning
G4			Process FTA
G5			Process FMEA
G6			Process QC

Figure 19-5   Matrix of Matrices (Summary)

- Quality-assurance planning by part
- Supplier versus manufacturing parts/materials
- Process-failure analysis

Activity	Matrix
Marketing/Planning	Voice of Customer Tables
System-Level Design/Planning	VOC/SQCs VOC/Functions SQCs/Functions
System-Level Design/Checking	Key SQCs/Methodology
	Key Functions/Methodology
	Key VOC/Failure Modes
	Key Functions/Failure Modes
Concept-Level Design/Doing	Key SQCs/Concepts
	Key SQCs/Top Concepts
Concept-Level Design/Checking	Key Concepts/Methodology
	Key Concepts/Failure Modes
	Key Concepts/Cost
	Key Concepts/Safety
Concept-Level Design/Acting	Select Best Concepts
Element-Level Design/Doing	Key SQCs/Elements of Best Concept
Element-Level Design/Checking	Key Elements/Methodology
	Key Elements/Element Failure Modes
	Key Elements/Cost

Figure 19-6 Comprehensive QFD Matrices (Summary)

The entire system encompasses several phases of product development, with a strong continuous-improvement emphasis. The phases have been labeled as follows:

- Gather VOC
- Define product and identify bottlenecks and breakthrough opportunities
- Develop design breakthroughs
- Develop process breakthroughs

The Akao QFD model is not intended to be used exactly as presented. Rather, it is intended to open up possibilities to a development team. The team is expected to create its own QFD model, because no two organizations and no two development projects have the same needs.

As indicated in Section 1.3 (Chapter 1), the Akao model was introduced in the U.S. by Bob King around 1984. In the early 1990s, King and others proposed a subset of the 30 matrices for those developers who were overwhelmed by the matrix of matrices—so overwhelmed that they were unable to realize any benefit from QFD at all.

The subset, called Comprehensive QFD, included 17 matrices generally based on the original 30, and it added the Voice of the Customer Table and a concept-selection activity. The matrices were presented within a grid that had product-development phases along the left (moving from VOC acquisition and proceeding to detailed design, as in the Four-Phase Model), and the elements of continuous improvement (Plan-Do-Check-Act) along the top. "Plan-Do-Check-Act" refers to the Deming-Shewhart cycle of continuous improvement,[6] a fundamental TQM concept. This formulation of QFD creates a continuous-improvement context for product development. The elements of Comprehensive QFD are listed in Figure 19-6.

The use of Plan-Do-Check-Act for structuring the quality tables is consistent with the notion that strategic corporate quality planning, the PDCA continuous improvement model, and QFD can be combined to provide an overall TQM model.

19.3 SUMMARY

In this chapter, we have looked at the Four-Phase Model and the Matrix of Matrices model of QFD. We've compared them and concluded that the more-elaborate model, the Matrix of Matrices, makes explicit activities that are implicit or optional in the Four-Phase Model. The two models are not in conflict.

The Four-Phase Model identifies a series of translations from the Voice of the Customer through to production-process steps. The Matrix of Matrices does the same, but also makes explicit various forms of reliability engineering, cost analysis, value engineering, and quality control.

QFD is a system of matrices that provides a backbone for the entire development process. Various models of QFD as a development backbone exist. The best known of these are the two described in this chapter (the Four-Phase Model and the Matrix of Matrices). Neither model is intended to be used blindly as presented. Instead, they are intended to provide templates or examples of the possibilities for systematic planning throughout the development cycle. Both models suggest that the VOC can be carried through the development cycle to the very details of implementation. The more-elaborate models suggest that notions of continuous improvement can be integrated with QFD.

We mentioned in the introduction to this book, and in many other places, that QFD helps any team relate its Whats to its Hows. A great deal of creativity has already been

---

6. W. Edwards Deming, *Quality, Productivity, and Competitive Position* (Cambridge, Mass.: MIT Center for Advanced Engineering Study, 1982).

applied to extending QFD's initial purpose of product development. In the next chapter, we'll look at some ingenious ways QFD has been used. Perhaps readers of this book will be inspired to apply QFD in yet other creative ways.

19.4 DISCUSSION QUESTIONS

- What does "Plan-Do-Check-Act" mean in your organization? Is it integrated with product or service development?
- Can you conceive of a multilevel QFD structure that would support and improve your development process? What would it look like? What do your colleagues think of it?

Special Applications of QFD

In this chapter, we look at some of the ways QFD has been extended beyond its initial conception. The specific areas we cover are QFD for Design for Six Sigma (DFSS), Total Quality Management, Strategic Product Planning, Organizational Planning, Cost Deployment, Software, and Service.

QFD's primary application has been for planning and managing product development. The dominant QFD models assume hardware product development activities. As we have seen, however, the QFD model can easily be applied to development of almost any type of product or service. My own initial experiences with QFD were in the area of software development. Very quickly afterward, I used QFD to help an internal service group develop its strategy for satisfaction of its internal customers.

Moreover, the general idea of matching Whats against Hows to understand their linkage and importance has even wider application than in product development. Let's begin our survey of these applications by considering QFD and DFSS.

20.1 QFD IN DFSS ENVIRONMENTS

There are many different DFSS approaches, and each has an associated roadmap. In fact, it seems as though each of the major Six Sigma consulting groups developed one of its own! Some are more focused on product development, while others are more focused on service development. The most popular roadmaps, which we will explore in detail, are DMADV, DMEDI, IDOV, CDOV and CDOC. Compounding this alphabet soup is the fact that the letters stand for different words in the different acronyms: for example, "D" for Define or Design, "I" for Identify or Implement, and "V" for Validate or Verify. If this makes the reader confused, one can only imagine how it is interpreted by those trying to purchase DFSS consulting services! And if this isn't enough, each DFSS consulting effort in which I have been involved has required some customization of the DFSS approach to align with the client's New Product Development (NPD) process. (Failure to customize in this way is a sure recipe for confusion and frustration in the design community.)

Notwithstanding the need for customization, most DFSS roadmaps offer strong support of QFD as a required element of the DFSS implementation. We will now explore each of the most common DFSS approaches and discuss where QFD fits into each specific roadmap.

20.1.1 DMEDI—DEFINE, MEASURE, EXPLORE, DEVELOP, IMPLEMENT[1]

The DMEDI roadmap appears most often when there is a new process to be developed and implemented. In this approach, QFD resides in the Measure phase, with the expectation that the Measure phase begins the process of measuring the VOC and ends with requirements for the new process. It is followed by the Explore phase, where new concepts are explored to meet the needs of the customers. This approach has been used in the financial services market segment, and also at Xerox.

20.1.2 DMADV—DEFINE, MEASURE, ANALYZE, DESIGN, VERIFY[2]

The DMADV roadmap appears most often when there is a new product or process to be developed and then verified as meeting customer needs. In the product-design area, I have seen it used mostly when doing product "tweaks" or redesign where there is more to measure. It has been used often in process design, without the tolerancing or transfer functions used in product design. In this approach, QFD also resides in the Measure phase, with the expectation that the Measure phase begins the process of measuring the VOC and ends with requirements for the new product or process. It is followed by the Analyze phase, where new concept options are analyzed to determine how well they will meet the needs of the customers. DMADV is generally credited as having been developed at General Electric. It later evolved into DMADOV, with "O" being the Optimize phase.

---

1. See Steven H. Jones, "To Use DMEDI or to Use DMAIC? That Is the Question," *iSixSigma*, available online at http://www.isixsigma.com/library/content/c060313a.asp.

2. See Kerri Simon, "DMAIC Versus DMADV," iSixSigma, available online at http://www.*isixsigma*.com/library/content/c001211a.asp.

20.1.3 IDOV—Identify, Design, Optimize, Verify[3]

The IDOV roadmap appears most often when there is a new product to be developed and verified as meeting customer needs. The Identify phase contains the Define-phase aspects found in DMAIC, DMEDI, and DMADV. The IDOV roadmap also contains the Optimize phase, and first gained acceptance in DFSS because of this key phase. In this approach, QFD resides in the Identify phase, with the expectation that the Identify phase begins the process of identifying the measures and deeds of the Customer, and ends with requirements for the new product. It is followed by the Design phase, where new concepts are designed according to the needs of the customers. The origins of IDOV are unclear, as it appears in two forms: one where the V stands for Verify as described previously, and one where it stands for Validate.[4] Several consulting groups use IDOV, but the most-recognized origin within the Six Sigma community is from Brue.[5]

20.1.4 CDOV—Concept, Design, Optimize, Validate[6]

The CDOV roadmap also appears most often when there is a new product to be developed and verified as meeting customer needs. This roadmap first appeared in 2002 in *Design for Six Sigma in Technology and Product Development.*[7] It is a derivative of the CDOC methodology originally created at SBTI, which is covered next. The Concept phase contains the Define aspects found in DMAIC. In this approach, QFD begins in the Concept phase, and the HOQ is completed there; this phase ends with requirements for the new product. It is followed by the Design phase, where the next few matrices in QFD help manage the design.

20.1.5 CDOC—Concept, Design, Optimize, Control[8]

The CDOC roadmap is utilized when there is a new product to be developed and verified as meeting customer needs. A high-level view of the CDOC roadmap is shown in

---

3. See Dr. David Woodford, "Design for Six Sigma—IDOV Methodology," *iSixSigma*, available online at http://www.isixsigma.com/library/content/c020819a.asp.

4. See "IDOV," *iSixSigma*, available online at http://software.isixsigma.com/dictionary/IDOV-426.htm.

5. Greg Brue and Robert G. Launsby, *Design for Six Sigma* (New York: McGraw-Hill Professional, 2003).

6. Clyde Creveling, Jeff Slutsky, and David Antis, Jr., *Design for Six Sigma in Technology and Product Development* (Upper Saddle River, N.J.: Prentice-Hall, 2002).

7. Creveling *et al., Design for Six Sigma.*

8. Randy C. Perry and David W. Bacon, *Commercializing Great Products with Design for Six Sigma* (Upper Saddle River, N.J.: Prentice-Hall, 2007).

Figure 20-1. It has been in constant use by the management consulting group SBTI since 1997 in one form or another. A process patent has been sought to cover details that now also combine lean precepts into the CDOC roadmap. The Concept phase in CDOC also contains the Define aspects found in DMAIC. In the CDOC approach, QFD begins in the Concept phase, and the HOQ is completed there; this phase ends with requirements for the new product. It is followed by the Design phase, where the next few matrices in QFD help manage the design.

The QFD practitioner may come across any one of these roadmaps (DMEDI, DMADV, IDOV, etc.) when working with a company that is employing DFSS, but in an ideal world these roadmap phases would not exist. They should be replaced by the company's own NPD-process phases, and the tools of DFSS should then be integrated into the NPD process itself. Otherwise, there is cause for confusion and consternation about which process is being followed. I have often followed this procedure when deploying DFSS: The roadmap gets replaced with a revised NPD phase gate process incorporating the sequenced tools of DFSS.

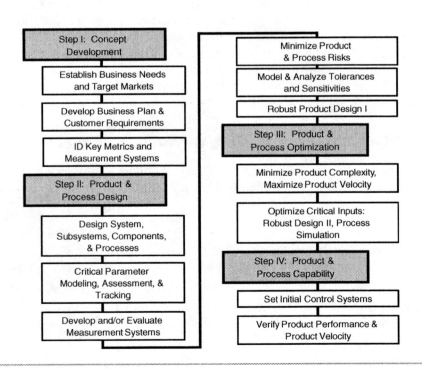

Figure 20-1 SBTI CDOC Roadmap

20.2 TOTAL QUALITY MANAGEMENT

There are many different definitions of Total Quality Management (TQM). All approaches to TQM contain a strong element of aiming for customer satisfaction. Putting this into QFD terminology, a common pursuit in TQM is finding a cost-effective way to link or align the organization's activities to best meet the needs of the customer. (While TQM has been displaced by Six Sigma in many Fortune 100 companies, the focus on customer satisfaction and the need to align activities to that goal remain.) A great deal of the acceptance of Lean Six Sigma was due in part to the successes of TQM.

One very powerful and widely understood formulation of this in the USA is the application criteria for the Malcolm Baldrige National Quality Award. The Baldrige criteria define seven major aspects of TQM. The exact detailed formulation of these seven aspects varies slightly from year to year.

The seven areas are these:

1. Leadership
2. Strategic Planning
3. Customer and Market Focus
4. Measurement Analysis and Knowledge Management
5. Workforce Focus
6. Process Management
7. Results

In any TQM model, these general areas must be linked to customer needs and to each other to form a coherent program. Typically, Baldrige applicants must address detailed questions in their written applications like these (paraphrased):

- How are the company's quality objectives translated into work units' operations plans?
- How are quality values integrated into the company's management system?
- What is the approach for each type of competitive comparison and benchmarking?
- How are performance-related results and quality feedback analyzed and translated into actionable information for developing priorities in operations?
- How does the planning process create a framework for customer satisfaction and leadership?

- How does the planning process drive improvement in operations and processes?
- How are plans communicated to work units and suppliers?

While the Baldrige criteria prescribe no answers for such questions, and while there may be other ways to establish the linkages, more and more U.S. companies are now using QFD in their efforts to achieve these linkages. Customer needs are formulated at a very high level and deployed into corporate objectives. These objectives are deployed through the levels of the corporation (and the number of levels is being reduced in many of these companies). QFD is becoming the vehicle of choice for these activities.

20.3 STRATEGIC PRODUCT PLANNING

20.3.1 PLANNING FOR FAMILIES OF PRODUCTS

Many business groups sell families of products into a market. Examples are

- Systems of computers and software
- Families of chemicals or drugs
- Telephones and accessories
- Financial services

Some management teams are using QFD to obtain a better understanding of how the members of a family of products work together to meet the market's needs. A typical approach is shown in Figure 20-2.

The planning matrix is used for strategic goal setting. The impact of each product family member on meeting needs determines the relative priorities of the product family members. The impacts also indicate whether certain products are redundant, and whether certain needs are not being properly met.

20.3.2 PLANNING MULTIPLE PRODUCTS FOR MULTIPLE MARKET SEGMENTS

A large company recently embarked on a new product strategy. In the past, it had developed a limited number of product lines. The company had been using a flexible pricing strategy to sell the same products into various markets. Company leaders recognized, however, that by lowering their prices to cost-conscious customers, they were lowering margins substantially, and could not stay competitive indefinitely. Most of

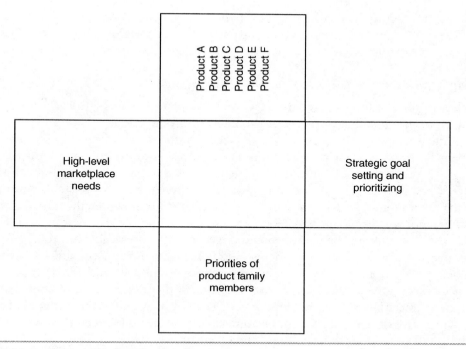

Figure 20-2 Strategic Product Planning

their products were too rugged and durable to allow them to compete in cost-conscious markets.

The company employed VOCALYST (see Section 17.5 of Chapter 17) to acquire the VOC for four different *types* of customers or users, all of whom were involved in purchasing, installing, or using a product. The types were

- **Recommenders:** People who recommended products for purchase, such as consultants
- **Managers:** People who made purchase decisions
- **Maintainers:** People who installed the products and provided ongoing maintenance of them
- **End Users:** People who actually used the products on a day-to-day basis

Note that these types do not represent alternative market segments into which the same product would be sold. Rather, they are all in the same market, and even within the same customer entity. They simply have different perspectives and needs for the product at different points in the product's life. The partial list of secondary and tertiary customer needs (attributes) looked similar to those in Figure 20-3.

Using the VOC to Develop a "Master HOQ"

The development team that used the VOC data was a ten-person, high-level multifunctional team, with members representing marketing, product design, sales, and manufacturing functions. The team members recognized that the strategic direction of their company lay in positioning products in much smaller and more-uniquely defined market segments than they had hitherto identified. In this way, product capabilities and pricing could be better targeted, more competitive, and therefore more profitable within each market segment. Even by the time of the market study, however, these market segments had not yet been identified. Furthermore, it was expected that new market segments would continue to be identified and targeted in the foreseeable future.

Because the strategic market segments were to be defined on an ongoing basis, the developers could not predict the relative importance of the customer needs for any particular market segment, although they were confident that the needs would not change much over time. Therefore, although the VOCALYST process delivered both qualitative and quantitative data about the four customer types, the team felt it could use only the qualitative data, i.e., the tertiary and secondary customer attributes. Team members felt comfortable that the VOC process had uncovered universal secondaries that constituted a superset of the secondaries needed to describe each as-yet-undefined market segment.

The team's strategy for using QFD was to create a "master" HOQ that listed all secondary attributes for all four customer types on the left. A successful product would have to satisfy all four customer types, regardless of the market segment. However, the relative

Secondary Attribute	Tertiary Attribute
Easy to Order	• Can get options I want quickly • Can understand prices easily • Equipment arrives in good condition
Easy to Explain	• Simple, intuitive controls • Controls within easy reach • Can learn adjustments without a manual
No Data Loss	• Data is safe even when there is a power failure • System continues to run even if a component fails • Can always retrace my processing steps

Figure 20-3   Examples of Affinity Groupings

importance of the four customer types, and the relative importance of each secondary attribute within each customer type, was expected to vary with the market segment. The team decided that the company would have to determine the relative importance of the needs for each market segment in the future.

Accounting for some duplicate secondary attributes shared among different customer types, the team ended up with a list of approximately 50 secondary attributes. Members then generated performance measures (Substitute Quality Characteristics) as the column headings in the HOQ for each secondary customer need. As is common in generating SQCs, two or three performance measures were required for each secondary customer need. The result was approximately 150 performance measures. This resulted in a Relationship section of about 7,500 cells—a formidable HOQ by any standards.

Nevertheless, the team committed itself to filling in the entire matrix. The result was a master HOQ that defined customer needs, performance measures, and their linkages applicable to many market segments, far into the future. Given that the secondary customer attributes, the performance measures, and their linkage to customer attributes were universal, the only missing data, to be supplied by individual development teams for each market segment in the future, were the relative importance values of the four customer types and the relative importance of the secondary customer needs.

Next, the team created a handbook that explained to future product developers in the company how to customize the master HOQ for their specific market segments. Each future product team will determine the relative importance of the four customer types and the secondary customer attributes for *its* market. (An example is shown in Figure 20-4.) The future team will then plug those numbers into the master HOQ, and thereby prioritize the performance measures for its market segment.

By experimenting with some hypothetical importance values, the current team determined that it would be possible for most future development teams to identify a small handful of performance measures out of the master list of 150 that would be critical for success in a particular market segment. The master HOQ will be the source for many market-specific HOQs to be created by future market segment teams. Example schematics for such market-specific HOQs are shown in Figure 20-5.

The process was highly publicized throughout the company, and it generated a great deal of interest. Even before the master HOQ was completed, a few development teams had begun using the preliminary results.

This case study demonstrates the use of an excellent VOC technique combined with QFD to create a strategic methodology that will influence product development for years. The resulting master HOQ provides a company-wide statement of both customer needs and top-level corporate responses to meeting those needs that can be customized by product-development teams throughout the organization.

Customer Segment & Segment Importance	Secondary Attribute	Importance within Segment	Overall Importance
Recommenders .40	Easy to order	100	40
	Easy to explain	70	28
	Parts available	60	24
	On-time delivery	60	24
Managers .30	Durable products	100	30
	Good looking	50	15
	No safety hazards	90	27
Maintainers .15	No data loss	100	15
	Easy cabling	70	11
	Protected from damage	60	9
End Users .15	Equip. at right height	80	12
	Reliable power source	60	9
	Organize work easily	100	15

Figure 20-4   Customer Types, Attributes, and Importance Values for a Market Segment

20.4 ORGANIZATIONAL PLANNING

20.4.1 SELECTING AN ORGANIZATIONAL SCHEMA

Often (perhaps too often!) it becomes necessary to change the structure of all or part of a company's organization. Many choices must be made, such as:

How many new organizations will replace the old?

Who should manage each new organization?

How should organizational objectives be assigned?

How will people be reassigned from the old organization to the new organization?

There may be several plausible options for each of these choices, making the combinations of plausible sets of options difficult to choose among.

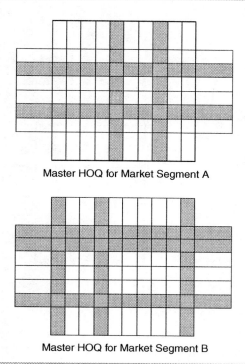

Master HOQ for Market Segment A

Master HOQ for Market Segment B

Figure 20-5   Examples of Market-Specific Houses of Quality

A combinatorial approach combined with the Pugh Concept Selection Process can help.

The most promising combinations of answers, plus the structure of today's organiza-tion, are selected as alternative concepts and labeled "Today's org." and Proposals A, B, and C. Today's org. is selected as the datum. The datum and the proposals are thus the initial concepts and are shown above the double horizontal line in Figure 20-6.

The objectives it is hoped the reorganization will achieve are placed on the left of the matrix. They are the criteria management hopes to optimize. The proposals are compared to the datum. As the management team evaluates the proposals against the datum, it will discover advantages and disadvantages, just as any in any other well-run concept-selection exercise. The possibility of hybrids and other variations of the initial concepts exists.

The opportunity for finding the best plan is enhanced through this process.

20.4.2 MATCHING THE ORGANIZATION'S WORK TO ITS OBJECTIVES

Every organization has customers. As in all VOC activities, it is possible to define these customers, determine the attributes of customer satisfaction, and measure current Customer Satisfaction Performance levels.

	Today's org.	Proposal A	Proposal B	Proposal C
Number of organizations	3	4	2	1
Who should manage?	L.C.	R.L.	R.L.	J.O.
Org. objectives	Plan A	Plan B	Plan C	Plan C
Reassignment plan	Plan X	Plan Y	Plan X	Plan Y
Better cust. service	DATUM	+	-	-
Reduce org. levels		-	-	+
Reduce headcount		S	+	S
Higher morale		S	-	+

Figure 20-6   Organizational Concept Selection

All organizations have activities that they normally perform. QFD can be used to evaluate the impact of these activities on customer satisfaction, as shown in Figure 20-7. In this figure, the entries on the left of the matrix are examples of customer needs (the VOC). The processes along the top, the Substitute Quality Characteristics, are a list of the organization's primary functions. These can be determined either by the organization's strategic plan or by a team brainstorming process. In the case of brainstorming, the team would identify all the jobs done by anyone in the organization, and then develop an Affinity Diagram of all the jobs brainstormed. The processes at the top of the matrix in Figure 20-7 would be the primary or secondary level of the Affinity Diagram.

	Process A	Process B	Process C	Process D	Process E
Quicker access	○		◎		△
Lower defects		◎	◎		○
Accurate answers		○		○	△
Easier forms			△		

Figure 20-7   Organizational Process Planning

Once the Relationships have been evaluated, the team members will understand how each of the processes affect the customers' needs. They will then see which processes have the greatest impact on meeting the needs. With this information, they can decide which processes must be delivered most efficiently.

20.5 COST DEPLOYMENT

Cost Deployment is used to allocate known development costs of any type to the customer needs or Whats they support. Cost Deployment can show developers what they are paying to support each function or customer need. The developers can determine whether they are paying too much to support unimportant needs, and therefore whether new concepts are needed.

The costs of the Hows are arrayed along the top of the Cost Deployment matrix. In Figure 20-8, a tree diagram representing a product design is shown at the top of the matrix. The product parts are at the lowest level of the tree, and below each part is the cost of that part.

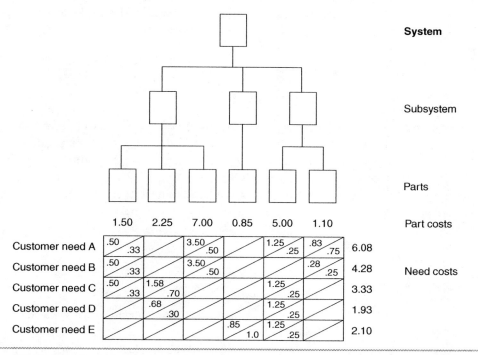

Figure 20-8   Cost Deployment

The team places the Whats along the left side. In Figure 20-8, the Whats are customer needs. The team then estimates the contribution of each How to each What as a fraction. In the figure, these are the numbers below the diagonal line in each cell. The fractions in each column should add to one. The team multiplies the cost of each How by the fractions and puts the resulting contribution in the same cell (above the diagonal line). The contribution represents the estimated cost that the How contributes to meeting the customer need.

The sum of the contributions in a row is the total cost to meet the What represented by that row. The total costs can be compared to the raw weights from the planning matrix, or to the priorities of these Whats as the result of any other matrix analysis. A serious misalignment of costs to priorities is a good indication of the need to redesign.

In the development of complex products such as computer systems, the costs of final testing are usually very high. Cost Deployment is an excellent way of determining whether the cost of each test is appropriate to its purpose. The VOC is placed on the left of the matrix, and the final tests are placed at the top. The fractional values below the diagonals are the extent that the test contributes to verifying that the product meets the Customer Satisfaction Performance targets. The cost of each test is at the top of each column. The completed matrix can then indicate the total cost of verifying success in meeting each need.

20.6 SOFTWARE DEVELOPMENT

The Four-Phase QFD model is based on the paradigm of designing and manufacturing physical objects—hardware. Software paradigms are different enough that software engineers cannot use this model beyond the HOQ. In fact, most QFD applications for hardware, software, and services use only the HOQ.

I have applied QFD to software development by making extensive use of the HOQ, using software functions as the SQCs, and deploying these functions down more than one level.[9,10,11]

9. Lou Cohen, "QFD: The House of Quality," *American Programmer* (June 1993).

10. Lou Cohen, "Quality Function Deployment: An Application Perspective from Digital Equipment Corporation," *National Productivity Review* (Summer 1988).

11. Lou Cohen, "Insights into QFD at Digital Equipment Corporation," *Proceedings of National Electronic Packaging and Production Conference* (June 1992).

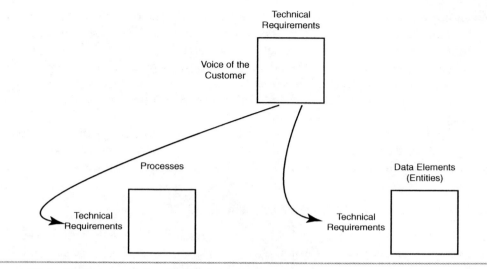

Figure 20-9   Deployment to Data and Processes

Two important software-analysis and design paradigms have been reconciled with QFD. Richard Zultner has developed a QFD model that incorporates elements of the **structured analysis** paradigm.[12] Prof. Hisakazu Shindo has developed a QFD model that incorporates elements of object-oriented analysis and design.[13]

20.6.1 STRUCTURED ANALYSIS AND QFD

Structured analysis is a general term referring to a number of graphical techniques that help software analysts translate customer requirements into technical requirements. Several structured-analysis systems have been developed since the early 1970s.[14,15] In general, these methods transform the technical requirements into data elements and procedural elements.

12. Richard Zultner, "Quality Function Deployment for Software," *American Programmer* (February 1992).

13. An English-language description of this approach can be found in Takami Kihara and Malik Mamdani, "Quality Function Deployment Applied to Software at Digital Equipment Corporation," working paper, Thayer School of Engineering, Dartmouth College, Hanover, N.H., May 1991.

14. Edward Yourdon and Larry L. Constantine, *Structured Design: Fundamentals of a Discipline of Computer Program and System Design* (Upper Saddle River, N.J.: Yourdon Press, a Prentice-Hall Company, 1979).

15. Chris Gane and Trish Sarson, *Structured Systems Analysis: Tools and Techniques* (St. Louis, Mo.: McDonnell Douglas Systems Integration Company, 1977).

In Zultner's QFD model, the VOC is deployed into technical requirements, the highest level of SQCs. The language of technical requirements refers to high-level functions of the software product, such as "provides graphic editing capability."

The technical requirements are then deployed into two separate sets of lower-level SQCs: data elements and procedural elements, as in Figure 20-9. Procedural elements might include "provides editing of circles" and "provides editing of rectangles." Data elements might include "circle" and "rectangle."

The procedural elements or processes and the data elements are prioritized independently against the technical requirements. These prioritizations can then be used to guide resource allocation, technical-breakthrough focusing, and testing resources, for example.

20.6.2 OBJECT-ORIENTED ANALYSIS AND QFD

Object-oriented analysis and design brings procedures and data together into objects. A software object is a description of some physical or conceptual object, together with the procedures that apply to that object. For example, the object "circle" might include the data that represents the circle, along with the operations that the software can perform on a circle, such as Create, Move, Change Size, and Delete.

A primary advantage of objects is that they can be conveniently reused and modified. They are normally placed in large **class libraries** that software engineers can select from. However, defining the objects incorrectly renders them difficult to use and reuse. One of the challenges of object-oriented analysis and design is to "discover the objects." Defining just what data and procedures should be encapsulated into a single object has proven difficult.

In Shindo's approach, the VOC is deployed into technical requirements. The technical requirements are deployed into processes and data elements. Then the processes and data elements are placed on the left and top of a new matrix (Figure 20-10). The matrix is input to a matrix-reduction and -manipulation process known as Quantitative Method Type III, which interchanges rows and columns to best arrange the nonzero matrix elements along the diagonal. This gathers those processes and data together that would appear to be best encapsulated into objects.

While Quantitative Method Type III provides an automatic method for clustering processes and data, it is possible to use judgment and observation to perform the process without automation.

20.7 QFD FOR THE SERVICE INDUSTRY

QFD has been used in the service industry for quite awhile. In Chapter 21, two case examples from the service industry will be provided. The interested reader may find

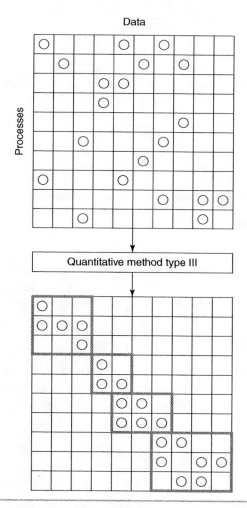

Figure 20-10 Quantitative Method Type III

these to be quite distinct from the product development examples, which are the primary focus of this book. References to service QFD date from as early as 1988 in Japan, in Yoji Akao's collection of Japanese papers on QFD.[16] I used QFD for service applications in 1987, without having seen any published case studies for service QFD.

---

16. Teiichiro Noda and Junji Ogino, "Quality Function Deployment for the Service Industry," translated into English in Yoji Akao, ed., *Quality Function Deployment: Integrating Customer Requirements into Product Design* (New York: Productivity Press, 1990), a translation of *Hinshitsu Tenkai Katsuyo No Jisai* (Tokyo: Japan Standards Association, 1988).

Step ID	What	Where	Who	When	Why
1	Ask customer how you can be of service	Customer counter	Customer Service Rep.	Business hours, 8:30 A.M. to 5:00 P.M.	
2	If customer has existing account, get account	Customer counter	Customer Service Rep.	Business hours, 8:30 A.M. to 5:00 P.M.	Verifies customer really has account, brings up account information on terminal
3	If customer wants account balance, get balance from terminal, write it on "Acct. Balance" form and hand it to customer	Customer counter	Customer Service Rep.	Business hours, 8:30 A.M. to 5:00 P.M.	By writing balance instead of saying it, customer is given privacy

Figure 20-11   Detailed Process Steps

A good formulation for service QFD helps the development team identify the detailed process steps required to perform a service. As described in Section 8.3, performance measures or process functions can be used as SQCs. If performance measures are used, then deployment of those measures to processes or subprocesses is a logical next step. Once these are deployed to the process or subprocess step, some teams next develop detailed process steps, using a What/Where/Who/When/Why format that resembles Figure 20-11.

Once the information is in this format, and especially if the format is input to a computer database, the steps can be sorted by the person responsible by time, or by any of the columns, to manage responsibilities, timing, and even space.

20.8 QFD AND TRIZ ENVIRONMENTS

QFD, in particular at the HOQ level, has synergy with environments where TRIZ (Section 3.4, Chapter 3) is being applied.[17] QFD practitioners are strongly encouraged to invest in their continuing education by doing additional research into TRIZ methods.[18] The most popular application synergy exists at the roof of the HOQ, identified as Item 6 in Figure 20-12. Here, conflicts are typically identified between engineering characteristics. TRIZ utilizes a contradiction matrix to suggest alternative approaches when these

---

17. Ellen Domb, personal communication, 2008.

18. Genrich Altshuller, *And Suddenly the Inventor Appeared: TRIZ, the Theory of Inventive Problem Solving*, trans. Lev Shulyak (Worcester, Mass.: Technical Information Center, 1996).

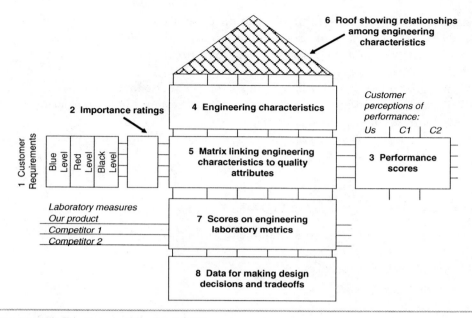

Figure 20-12   The House of Quality (HOQ)

types of trade-offs have happened in previous situations. In Item 5 of Figure 20-12, the correlation matrix between quality attributes and engineering characteristics, empty rows may be explored with TRIZ as well. Potential solutions can be developed with TRIZ principles and/or other methods. Empty columns in this area indicate potential opportunities for functional simplification, unless required for other reasons, such as regulatory needs, not expressed in Item 1 of the figure. In addition, there are other possible synergies between QFD and TRIZ for Items 3, 4, and 7 in Figure 20-12 as well. They apply to developing performance measures and methods, and in dealing with management problems associated with conducting customer interviews.

20.9 SUMMARY

In this chapter we have seen how QFD can be used for many applications far beyond the initial concepts of QFD. The application areas ranged from DFSS, TRIZ, and TQM to object-oriented design of software.

The better we understand QFD, the easier it is to see how to use it in new situations. The inescapable conclusion is that QFD is a very robust tool with as yet unbounded application. Suppose we restrict ourselves to the planning and development of products and services. We need only look around us at products that don't work well or services that aren't worth the money we paid for them to realize how many more routine applications of QFD are needed.

I encourage you to think of products and services you use that might benefit from QFD. Then consider products or services that you are responsible for delivering. Could QFD help to make them better?

Beyond that, we can hope to see systematic analysis of Whats and Hows applied in many untapped areas.

20.10 DISCUSSION QUESTIONS

How could or should QFD be applied to the following:

- Personal investment strategy?
- Selection of a new car?
- Deciding who to vote for in an election?
- The process of legislation?
- Formulation of public policy?
- International diplomacy?

QFD in Service Businesses

This chapter explores some of the many and varied challenges seen in QFD and DFSS for service businesses. An extended case example is presented, including key Voice of the Customer (VOC) and Voice of the Business (VOB) details for a residential solar-power installation. An additional case, for an urban coffeehouse renewal, is also presented. These cases are meant to illustrate the varied manners in which the tools can be applied to service businesses, and will be helpful to sole proprietors. Remember that fantastic product and service results are always possible, with the right methods, tools, and deployment of customer voices into development. The keys to business success and a growth strategy need to be unearthed for each company.

QFD can play a role no matter what services your company provides. The starting and ending points are always about working with the customers of your business or service.

For the reader who has jumped ahead to this chapter, some introductory comments are included here. Those who have been following along may wish to skip the rest of this introduction and peruse the examples that follow.

Whatever size or type of business you are in, you need customers to stay in business. All successful businesses have them, and most growing businesses listen and respond to them in some manner. Someone, or some group, receives product or service outputs from your company. Since the "quality" part of QFD begins with the customers, it is important that you find out what they value from you, and which aspects meet, exceed, or fall short of their needs. A four-step VOC process is shown in Figure 21-1. These steps can be found in the case study that follows as well.

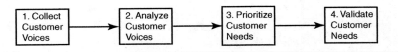

Figure 21-1   A Four-Step VOC Process

Chapter 17 provides a more extensive look at VOC. How you perform these VOC tasks can vary based on your budget and individual situation. What is critical is that you accomplish them, not necessarily which methods you choose. Whichever methods you choose, as long as you have open-ended questions and solicit feedback from all potential areas, you are beginning Quality Function Deployment.

Customers who purchase regularly should be your first priority in collecting voices. You can obtain customer feedback in several ways:

- solicit comments verbally
- drop off or send customer comment postcards
- conduct follow up phone calls with frequent customers
- interview your best customers
- create a website address for them to post their comments

Ask as many customers for feedback as are willing to provide it within the practical constraints of time and resources. However, there are some groups whose feedback may provide the greatest potential to improve your company in their eyes:

New Customers:

- First time buying with you
- Returning customer
- Switching from competitor
- Large nessw purchaser

Existing Customers:

- Regularly purchase from you
- Stopped purchasing from you
- Unhappy customers
- Very happy customers

It is useful to set a target of acquiring between 30 and 100 distinct customer voices in order to proceed effectively with the next steps. You should try to cover as many customer categories as practical. Diversity of perspectives is important when making service design or redesign decisions.

Once you have collected customers' feedback, it is important to analyze the feedback with some grouping based on similarity. (See Sections 4.2 and 4.3 in Chapter 4 for the Affinity Diagram and tree-diagram tools, with explanations of how to analyze the VOC. For a more-detailed explanation, see Chapter 17.) You may find, as you place the customer voices into the diagrams, that you want to obtain more customer information. If that happens, do not hesitate to go back and collect a little more.

After applying some diagramming tools to understand the hierarchy of customer needs, the service designer must validate the perspectives contained in the diagram. This may be done in several ways, but is usually done with a survey. The survey can be made available on paper or over the Internet. Survey Monkey[1] is one Web site that makes it very easy to design and deliver an electronic survey.

At a bare minimum, the following general areas should be covered in the survey:

- How did you first hear of our service? Our company?
- What factored into your decision to choose us?
- What aspects of the service did you like, and why?
- Which features or areas of the service could be improved?
- Which parts of the service, if any, did you not like, and why?
- Will you repeat your business with us? Would you recommend us?
- How do you rate your overall satisfaction with this service?

This list of bullet points should be customized and expanded to your specific situation.

The important factor in successful VOC collection and processing is the performance of all of the key steps. The work within each step need not be complicated. In fact, you are advised to do what is practical within the constraints of the business, starting small if necessary. Most often, key steps are skipped because they are deemed to be too much work. The truth is that this VOC work must take priority above all else before designing or redesigning a service. The example of a service application that follows should help illustrate how to perform these key steps without creating too much complexity while still obtaining valuable results.

---

1. www.surveymonkey.com.

DFSS, SERVICE DESIGN, AND QFD

In Section 3.2 (Chapter 3), we introduced the key steps of DFSS for Service Design or Six Sigma Process Design (SSPD) shown below. This roadmap helps guide the development of new services or processes when an existing process or service already exists. The first few steps deal with the existing situation or process.

- **Define** the process or service challenges, determining key stakeholders and project outcomes, and developing and validating key customer needs
- **Measure** current process or service performance versus defined needs and outcomes
- **Analyze** existing process or service data for improvement opportunities
- **Conceptualize** ways to meet customer and business needs with the right value proposition
- **Design** a process or service that meets customer, stakeholder, and business needs
- **Optimize** the process or service design to be effective, efficient, and flexible, with the right metrics, targets, and specifications to:
- **Control** the processes or service to always deliver the value (or capability) expected

In the solar-power case study that follows—where a service company designed and installed a new solar-power energy system—Phase I involved the Measure and Analyze steps, while Phase II illustrates several of the later steps (Conceptualize, Design, Optimize, and Control). The case study shows how QFD played a role in deploying and managing the customer needs.

21.1 QFD IN A RESIDENTIAL SOLAR-POWER INSTALLATION

21.1.1 CASE STUDY #1: INTRODUCTION—DEFINING PROCESS OR SERVICE CHALLENGES

This case study highlights in detail the QFD aspects of a DFSS project, namely the use of DFSS tools and methods for a residential solar-power installation. The first few aspects covered are normally found in the Define phase of the Six Sigma Process Design roadmap for DFSS used by SBTI.[2] Phase I of the case study covers the Measure and Analyze bullet points for current-improvement opportunities. Phase II of the case study covers the more-complex aspects Conceptualize, Design, Optimize, and Control for the

---

2. SBTI is Sigma Breakthrough Technologies, Inc., a management consulting firm that employs the author.

new system, where QFD plays a large role. The four-step VOC process at the beginning of this chapter runs throughout both phases of the case study.

In this example, we show how the use of QFD and DFSS in a new service led to success for a solar-power installation in a rapidly growing new market segment, residential solar. The installer had previously installed solar power on large building rooftops in the commercial market segment. However, the State of New Jersey had recently introduced incentives to encourage smaller-scale installations for homeowners, making a new market segment attractive.

In the contractor's prior commercial-rooftop work, formal requests for quotations or proposals were issued, and then either won or lost. VOC work was not done, nor was any QFD work; most designs were either largely preset by the commercial customers or left to the installers to determine. In the latter situation, the installers were held to performance standards alone. In addition, the financial payback and justification or VOB efforts from the customer side were typically done by the commercial customers. The installers each had their own financial considerations or VOB internally, but that is not the focus of this case study as regards VOB.

Instead, we focus on the *customer's* financial considerations or VOB. To sell a solar-power generation system to a homeowner, justification of the financial benefits needs to be detailed before proceeding. Many questions arose in the VOC, as customers sought help in understanding the financial aspects or VOB, the design, and the installation. Thus the selling, designing, and installation processes for the residential segment were not similar to those considerations in the commercial segment, due to differing needs and activities required to install at someone's personal residence.

In our example, QFD and DFSS tools played a vital role in the design of this service. Indeed, since every home would be different, using these tools created a competitive advantage for the solar-power installer when designing systems in the new and rapidly growing residential segment.

21.1.2 VOC PROCESS STEP 1: COLLECT CUSTOMER VOICES

Here are some VOC/VOB samples encountered during the first dialog with the customer:

- We are interested in "going solar"…
- What will it cost?
- What about rebates, state and federal?
- What's the payback period? Solar Renewable Energy Credits (SREC) market pricing?

- How many installers will quote?
- Does it add value or decrease the home value?
- How much maintenance is involved?
- How long will everything last?
- Do we mount it on the roof or on the ground?
- How does it look on the roof?
- Will it damage the roof?
- What will it look like when finished?

In this instance, the key questions asked could be viewed as the customer voices. Each one represents a need for general information, or for information on some key performance or quality aspect of the physical system. Both types of information are important to provide in order to have satisfied customers in the service business. More-detailed VOC work and analysis was done, and will be described later in this example.

21.1.3 CASE STUDY PHASE I: MEASURE AND ANALYZE FOR IMPROVEMENT OPPORTUNITIES

Since there was no request for a quote in this residential market segment, nor was there a previously defined business case driving these early discussions—as would normally exist in the commercial segment—it was decided to work with the homeowner to understand the customer's interests in "going solar." Working backwards to the customer's "business problem" in the first discussion, it was discovered the customer had seen a rise in the total cost of electricity in the current year and was concerned that an increase in electricity rates was looming on the horizon. A simple plot of monthly electric bills, as seen in Figure 21-2, provided evidence of this issue. The homeowner felt a trend was evident, but large variability existed both seasonally and month-to-month within a season, invalidating any statistical conclusions one might draw from a linear trend fit. The homeowner was somewhat well-informed on conservation issues, understanding that some benefits obtained through consumption management could shrink the household cost, and that these benefits would continue after the solar array was installed, since excess electricity could be sold back to the power utility. A seasonal model of the kilowatt hours consumed and associated costs was created after a tabulation of all major electrical appliances was made. This model is illustrated in Figure 21-3. A range of between 1,500 and 1,900 kilowatt hours (kWh) per month was set, except during the summer months when consumption would be higher due to air-conditioning use. The homeowner did not wish to see any monthly electricity bills with amounts exceeding $500 in the summer months. This was viewed as an upper specification limit for planning purposes.

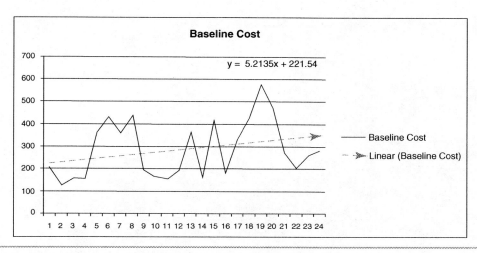

Figure 21-2 Measuring the Cost Problem, as Seen By the Homeowner

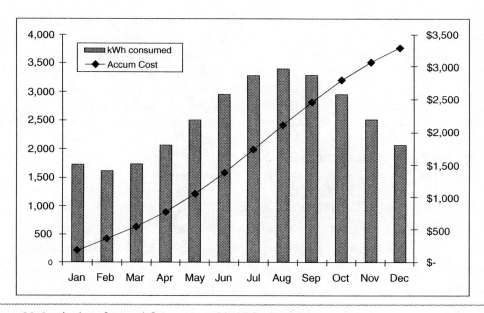

Figure 21-3 Analysis: Seasonal Consumption Model for Residence

The homeowner was grateful for the help provided, and commented that the solar installation salesman was viewed as part of his "personal problem-solving team" instead of just a sales resource. An illustration of the homeowner's aggressive cost target versus

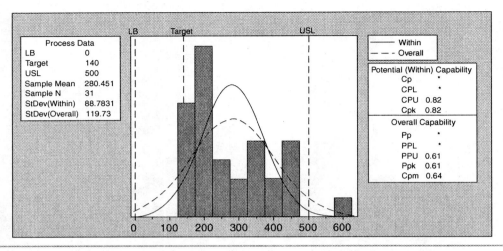

Figure 21-4 Capability Analysis of Monthly Costs, as Seen by the Homeowner

the 31 prior monthly bills is shown in a Process Capability Analysis in Figure 21-4. In the Six Sigma world, setting a goal is often an aggressive procedure. Typically, the use of the word 'entitlement' is done to help create a vision of what is possible. In this case, the homeowner was encouraged to look by asking, "How low has it ever been recently, and could you imagine if it was like that all the time?" By doing this, the target of $140 per month was selected. The savings potential achieved by hitting this target is over one thousand dollars in annual electricity costs at the current electric rates, which will likely increase over time.

Efforts that could reduce electricity consumption included both high- and low-con-sumption seasonal suggestions. These suggestions included:

- Room-darkening shades for south-facing windows in the summer months
- Setback thermostat time and temperature changes for the HVAC system
- Elimination of a second refrigerator
- Reduced well-pump operation through water conservation via:
  - o Insulating hot-water pipes to reduce wait times for hot water
  - o Flow reducers in showerheads
- Installation of Compact Fluorescent Lamps (CFLs) in living areas
- Installation of ceiling fans to better circulate cool air in the summer
- Timers on lights, and on all Christmas lighting during the holiday season

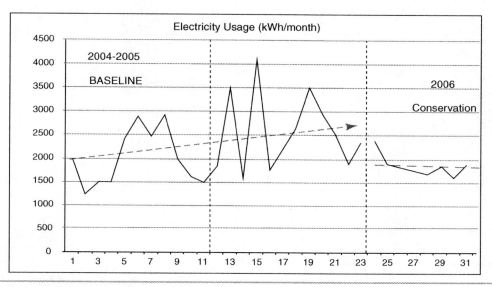

Figure 21-5 Consumption Baseline and Conservation Improvement Results

Figure 21-5 illustrates the timeline result and comparison to baseline of these conservation efforts during seven months after the changes were made. During this time period, the VOC efforts, QFD efforts, installation planning, requests for permits, and design and installation of the solar array were also done, as described next in this chapter. A histogram of a process-capability analysis of electricity consumption was created with this limited data set to illustrate the effect of the conservation efforts relative to the aggressive lower-consumption target. Although such an analysis normally requires 30 data points or more for statistical validity, it is shown in Figure 21-6 for illustration purposes. The upper specification of 2,500 kWh was based upon the average of the baseline electricity consumption in the prior year. Figure 21-6 illustrates how well the conservation efforts worked in transitioning from the low seasonal use in the winter months to the higher use season of July. Average consumption was below the 1,900-kWh target, coming in at 1,778 kWh. This compares favorably with the average use in the prior year of 2,500 kWh. The end result was a savings in six months of nearly 4,500 kWh, which translates to approximately $450 in electricity costs over these seven months, approaching the annual target of $1,000.

21.1.4 Case Study Phase II: Solar-Power Generation

The homeowner was interviewed on multiple occasions. These comprised an initial phone call, the initial meeting at the home, a follow-up phone call, and a second follow-up visit.

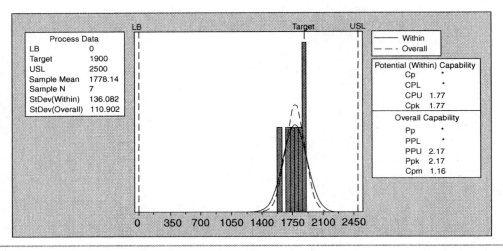

Figure 21-6 Consumption Capability Analysis of Conservation Improvements

Samples of the VOC thus collected, in the form of questions, were listed in Section 21.1.1. Records were kept of the owner's questions and comments. These were all listed in one location, and the business owner, salesmen, and installers were gathered to work through VOC Process Step 2, Analyze Customer Voices, by analyzing the comments for key questions or thoughts. Once the preliminary screening was done, the team agreed these points were all that they had collectively heard that seemed important.

Sometimes a step is inserted to translate the VOC into a statement of need or requirement. That step was not done here, as the team elected to use verbatim statements or questions as much as possible.

Figure 21-7 shows the output of VOC Step 2, after Affinitization of the 30+ individual statements and questions into five distinct columnar groups, titled with "I want..." statements in the grey-shaded second row of the figure. The titles are shown in the first-person voice to represent the customer's needs. They are at a higher level of abstraction, necessary to encompass all the statements grouped below them in the smallest type. In addition to these five "I want..." groups, four of them were further grouped by the team into two higher-abstraction groups, labeled across the top row in larger type. The five shaded groups were then discussed as being generally representative of the homeowner's wants that were to guide design and service decisions, and then the team ranked them for priority .

That priority order is reflected in the organization of the table as shown, from left to right. The leftmost column is the highest-priority group, and the rightmost column is ranked the lowest priority. The homeowner was also given a sheet with all 30+ statements and questions and asked to weight them on a 10-point scale as input for the HOQ.

A	B	C	D	E
I want to maximize the system's value			I want to maximize the operating time of the system	
I want the highest return on my investment that I call get	I want to maximize the energy produced	I want the system to look pleasing to the home owner	I want the system to last as long as possible through any weather	I want the system to be operational as soon as possible
What's the payback period?	Is there enough sunlight with the trees behind us?	Where will the array mount on the roof?	How much maintenance is involved?	Who has to get the permits?
What will it cost?	Is there enough sunlight in the winter?	What will it look like? The appearance is a concern.	How long will the whole system last?	How long will the installation take?
How many installers will quote?	How much power can be generated?	Do any of our roofs face the right direction?	How long will the solar panels last?	How long do we wait for state approval?
How do we sell the extra energy generated?	How do we sell the extra energy generated?	Where do I mount it on the ground?	Does the roof require extra maintenance?	When can you begin?
What about rebates, State and Federal?	How many panels do we need?	Will you be able to see it from the front of the house?	Do we need to reinforce the roof?	
How does it change the home value?	Do any of our roofs face the right direction?	How does it change the home's value?	Will ice/snow damage the system?	
How long will the whole system last?	Hosw efficient is the DC/AC convesion?	Where will the inverters be mounted?	How can it be damaged by lightning?	
How does it change the home's value?		How much of the wiring will show?	Will hurricanes damage the system?	
		Where will the wires be routed in the house/garage?	Will the roof leak? How will you mount everything?	

Figure 21-7 VOC Process Step 2—Analyze Customer Voices: Affinitization Table

Another way of looking at the grouped voices and titles which has further utility towards customer satisfaction is a tree diagram. Figure 21-8 shows a tree diagram for one branch which was of particular interest to the solar-power installer. By looking at each of the boxes to the far right, the installer can see which questions need to be addressed in detail to make the customer happy with regard to power generation. By understanding the hierarchy of needs, starting with the shaded "I want..." text to the left, as installers and designers make design decisions along the way, they can address the underlying need for maximum power generation.

In addition, the key questions were used to create a checklist of measures that could be presented to the customer to show that answers to his questions were being provided. (See Figure 21-29 .)

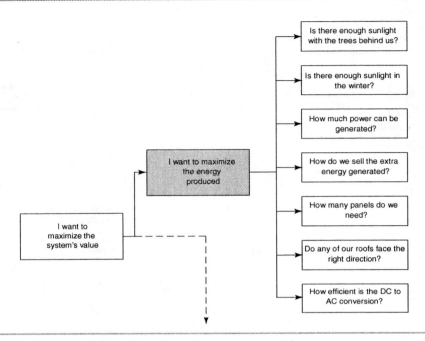

Figure 21-8   VOC Tree Diagram Illustration

Each column of issues in Figure 21-7 was also given a letter, from A through E, and these were plotted on a Kano plot (see Section 4.11, Chapter 4), as shown in Figure 21-9, to illustrate how these higher-abstraction needs were viewed by the customer relative to overall customer satisfaction and to provide the teams some guidance during planning and installation of the solar-power system. Groups A, B, and D represent competitive items where the more of the item provided, the greater the satisfaction. These were emphasized in the final proposal to the customer. Group E is stated that way as well, but with further customer discussion it became evident that "as soon as possible" actually meant "in time for the air-conditioning season." Group E really was a need that was more binary in nature: summer operation or not. So E was plotted as taken for granted given that the proposed summertime installation date would be met. Finally, Group C indicated that a pleasing (in this case, hidden) appearance would be very satisfying to the customer. It was this item that identified the need to use the back roof for the panels as much as possible in the layout proposal, to illustrate that the team would hide the wiring, and to lay out the panels in an aesthetically acceptable way whenever visible. Every aspect of Group C contributed to increased customer satisfaction, and so it was plotted as a delighter on the Kano plot.

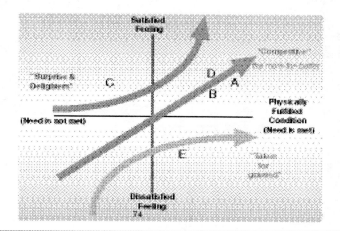

Figure 21-9 Kano Plot of Abstractions from Shaded Row of Figure 21-7

With the voices well in hand, and validated through a short survey that provided weighting, filling out of the HOQ could begin. Figure 21-10 illustrates the outcome of VOC Process Steps 3 and 4, Prioritize Customer Needs and Validate Customer Needs. It shows the entry on the left-hand side of the HOQ, including the customer weights. By retaining the top two rows of groupings from Figure 21-7 on the left-hand side, the team could see why certain voices were more important to the customer, and could aggregate the scores and provide an average importance to each of the shaded attributes. A tabulation of these scores is shown in Figure 21-11.

21.1.5 VOC Steps 3 and 4: Prioritize and Validate Customer Needs

The installer was a little surprised by the scoring on C, and returned to the homeowner to investigate. After some brief discussion, it was found that the customer in this case wanted the solar array to be as hidden as possible, while still generating the most power practical for this constraint. This dialog led to what became the final system concept, placing as much of the array as possible on the back roof, and mounting the inverters in the interior of the garage, hidden from view.

Now the team was ready to begin filling in the top section of the HOQ. Solution-independent solar-power system attributes, or what we call Substitute Quality Characteristics or SQCs (see Chapter 8), were identified and listed along the top, as shown in Figure 21-12. For each SQC, the direction of improvement was identified

KANO QUALITY	Blue Level Cust Req. (Function)	Red Level Cust Req. (Function)		Customer Needs	Customer weights
			1	What's the payback period?	10
			2	What will it cost?	9
		I want the highest return on my investment that I can get	3	How many installers will quote?	5
			4	What about rebates, State and Federal?	7
	I want to maximize the system's value		5	How does it change the home's value?	9
			6	How long will the whole system last?	9
			8		
			9		
		I want to maximize the energy produced	11	Is there enough sunlight with the trees behind us?	5
			12	Is there enough sunlight in the winter?	6
			13	How much power can be generated?	9
			14	How do we sell the extra energy generated?	10
			15	How many panels do we need?	7
			16	Do any of our roofs face the right direction?	3
			17	How efficient is the DC/AC conversion?	3
			19		
		I want the system to look pleasing to the home owner	20	Where will the array mount on the roof?	5
			21	What will it look like? The appearance is a concern.	6
			22	Do any of our roofs face the right direction?	3
			23	Where do I mount it on the ground?	5
			24	Will you be able to see it from the front of the house?	7
			25	How does it change the home value?	10
			26	Where will the inverters be mounted?	2
			27	How much of the wiring will show?	3
			28	Where will the wires be routed in the house/garage?	1
			29		
		I want the system to last as long as possible through any weather	30	How much maintenance is involved?	3
			31	How long will the whole system last?	7
			32	How long will the solar panels last?	7
			33	Does the roof require extra maintenance?	8
			34	Do we need to reinforce the roof?	8
	I want to maximize the operating time of the system		35	Will ice/snow damage the system?	7
			36	How can it be damaged by lightning?	4
			37	Will hurricanes damage the system?	4
			39		
		I want the system to be operational as soon as possible	40	Who has to get the permits?	3
			41	How long will the installation take?	5
			42	How long do we wait for state approval?	7
			43	When can you begin?	4
			44		

Figure 21-10 VOC Steps 3 and 4—Prioritize and Validate Customer Needs, Left-hand Side of the HOQ

Attribute	Average item score
A. I want the highest return on my investment	8.2
B. I want to maximize the energy produced	5.1
C. I want the system to look pleasing	4.7
D. I want the system to last as long as possible	6.0
E. I want the system to be operational as soon as possible	4.8

Figure 21-11 Average Item Scores for Attributes from Shaded Row of Figure 21-7

relative to increasing customer satisfaction. These can be seen in Figure 21-13. Once the top part of the HOQ was completed, the interior of the matrix was filled in. With the customer weighting and the interior filled in, the scores of the technical attributes or SQCs could be graphed, as shown in Figure 21-14. In doing this step, it became apparent that the top five SQCs considered important for system design leading to customer satisfaction were:

- The number of solar panels and **wiring strings** (groups of wiring)
- The number of roofs utilized
- The overall solar-power system simulation
- The solar panel power rating
- The system total power

Of somewhat lesser importance, but still not insignificant, were these seven SQCs:

- The layout of the solar panels on the roof(s)
- Federal and state rebate plans and SREC sales
- Locations of the inverters
- System price
- Inverter types and brands
- Inverter locations
- Inverter capacity

While price is not typically included, it is always important—and as the sale was not guaranteed yet by the consumer, finding the right price and system design were both crucial for the installer to make the sale. Since each residential solar-power system is

Customer Needs	Customer Weights	System price	System Power Total	Number of Inverters	Number of panels, strings	Layout of solar panels	Overall System Simulation	Solar Panel Power Rating	Wiring Layout Complexity	Inverter types, brands	Inverter Locations	Inverter Capacity	Township Permitting Length of Time, $ of steps	Fed. & State Rebates SREC Plans	Installation Crew size	$ of Roofs utilized	Installation Length of Time	Electrician Length of Time	Component Guarantees	System Guarantee	State Rebate Timing
1. What's the payback period?	10	9	9	3	9		3	6		3		1	3	9	3	9	3	3			1
2. What will it cost?	9	9	9	6	9		3	3	3	3		1	3	9	3	9		6		3	
3. How many installers will quote?	5	9	9	3	3	3	6	3	3			3	3	9		3	3	6	3	3	3
4. What about rebates, State and Federal?	7	9	9	3	6		6	6	3	3	6	3				6	6		3	6	9
5. How does it change the home value?	9	9	9		6	9	3	9	3	9	9	3		3		6				3	
6. How long will the whole system last?	9		6		6		9	9		9	9	3									
8.																					
9.																					
11. Is there enough sunlight with the trees behind us?	5		6		9	6	9	9								9					
12. Is there enough sunlight in the winter?	6		6		9	6	9	9								9					
13. How much power can be generated?	9		9	6	9	6	9	9		6	6	3		9		9					
14. How do we sell the extra energy generated?	10		9		9		9	9		3		6	3			9	3	6			
15. How many panels do we need?	7	3	9	6	6	9	9	9		3		6				9					
16. Do any of our roofs face the right direction?	3		9	6	6	9	9	9		9		3				9	3	3			
17. How efficient is the DC/AC conversion?	3		6	9	3	6	6	3		9	9	6				9					
19.																					
20. Where will the array mount on the roof?	5	3	6		6	9	9	9	9	9	3	6	3			9					
21. What will it look like? The appearance is a concern.	6				6	9	6	9	9	9	3	3				9					
22. Do any of our roofs face the right direction.	3	5	6		6	6	6	9	9	6	9	6	3			9					
23. Where do I mount it on the ground?	5		6		6	9	9	3	9	3	9	3	3			9		3			
24. Will you be able to see it from the front of the house?	7				6	6	9														
25. How does it change the home value?	10	3		6	6	6	3	9	3	9	9	3		9		9	3	6	3	6	
26. Where will the inverters be mounted?	2	3		3	6	6	3		3	3	9	9				9		6	3	3	
27. How much of the wiring will show?	3	3		3	6	6			3	3	6	9	3			9		6	3	3	
28. Where will the wires be routed in the house/garage?	1	3		3	6	9	6			9	9	3	3			9		6			
29.																					
30. How much maintenance is involved?	3	3		6	6	3	6	3		6	6	6	3			3		6	3	3	
31. How long will the whole system last?	7	3		3	6	3	6	3	6	6	6	3	3					6	3	3	
32. How long will the solar panels last?	7		3		6		6	3	3				3			3					
33. Does the roof require extra maintenance?	8			6	6	6		3	3				3			3					
34. Do we need to reinforce the roof?	8					6	6		3	6			3								
35. Will ice/snow damage the system?	7					3	3			3	9		3			3			6	3	
36. How can it be damaged by lightning?	4					3	3			3	6	9	3			3			6	3	
37. Will hurricanes damage the system?	4					3	3			3	3	9	3			3			6	3	
39.																					
40. Who has to get the permits?	3	3						6	6					9	9	9	6	3	3		3	3
41. How long will the installation take?	5	3		6	6	9	6	9		6			9	9	3		9	9	6	6	9
42. How long do we wait for state approval?	7			3	3			6					9	6			6	3	3	3	9
43. When can you begin?	4	3	3				3	3		3			9				9	9			9
44.																					

Figure 21-12   HOQ Body Illustrating Correlation Scores between Needs and SQCs

Direction of Increasing Satisfaction		↓	↑	↓	↓		↑	↓		↑	↓		↓	↑	↓	↓	↑	↑	↓

Functional Performance Requirements/Substitute Quality Characteristics

Needs	Customer Weights	System Price	System Power Total	Number of Inverters	Number of Panels, Strings	Layout of Solar Panels	Overall System Simulation	Solar Panel Power Rating	Wiring Layout Complexity	Inverter Types, Brands	Inverter Locations	Inverter Capacity	Township Permitting Length of Times, $ of Steps	Fed. & State Rebates SREC Plans	Installation Crew Size	$ of Roofs Used	Installation Length of Time	Electrician Length of Time	Component Guarantees	System Guarantee	State Rebate Timing
	10	9	9	3	9		3	6		3		1	3	9	3	9	3	3			1

Figure 21-13   HOQ Top, with SQC Direction of Satisfaction Impact at the Top

customized to the home receiving it, it was felt that price was an important attribute to include in the overall QFD approach.

Before proceeding any further, the installer wrote a short proposal based upon the VOC work, with a standard system concept and a technical description of the proposed system. The homeowner had already favored the installer and agreed to proceed a little further, but not much more until a final system could be shown with a simulation. At this point, sensing the customer's hesitation, the team decided to review the feedback on the VOC questions. The row highlighted in Figure 21-15 had a 10 weight from the customer, with a score on the right as "1" or poor. That customer voice was, "How do we sell the extra energy generated?" So the installer came back and showed the customer on his laptop PC how the energy credits would be accumulated and how the customer could indeed sell these credits from his home PC. This allowed the customer to accept the initial proposal and move forward with confidence that his voice would be heard and addressed as the system design went forward.

Figure 21-16 illustrates the roof of the HOQ, showing areas of SQC correlations, both positive and negative. Negative correlations highlight system technical areas that require trade-off to achieve multiple customer goals. The strongest technical trade-offs identified between SQCs were:

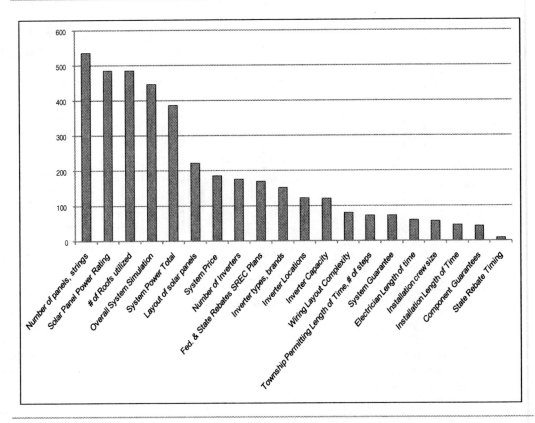

Figure 21-14   Technical Parameter (SQC) Scores from the HOQ

- System price versus system power
- Installation time versus number of roofs used and crew size

To address the price-versus-power issues, some financial return-on-investment calculations and simulations were performed, as described in the next section. One key deciding factor was that the state rebate was limited to the first 10 kilowatts of a residential solar-power array. Any system concept beyond 10 kilowatts was less financially attractive. The installation time versus number of roofs used and crew size was also a concern. As shown in Figure 21-3, seasonal consumption was a concern here as well. With the summer air-conditioning season falling at the tail end of the installation, the customer advised that perhaps the number of roofs should be reduced. However, using fewer roofs negatively impacted total system power, something highly desirable to maximize. The solar-power installer assured the customer that the schedule would be met, and system operations would commence by the middle of the warm summer months, in time to positively impact the peak-season electricity bills.

21.1.6 Technical Feasibility During the Conceptualize and Design Phases

Part of the VOC/VOB heard from the customer revolved around several key technical issues. From the DFSS aspects of service and product development, these SQCs needed to be addressed by the solar-power installer. Key answers to customer questions revolved around payback time, roof directions, and sunlight hours, plus the actual system design. This section will concentrate on the technical-feasibility aspects, included as reference for the interested reader.

One key aspect to any renewable-energy system is to understand how much energy is available from the natural resource. This is part of the Critical Parameter Management (CPM) methodology described in Chapter 4, Section 4.6. In this instance, the CPM scorecard would list monthly power generation of the solar-power system as the critical parameter. In order to calculate the parameter of interest, we begin with the energy available. Shown in Figure 21-17 is the amount of solar energy—called insolation—available at the latitude of the home residence in New Jersey. The line function plotted is the amount of solar energy from month to month for Trenton, NJ. The average for the year is approximately 3.6 kilowatt hours per square meter per day, which is shown by the central dashed line and corresponds to what the homeowner will experience in an average year. Included for reference is an upper dotted line corresponding to Miami, Fla., which at 26 degrees north latitude has a large solar influx. A lower-reference dotted line is shown for Oslo, Norway, at 60 degrees north latitude, considered to be the northernmost limit for any practical solar power. This graph shows that the central portion of New Jersey is feasible for solar-generated electricity.

In addition to where the homeowner's residence is located geographically, a second component of technical feasibility are the directions of the residence roofs, affecting the ability of a solar photovoltaic power system to capture the sunlight available. Essentially, the three directions east, south, and west work reasonably well, with south-facing roofs tilted at the latitude angle working best. This residence had a "hip" roof, which has trapezoidal-shaped front and back roofs, met on the side with triangular-shaped side roofs. The front of the residence faces nearly due west, at an angle of approximately 280 degrees from due north.

The aerial illustration of the residence shown in Figure 21-18 illustrates the sunlight that falls upon the back and south-facing roofs of the residence. These two roofs have directions of 100 and 190 degrees from north, respectively. The south-facing roofs of

House of Quality (QFD) Matrix — Solar Installation

#	Customer Needs	Customer Weights	System Price	System Power Total	Number of Inverters	Number of Panels, Strings	Layout of Solar Panels	Overall System Stimulation	Solar Panel Power Rating	Wiring Layout Complexity	Inverter Types, Brands	Inverter Locations	Inverter Capacity	Township Permitting Length of Time, of Stops	Fed.& State Rebates SREC Plans	Installation Crew Size	$ of Roofs Used	Installation Length of Time	Electrician Length of Time	Component Guarantees	System Guarantee	State Rebate Timing	Customer Assessment (1.Poor – 5.Excellent)	Perceived Quality Performance / Improvement Ratio	Sales Quality Performance Point Ratio
1	What's the payback period?	10	9	9	3	9		3	6		3		1	3	9	3	9	3	3			1	3 Acceptable	1	1
2	What will it cost?	9	9	9	6	9		3	3	3	3		1	3	9	3	9	3	6			3	3 Acceptable	1	1
3	How many installers will quote?	5		9	3	3	3	6	3	3				3			3	3	6			9	3 Acceptable		
4	What about rebates, State and Federal?	7	9	9		6		6	6				3		9			6		3	6		5 Excellent		
5	How does it change the home's value?	9	9	9	3	6	9	3	3	3	3	6			3		6			3	3		3 Acceptable		
6	How long will the whole system last?	9		6	3	6		9	9	3	9	9	3								3		4–5		
8																									
9																									
11	Is there enough sunlight with the trees behind us?	5		6		9	6	9	9				3				9						3 Acceptable	1	1
12	Is there enough sunlight in the winter?	6		6		9	9	9	9				6				9						3 Acceptable	1	1
13	How much power can be generated?	9		9	6	9		9	9		6						9						1 Poor / 3	1	1
14	How do we sell the extra energy generated?	10		9		9		9	9		3		6		9		9						1 Poor		
15	How many panels do we need?	7	3	9	6	9	9	9	9								9						3 Acceptable		
16	Do any of our roofs face the right direction?	3		9		6	9	9	9								9						3 Acceptable		
17	How efficient is the DC/AC conversion?	3		6	9	3		6	9		9		3										4–5		
19																									

#	Customer requirement																					
20	Where will the array mount on the roof?	5	3	6	6	9	9	9	3	3			9	3	6	6				◯	1	1
21	What it will look like? The appearance is a concern.	6			6	6	6	9	3	3			9							◯—◯	1	1
22	Do any of our roofs face the right direction?	3		6	6	6	9	9	3	3			9	3	6						1	1
23	Where do I mount it on the ground?	5	5	6	6	3	9	3	3	3			3	3	3	3		6		◯—◯		
24	Will you be able to see it from the front of the house?	7			6	9	3						9									
25	How does it change the home's value?	10			6	3	9				9		9	3	6	3 6				◯—◯	1	1
26	Where will the inverters be mounted?	2	3	6	6	6	3	3	9	9	6		9		6							
27	How much of the wiring will show?	3	3	3	6	6	3	3	6	3	3	6	9		6					◯		
28	Where will the wires be routed in the house/garage?	1	3	3	6	9	9	3	3	9	6	6	9		6					◯		
29																						
30	How much maintenance is involved?	3		6	6	6	3	6	3			3	3							◯	1	1
31	How long will the whole system last?	7		3	3	6	6	3	9	3						3	3			◯—◯	1	1
32	How long will the solar panels last?	7	3		6	6	9	6	3							3	3			◯—◯	1	1
33	Does the roof require extra maintenance?	8			6			6					3		3					◯	1	1
34	Do we need to reinforce the roof?	8			6	6	3	3				3	3			6	3			◯—◯		
35	Will ice/snow damage the system?	7				3	9	3				3	3			6	3			◯		
36	How can it be damaged by lightning?	4		3	3	6	9	3				3	3			6	3			◯—◯		
37	Will hurricanes damage the system?	4		3	3	3	9	3				3	3						3			
39																						
40	Who has to get the permits?	3	3									9	9	3	3					◯		
41	How long will the installation take?	5	3	6	9	6	6				9	9	6	9	9			3		◯—◯		
42	How long do we wait for state approval?	7	3		6						3	3	6	3	6	3	3	9	3	◯—◯	1	
43	When can you begin?	4			3	3					6	3	9	9	9	3		9	6			
44																						

Figure 21-15 Customer Ratings on the Right Side of the HOQ

PHASE I QFD

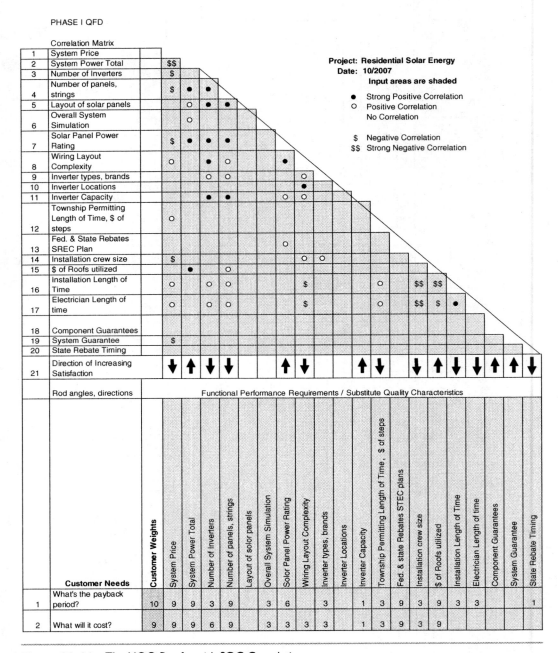

Figure 21-16 The HOQ Roof—with SQC Correlations

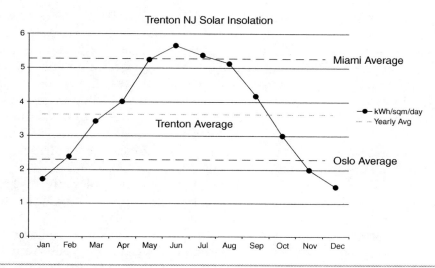

Figure 21-17   Solar Insolation

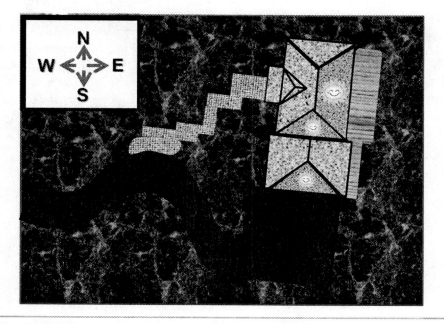

Figure 21-18   Aerial Illustration Showing Three Roofs with Solar Potential

the main residence and garage are tilted at nearly 30 degrees, but have limited surface area, while the east-southeast-facing back roof has a much larger surface area. At first glance, solar-powered electricity generation seems potentially feasible for this residence.

Following this check of feasibility relative to roof directions, some initial estimates were made of power generation across a day and across the year. Depending upon the roof selected and the type of solar panel chosen (175-watt versus 200-watt panels), power generation could be predicted both daily and annually. The top six technical items (SQCs) from the HOQ work were factored into a power-generation model for thesresidence. An initial concept model of 18 panels placed onto one of the south-facing roofs was created first. This is illustrated in Figure 21-19.

Eighteen panels at 175 watts would generate approximately 3 kilowatts (kW) peak power on a sunny day. This equates to about 20 kWh generated on a sunny day during the year's peak solar-insolation months. Assuming 20 sunny days in a month, this projects to 400 kWh monthly generation in the summer. The homeowner could then compare this against the 1,900 kWh monthly consumption target—post conservation efforts—and draw some conclusions about technical feasibility. In this case, the homeowner would see only about 20 percent of the monthly consumption covered by this concept, which was considered insufficient. As each concept was developed, advantages and disadvantages were also listed in the concept description. They will be omitted here for the sake of brevity and keeping to a focus of illustrating the methodology.

From here, a series of concepts was created covering the various available solar panels, layouts on the individual roofs, and combinations of two or three roofs. Figure 21-20 shows the first round of concept development using the Pugh Concept Selection Matrix. In this early concept work, the initial concept of one south roof was compared to a two-roof, south-facing, 4.5-kW concept and an 8-kW east-southeast-facing, back-roof concept. The highest-ranked technical factors from the HOQ were listed on the

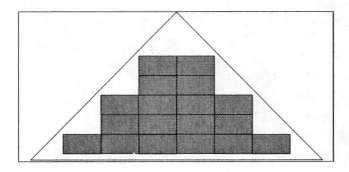

Figure 21-19   Lower South Roof Concept

Concept	3 kW S Concept	4.5 kW S Concept	8 kW ESE Concept		
Criteria	1	2	3	4	5
System Power Total		+	+		
Number of Panels, Strings		+	+		
Layout of Solar Panels		−	+		
Overall System Simulation		S	S		
Solar Panel Power Rating		S	+		
# Roofs Used	Datum	−	S		
Inverter Locations		S	S		
Fed. & State Rebate SREC Plans		+	+		
System Price		−	−		
Inverter Types, Brands		S	S		
Number of Inverters		S	−		
Total (+)'s		3	5	0	0
Total (−)'s		3	2	0	0
Total (S)'s		5	4	0	0
Rank			1		

Note: (+) = clearly better, S = about the same, (−) = clearly worse

Figure 21-20 Initial Concept Development Work

left. The first concept was considered the Datum or baseline, and the two larger arrays were considered against the Datum for each of the technical criteria. Each cell was filled in according to whether the concept was the same (S), clearly better (+), or clearly worse (-) in each technical criterion. The larger array on the back roof was rated the highest, so inclusion of the back roof seemed to be a desirable part of the overall concept. There were, however, two negative technical-criterion cells with this concept, namely system price and number of inverters. These were offset later by doing the financial feasibility calculations that follow later in this case description.

Another round of concept exploration followed the first round. In the second round, the 8-kW concept was used as the Datum. Figure 21-21 highlights the second round of concept work. In this new round, two concepts covering three roofs were compared to the single-roof, 8-kW concept. The 10-kW concept scored highest, due to clearly better performance on the state-rebate and inverter-location technical attributes. System price was rated nearly the same, due to using lower-cost and slightly lower-power solar panels. The state rebate was limited to 10-kW maximum, negatively impacting the 11.6-kW concept. The 11.6-kW concept also required an inverter location change, also negatively impacting this concept's rating.

Concept	8 kW ESE Concept	10 kW S, Concept	11.6 kW S, ESE Concept		
Criteria	1	2	3	4	5
System Power Total		+	+		
Number of Panels, Strings		+	+		
Layout of Solar Panels		−	−		
Overall System Simulation		+	+		
Solar Panel Power Rating	Datum	−	+		
# Roofs Used		−	−		
Inverter Locations		+	−		
Fed. & State Rebate SREC Plans		+	−		
System Price		S	−		
Inverter Types, Brands		S	S		
Number of Inverters		−	−		
Total (+)'s		5	4	0	0
Total (−)'s		4	6	0	0
Total (S)'s		2	1	0	0
Rank		1			

Note: (+) = clearly better, S = about the same, (−) = clearly worse

Figure 21-21   Second-Round Concept Development Work

The 10-kW plan thus became the highest-rated concept, in part because it produced the highest return on investment. This concept was conservatively rated at 900-kWh per month, against the homeowner's monthly (post-conservation) 1900-kWh consumption target, approaching 50 percent of monthly needs. The seasonal model for power generation of this final concept is shown in Figure 21-22, which can be compared to the original (pre-conservation) baseline power-consumption model. The shorter bars are the power generated, and the taller bars are the expected consumption from the baseline data. This final concept was considered to offer sufficient technical feasibility of power generation for the residence, and was chosen to move forward, especially in light of the conservation efforts, which would reduce the taller bars substantially. The line graphed shows the net expected cost of energy without conservation efforts for the original consumption model and the 10-kWh solar array. The expected annual cost for this situation shows almost a 50-percent reduction in annual electricity costs over the prior two-year baseline.

21.1.7 FEASIBILITY OPTIMIZATION (FINANCIAL)

The Optimize phase of the DFSS roadmap in this case example focuses on financial optimization per the VOB from the homeowner's perspective. Within a general DFSS framework, there are three typical value questions posed of any new project:

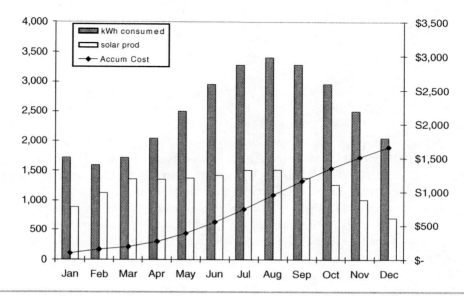

Figure 21-22 Power Consumption and Generation Models with Cumulative Cost

- Is it real? *(Are there real customers with real interest in this? VOC)*
- Can we win? *(Can this deliver more value than other alternatives?)*
- Is it worth it? *(Will the investment needed pay enough back? VOB)*

In this case example, we remind the reader that we view the VOB from the home-owner's perspective. In parallel with technical feasibility work per the VOC, questions were being answered about financial feasibility per the VOB. The focus discussed here is on the financial benefits of the solar-power system versus the planned costs.

When the three-roof concept came into full focus, a price proposal was delivered to the homeowner. This price proposal was quite detailed, except that it lacked sufficient detail in the payback model as proposed. This relates well to two pointers from the beginning of this case example. Firstly, the installer was previously bidding on commercial installations, where the customer would do the payback modeling, so the payback modeling was not very detailed due to the contractor's lack of experience in this area. Secondly, the customer was quoted in the "A" grouping titled "I want to get the highest return on investment possible," which includes the questions:

- What is the payback period?
- What will the total system cost?

- How will it change the home value?
- How do we sell the extra energy generated?
- What about the state and federal rebates?

The homeowner clearly wanted a very detailed and well-understood payback model from the outset. The original payback model assumed a single large outlay from the home-owner's savings account, and payback consisting of cost savings on the monthly electric bill plus sales of the extra energy produced, sold as Solar Renewable Energy Credits (SRECs). This model showed payback in seven years, assuming savings and SREC sales benefits of $4,300 annually after a $26,000 outlay, as can be seen in Figure 21-23. This scenario has an Internal Rate of Return (IRR) of 11.5 percent and a net present value near zero.

This initial payback model lacked a few key aspects of reality. The model lacked the cost of replacing the inverters, conservatively expected to last five years minimum. Figure 21-24 shows what happens to the payback with inverter replacement included. This changed the IRR to 3 percent, and the payback period became extended to 11 years, which was not a happy situation for the homeowner.

However, the homeowner would be borrowing the money, but the interest rate cost was not factored into the model, nor was the large initial outlay. The rising cost of electricity was not included in the savings, since the solar-power system would be effectively locking in the cost of electricity for half the consumption needs. The initial model also lacked a $1,000 federal energy credit that would be received in the next tax year.

An updated model with the inverter replacement and these other factors included is shown in Figure 21-25. This model contains the cost of the loan payments made over 10 years, a 7-percent annual electricity cost increase, and inverter-replacement costs. This model shows positive cash flows from Year 3 onward. The equivalent net present value is well over $8,000, the IRR is over 20 percent, and the payback is still positive in Year 7 with the expected series of inverter replacements. Should the inverters need replacing as soon as Year 5, there is still a net payback by year 7. These responses to VOB financial issues assured the customer that these matters would be addressed, and helped win the business for the service installer.

21.1.8 CASE STUDY WRAP-UP, CONTROL PHASE

In the control phase, key improvements are normally made permanent through the use of a control plan. The control plan is not shown here for the sake of brevity, as the major improvements are all permanent and have been previously described in Phase I (conservation) and Phase II (design and installation of the solar-power system). The net reduction in electricity usage in this case study is illustrated in Figure 21-26. The homeowner's electricity consumption is plotted in a grouped SPC chart, showing the reductions from

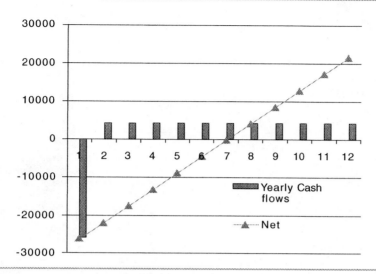

Figure 21-23 Initial Three-Roof Concept Payback Model

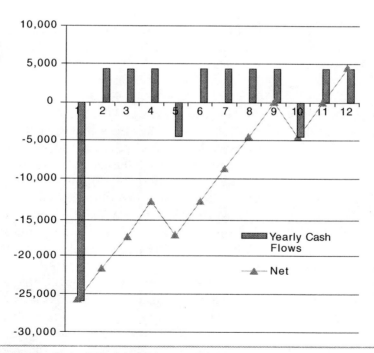

Figure 21-24 Payback Model with Inverter Replacement in Years 5 and 10

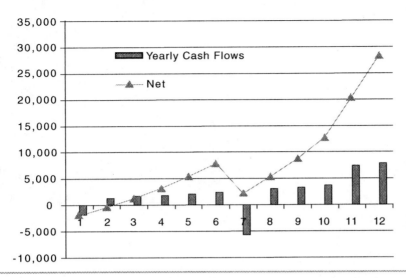

Figure 21-25   Complex Payback Model

Phase I, conservation, and then Phase II, with the solar-power generation system installed. Overall net consumption was reduced from an average of 2,300-kWh to 1,050-kWh, a 55-percent reduction in electricity usage.

During the time of study, the electric rates the consumer experienced rose by nearly 70 percent across 57 months of study, as shown in Figure 21-27. This is an increase of more than 1 percent per month in cost per kWh. The rates include additional delivery use charges, state and local taxes, and other small charges that just *show up* on an electricity bill. So while consumption was being decreased, the net effective rate to procure electricity was rapidly rising. A three-to-four-month bump occurs in the summer months, when the consumption and prices both increase while air conditioning is running.

If you recall, key goals of the customer were to save money and to yield the highest return on investment. Shown in Figure 21-28 is the final measurement of cost in a capability-analysis graph. While the average monthly cost is still not quite at the target of $140, it is close, at $171 per month. Without the rapid increases in electricity rates, the target would have been met and exceeded.

The primary use of QFD is to deploy the customer needs and desires into the design and delivery of new products and services. Each installation of solar power in a residence is unique. In this case study, the key voices were addressed during the design and installation of the solar-power generation system, and in the conservation efforts made prior. Throughout the process, the customer was involved and informed as to design progress, which is perhaps more involvement than is normal during *product* development efforts. This was a key to customer satisfaction: After all the customer is actually *living* with the final result!

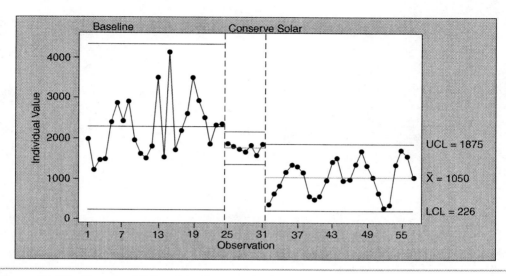

Figure 21-26   Phase I and II Electricity Consumption Improvements

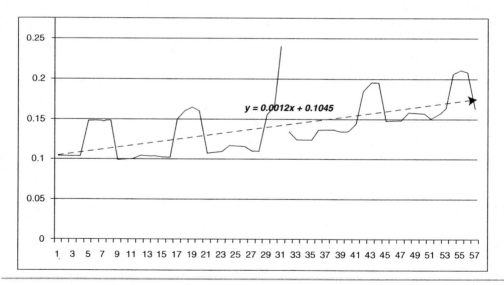

Figure 21-27   Net Electricity Rate Over 57 Months

As a final roundup to the control phase, the key voices from the lowest-level affiniti-zation of the VOC were listed, and appropriate measures were constructed to verify customer satisfaction. A sample of this is shown in Figure 21-29, a VOC Checklist Table. By identifying appropriate measures from the customer's perspective, we see the

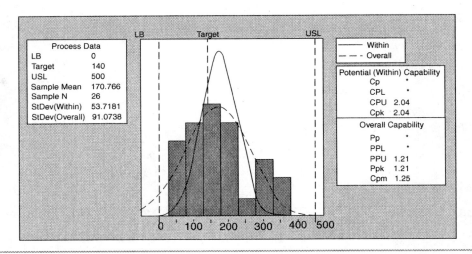

Figure 21-28 Electricity Cost Capability Analysis

VOC Phrase	Primary Metric	Target	Actual Measure
Is there enough sunlight with the trees behind us?	July 22nd system start time	6:00 AM	5:51 AM
Is there enough sunlight in the winter?	December kWh total	20 kWh	19.8 kWh
How much power can be generated?	Annual SRECs	10	12
How do we sell the extra energy generated?	SREC market training	2x per year	October, May
How many panels do we need?	System Power	10 kW	9.86 kW
Do any of our roofs face the right direction?	Compass Roof Angle from North	180	185
How efficient is the DC to AC power conversion?	Inverter efficiency	95%	96%

Figure 21-29 Extract from VOC Checklist Tabulation

service as measured from the customer's viewpoint. Follow-up work was done to show the customer after installation how the system was performing against the expectations. At that time, the customer felt comfortable acting as a reference for future customers as well.

This QFD case study included the use of those DFSS tools that relate specifically to the customer and the QFD focus of this book. Additional DFSS tools were utilized, but were not included here for brevity. These tools included a Design FMEA, Critical Parameter Management, the Design Scorecard, Tolerance Analysis, and Measurement Systems Analysis, to name a few. I have illustrated how a small service company can use QFD and DFSS to become more successful in a new market segment by focusing on the customer, which is always a key to success.

21.2 VOC AND QFD IN AN URBAN COFFEEHOUSE RENEWAL

With Bill Rodebaugh

21.2.1 CASE STUDY #2: INTRODUCTION

This case study highlights the initial months of startup and operation of a coffeehouse in the northeast section of Philadelphia, Pa. The name of this establishment is Great Awakenings. This case study was written by the owners of the establishment.

Some facts of note around this operation:

- Previously, a coffeehouse under two separate names between 1992 and 2008
- Purchased in June, 2008 after the property had been vacant for approximately 4 months
- Startup on Oct. 25, 2008—A four-month startup period was planned
- Author and spouse were frequent customers of both prior coffeehouses at this site
- Initial owner had begun a large transition from another business establishment to a coffeehouse, and much of that work remains
- Second owner made little physical change to the site; however, there was a noticeable difference in the offerings of the place
- Current customers remember the former shop, including its people and products

The vision for ownership of this particular shop is a civic-minded and people-oriented one. The goal is to use this upscale, downtown coffeehouse to be a catalyst for growth in a residential community far from the downtown scene. Also, the vision is to hire from the city and neighborhood, giving the business a chance to serve and grow as well. As a Master Black Belt and Operations and Supply Chain leader, the owner has an outstanding opportunity to put the axioms and algorithms into practice in the real world.

21.2.2 VOICES OF THE CUSTOMER, BUSINESS

Beyond the vision, there are specific guiding principles around this start-up operation. The following list is considered as the Voice of the Business (VOB):

- High and consistent quality of beverages and foods
- Consistent, friendly service
- Easy and speedy purchasing
- Desire to be the community coffeehouse—want neighbors to think of the shop as the extra room in their house

There are also operational strategies and tactics within the shop that are based upon Lean thinking. These points allow the guiding principles to be implemented successfully in an efficient manner.

- Sales taxes built into costs—all prices end with .00, .25, .50, or .75
- Standard sizes (12 oz. and 20 oz.), shared lid and drink sleeve sizes, and related efficiencies
- Most extra charges are avoided (for extra flavors, ice, etc.)
- Point-of-sale system purchased with inventory-tracking ability
- Ability to accept credit cards
- Development of a local supplier base for faster shipping
- Order at minimum order quantity—order more often with fewer items
- Some supplies to be purchased as they run out; other supplies are stocked in advance

All of the preceding are examples of guiding principles, strategies, and tactics that are necessary as a business is opened. These need to be sharp, well-defined, and well-executed; however, the part that is of paramount importance is the Voice of the Customer (VOC). In this case study, the VOC came in two parts—an early set of key comments acquired well before opening, and hardcopy surveys taken at open houses and during initial start-up.

21.2.3 VOC PROCESS STEP 1: COLLECT CUSTOMER VOICES

Shown below are some key pre-opening VOC comments gathered from speaking to past customers in the neighborhood.

- *"We didn't feel like we could come in and stay at the shop—just get our drink and leave"*
- *"It wasn't a friendly place"*
- *"We like our coffee strong around here—and a large cup of it"*

Upon recent review with other customers around "friendliness," there was some disagreement with the comment. Practically speaking, since current customers are treasured and new ones difficult to obtain, it is necessary to keep the friendliness comment at a high level of importance. This is a lot like offering a product with a sporadic problem that is difficult to measure. Even though it may not be seen by all customers, the issue must be designed out of the situation.

The owners felt it would be helpful to combine VOC and VOB into the table shown in Figure 21-30 to evaluate these "voices" further and begin identification of the key performance measures for the coffee shop. In this case example, Figure 21-30 represents a shortcut version of the HOQ, wherein the first column represents Whats and the second column represents Hows. In this case, perhaps more could have been learned with a complete HOQ, but time was short and the owners wished to be certain that all of the pre-opening customer and business voices were represented in the design phase of the coffeehouse.

21.2.4 Substitute Quality Characteristic (SQC) Analysis

To summarize Figure 21-30, there are four design solutions, two targets, and one need. It is important to see what each of these aspects mean with respect to measurement and performance. We call these Substitute Quality Characteristics (SQCs) throughout this book. (See Chapter 8 for details about SQCs). In Six Sigma, they are usually called Critical to Quality elements (CTQs). As the owner is a Six Sigma Master Black Belt, he utilized the CTQ terminology.

CTQs are helpful for determining whether the shop is on the right performance track to satisfy the pre-opening VOC and VOB. By looking at the CTQs at the current and a target state, we can think about how to best satisfy these voices, through an objective discussion of where we are and trying to understand where we can be.

It is acceptable to be inexact at this point in the analysis. The process of reviewing these CTQs is beneficial in itself. This is a form of our **VOC Process Step 2,** *Analyze*

Customer/Business Statement	Statement Translation for Design	Solution	Measure	Target	Need
'We didn't feel like we could come in and stay at the shop ... just get our drink and leave.'	Design a place where folks feel able to stay.	X			
'It wasn't a friendly place.'	Design a friendly place.	X			
'We like our coffee strong around here ... and a large cup of it.'	Prepare for high volume cups of bold coffee.				X
High and consistent quality of beverages and foods	Obtain good suppliers (what is good?).			X	
Consistent and friendly service	Hire the right people			X	
Ease and speed of purchase	Allow for credit and cash with tax included.	X			
Desire to be the community coffee house	Design a place where folks feel able to stay.	X			

Figure 21-30   VOC and VOB Application Table

Customer Voices, and is shown in Figure 21-31. In terms of the steps for Six Sigma Process Design (SSPD) listed in the introduction to this chapter, Figure 21-29 is an illustration of the first step combined with one later step:

- *Define* the process or service challenges, determining key stakeholders and project outcomes, and developing and validating key customer needs
- *Design* a process or service that meets customer, stakeholder, and business needs.

As the time to start the business was short, the owners moved through the design and development processes quickly in order to be ready in time.

VOC and VOB (Translated)	Comment Type	Key Issue	CTQ	Current State	Target
Design a place where folks feel able to stay.	Solution	Customer Satisfaction	Over time develop a group of regulars and see word of mouth communication on business. Sales is ultimate metric.	See approximately 50 orders per day on average. Have 43 members of our Facebook Group	Look to have 150 orders per day. Look to increase the Facebook Group membership 10-fold.
Design a friendly place.	Solution	Customer Satisfaction	See no commentary from our customer base on bad experiences. Have our internal management audit our performance with customers.	Only personal commentary, but we have positive feedback.	Quantify this feedback via on-going survey
Prepare for high volume cups of bold coffee.	Need	Product Availability	People have their product requests available for them.	Have rolling list of requests.	Continue process of taking product requests, but we need to study whether to bring product on.
Obtain good suppliers (what is good?)	Target	Product Price and Availability	On-time and right delivery (consistent). Prices are comparable.	Have developed 4 main suppliers (coffee, milk, food, paper). Need to develop scorecard for each.	Review scorecard. Develop feedback mechanism. Looking for 100% on-time, right deliveries.
Hire the right people	Target	Customer Satisfaction	Drink quality is good. All employees are equipped equally to serve. Good customer commentary all around.	Develop mechanism to measure out employee's performance	Hold quarterly reviews. Review customer feedback to add to review.
Allow for credit and cash with tax included	Solution	Processing Timeliness	Minimal wait times. (Potential for 60 secs in-and-out with no line).	No seemingly difficult issues with wait times, but need to capture data	Drink prep and pay shouldn't be more than 2 mins.
Design a place where folks feel able to stay.	Solution	Customer Satisfaction	Over time develop a group of regulars and see word of mouth communication on business.	See approximately 50 orders per day on average. Have 43 members of our Facebook Group	Look to have 150 orders per day. Look to increase the Facebook Group membership 10-fold.

Figure 21-31   VOC Process Step 2—Analyze Key Voices; VOC/VOB-to-CTQ Mapping

Key takeaways from the VOC/VOB mapping analysis:

- Assumption that customer satisfaction will drive sales
- Our Facebook and other Web presences are a good indication of our overall saturation
- A supplier scorecard will be a helpful tool to develop and maintain
- We need to determine what is an acceptable wait time for the customer. How quickly can a bagel be toasted and an espresso drink prepared? What about a simple cup of coffee only?
- We need to develop good methods of understanding how good our employees are and how happy our customers are with them. We think we are at a good level on these measures now.

21.2.5 START-UP VOICES OF THE CUSTOMER (EARLY SURVEYING)

The time from design and rebuilding to opening was relatively short. Clearance from the city to open the establishment came early in the week, and the coffeehouse opened later that same week. The shop was opened for a few hours on two successive evenings in order to let the two largest customer segments (neighborhood and church friends/family) see the shop and evaluate the establishment quickly. We had 150 people to potentially survey at the two open houses. We received more than 60 surveys. The survey questionnaire is shown in Figure 21-32. Response graphs for the first six survey questions are shown in Figures 21-33 through 21-38. Group 1 comprised 24 people from the church friends/family segment, while Group 2 contained 37 people from the neighborhood segment.

On Survey Questions 7 and 8, recommendation and additional usage, all of those who attended the open house and completed the survey indicated they would recommend the place to someone. Of all survey respondents, 72 percent said they would consider renting the back room.

On Survey Questions 9 and 10, open-ended questions regarding future offerings and characteristics of a good coffee shop, the highest response was no additional service offerings, while Friendly Customer Service and Coffee/Drink Quality were the overwhelming coffee shop characteristics desired. These were followed by Comfortable Environment, receiving half as many votes. All other characteristics appeared less frequently than these three.

Takeaways from the initial survey are as follows. (Please note that there was no charge for most drinks and customer levels were at an all-time high, so evaluating speed is not reasonable at this time.)

Which types of beverages would you have interest in choosing at Great Awakenings? (Check all that apply)
_Coffee _Tea _Espresso Drinks _Blended Coffee Drinks _Smoothies _Protein Shakes

Which food items are of interest to you? (Check all that apply)
_Bagels _Pastries _Muffins _Croissants _Dessert items _Sandwiches _Salads _Yogurt

Which time of the week would you most likely come to Great Awakenings?
_During the Morning Rush _Lunchtimes _After Dinner _Saturdays

Which features of the shop would be of interest to you? (Check all that apply)
_XM Radio _Flat Screen TV _Wireless Internet _Choice of Newspapers _Faxing Capability _Our back room seating area
_Our eat-in area in the front room _Our loyalty card program

Which offerings would be of interest to you on top of our normal coffee house food and features?
_Music nights _Local Artist Open Houses _Book and Poetry Readings _'Meet Your Politician' Nights

Our mission is to provide our guests an excellent quality product that is served in a friendly and efficient way. We want you to relax on a Saturday and get in-and-out on a Monday morning. Which values are important to you?
_Great Coffee and Beverages _Family Service _Quick Service _Payment Options (Credit/Cash)

After your initial visit would you recommend Great Awakenings to a friend? _Yes _No

Would you consider renting our back room for a private party or business event? _Yes _No

After your initial visit to Great Awakenings, what would you like to see as another service or offering of this shop?

What characteristics of a coffee shop would make you a loyal customer through the months and years?

Figure 21-32   Survey Questions for VOC Process Step 1—Collect Customer Voices

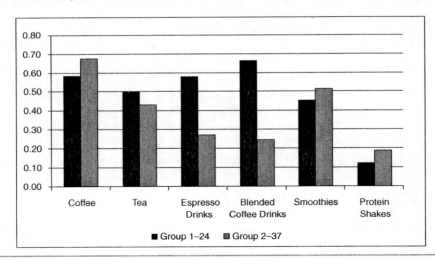

Figure 21-33   Survey Question 1: Identify Favorite Drinks (% Selected)

The two groups—church friends and family and neighborhood guests—had some commonalities, but some differences:

• Commonalities: Almost exact agreement on the types of special events desired. Coffee shop features like the back room and fax capabilities were appreciated equally between the two segments, while newspapers were more desired by the neighbors. There was also almost complete agreement regarding food types.

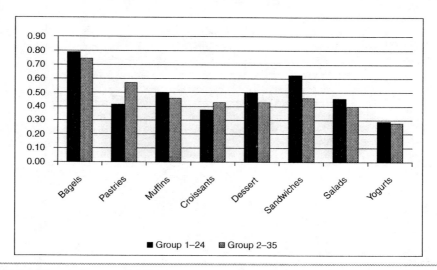

Figure 21-34   Survey Question 2: Identify Favorite Foods (% Selected)

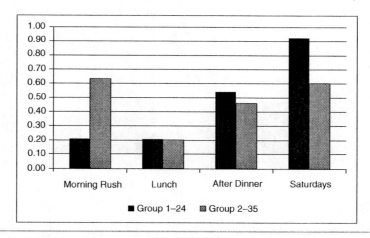

Figure 21-35   Survey Question 3—When Would You Come to the Shop? (% Selected)

- Differences: Neighbors had more of a desire for coffee, while the church friends and family group was far more into the espresso and blended drinks. The neighbors would come more in the morning, while the church friends and family would more heavily target Saturdays. The neighbors were more concerned with speed, while the church friends and family group was more interested in the variety of payment methods.

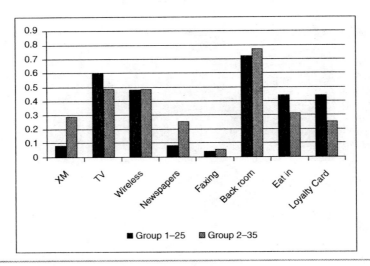

Figure 21-36   Survey Question 4—Features Desired (% Selected)

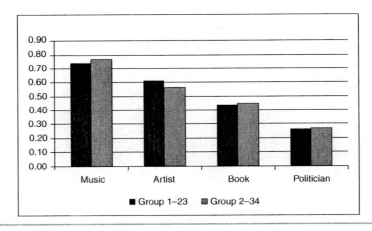

Figure 21-37   Survey Question 5—Special Events (% Selected)

With the coffeehouse up and running, there were now two sets of voices, pre-opening and initial launch, to compare and contrast. As shown in Figure 21-39, a correlation matrix was created between the pre-opening CTQs and the survey results from the open-house events. The CTQs from pre-launch are shown on the left-hand side, while the launch feedback is shown across the top. This can help in validating the initial design by observing positive correlations. All CTQs showed positive correlation with the initial launch feedback, except for offering credit and cash prices with tax included to add up to round numbers.

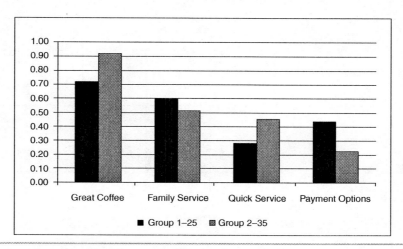

Figure 21-38 Survey Question 6—Why Would You Come? (% Selected)

Takeaways from this CTQ correlation and VOC analysis:

- There is a good understanding of the need for a comfortable environment and good customer service. This has been established, and can be maintained by having the right people and constantly looking for ways to make the shop even more comfortable.

- There were several comments on coffee, food, and price. There is already data that "coffee is king," especially in the morning hours. The question is whether we've done enough to satisfy the coffee desires of the neighborhood. Should we expand to some additional flavors? Should we make the aroma of coffee even more of a prevalent part of the shop?

- The CTQ on easy payment was important to owners, and it did get some votes for importance, but it didn't stand out as a key characteristic. Perhaps this is because there was no payment during the open house. Should we develop what is thought to be a highly capable part of the operation, and let the public know how easy it is to deal with us?

- There are three positive associations around suppliers. As mentioned previously in the discussion, what is a good supplier? Is it offering the best pricing and quality? Do we need to get a real understanding of supplier quality?

- We have looked to hiring the right people, but have we trained them, or even talked to them, regarding the best way to make the customer experience even better?

	Friendly Customer Service	Drink (Coffee) Quality	Comfortable Environment	Good Food	Good Prices	Owner-Related Comments
Design a place where folks feel able to stay.	+		++			+
Design a friendly place.	++		++			+
Prepare for high volume cups of bold coffee.		++		++		
Obtain good suppliers (what is good?).		++		++	+	
Hire the right people.	++		+			
Allow for credit and cash with tax included.	0	0	0	0	0	0
Design a place where folks feel able to stay.	+		++			

Figure 21-39   Comparison between Pre-Opening CTQS and Initial Launch Feedback

21.2.6 QUALITY FUNCTION DEPLOYMENT PART 2 (CRITICAL FUNCTIONS TO PROCESSES)

After translating the VOC/VOB into CTQs, and after having validated the pre-launch CTQs with the launch survey VOC data, we can deploy these CTQs into the critical process functions of the coffeehouse. After these analyses are complete, we can identify any gaps and understand where to best place further process efforts.

The critical process functions are shown in Figure 21-40 at a very high-level view.

The process functions are related to the CTQs from Figure 21-31, which were developed from the pre-launch VOC/VOB. Importance levels for the VOC/VOB were determined by ownership, with the thoughts of the customer in mind. Then the cross analysis was completed. The full matrix is shown in Figure 21-41.

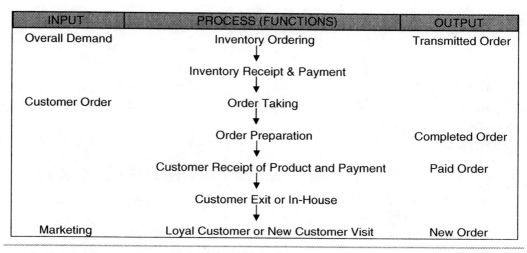

INPUT	PROCESS (FUNCTIONS)	OUTPUT
Overall Demand	Inventory Ordering	Transmitted Order
	↓	
	Inventory Receipt & Payment	
	↓	
Customer Order	Order Taking	
	↓	
	Order Preparation	Completed Order
	↓	
	Customer Receipt of Product and Payment	Paid Order
	↓	
	Customer Exit or In-House	
	↓	
Marketing	Loyal Customer or New Customer Visit	New Order

Figure 21-40 Top Level Process Functions of the Coffeehouse

Project: Customer(s):	Coffee House Re-Start Owners Decide Importance–Customer Reaction is Key			Relationship Ratings Blank = None, 1= Weak, 3 = Moderate, 9= Strong				
Key Question								
					Process Functions			
Critical to Quality (CTQ) Measures	Importance Rating	Inventory Ordering	Inventory Receipt & Payment	Order Taking	Order Prep	Customer Receipt & Payment of Product	Customer Exit or In-house	Loyal Customer or New Customer Visit
Growth of 'regulars' and word of mouth customers	10			1	3	3	9	9
No comments of bad experiences by customers along with positive survey results	7			1	3	3	9	9
Product requests are met	4	3		3	9			
On-time and correct inventory delivery (supplier)	6	9	3		1			
All workers make equally good quality beverages	8				9	1		
Minimal customer wait times	9			9	9	9		
Function Importance		66	18	110	246	140	153	153
Scaled 1–10		1.5	0.4	2.5	5.6	3.2	3.5	3.5

Figure 21-41 CTQ correlations with Top Level Process Functions of the Coffeehouse

Takeaways from the process functional level QFD matrix:

- Order preparation is the most important process with the greatest impact on the CTQss
- At the end of the day, the best way to get a happy customer, avoid complaints, and increase the business is to deliver a good, consistent order of coffee (or other menu item)
- The final two processes are the second and third most important parts of the process. These final two processes may not have the simplest structure and clearest sets of activities, but we know that sthe outcomes of these processes are very important.
- Although it is important to develop a good inventory system, the inventory processes do not have much impact on the CTQs.
- The payment process around inventories is the least critical process area (which makes good sense).

21.2.7 PERFORMANCE AND NEXT STEPS

The coffeehouse opened approximately two months before Christmas. Operationally speaking, we can characterize the first two weeks as a somewhat out-of-control experience! Generally without the customer's knowledge, we were attempting to understand sales choices while translating them into vendor orders. Data from the first month reveal that 32 percent of our total sales were in 12-oz. and 20-oz. coffees. Most of these were sold to the morning-rush crowd. The more-expensive beverages, like smoothies and lattes, were reserved for the evenings. The suppliers of food, milk, coffee, and paper each had issues with regard to over-purchase and expiration, lack of product, and inconvenient delivery times.

After two weeks, we became more in-control, and the operating costs were managed much more efficiently, while product margins looked to be reasonable at this time.

The Lean principles created general operating efficiency, and were now working better in a more controlled demand and operational environment. For example, 95 percent of the milk orders were three items. Certain types of milk were used less frequently, and did not require resupply on a regular schedule. It made better sense to purchase them in smaller increments per demand versus larger quantities when the other supplies needed reordering.

In another example of improved operations, the owner was able to have an excellent meeting with the paper supplier. During this meeting, it was discussed that there are needs which did not show on the traditional order form. Now, other items that were taking extra trips to purchase can be ordered. Also, we had been working with regard to certain minimum-order quantities. These were dictated by the vendor because of shipping requirements. The shop is a five-minute drive to the vendor's warehouse. The vendor will now load any item into the owner's vehicle, generally on the same day. This knowledge allows for spreading out paper-goods purchases as long as possible, conserving cash flow.

Gap analysis

With the various analyses shown previously, one can see that the initial design principles have been sound. We have potentially overdesigned some systems, like the inventory tracking, which are really invisible to the customer at this point. Issues that remain include the following:

- There is a need for some supplier rating at this point, which will help as we go forward. Developing at minimum a simple scorecard would be helpful.
- We need to fully develop the coffee process at the shop. Do we have all of the flavors and types necessary? Are we able to handle bulk sales easily, even for blends?
- How can we better leverage the friendliness and speed that are our core competencies? Are our employees living up to these at all times?
- As shown in the surveys, we should leverage the special events. The music events have begun and have been quite successful, but we need to begin our engagement within the art, book, and political arenas.

21.2.8 URBAN COFFEEHOUSE RENEWAL EXAMPLE: SUMMARY

In this example, we have focused on VOC/VOB, CTQ assessment, surveying, and matrix analysis of VOC over time. The DFSS tools of VOC and QFD are helpful for any business that is just getting started. Certainly, the functional processes of order taking, beverage preparation, and cash fulfillment are simple and well known in this case. The DFSS tools assisted in understanding the needs most important to the customers, ensuring overall business effectiveness and growth. Adding the Lean fundamentals to a customer-focused design assures efficiency of operations.

In this case, in four months of business start-up, the few VOC comments gathered early as well as the important VOB concepts were applied to the design and setup of the business. By reviewing the CTQ definitions, one can immediately realize that failure to properly measure and deploy critical matters can be a large gap.

If there were major gaps between customer expectations and the operating-process outputs, the proprietors would restart a CTQ flow-down and run some high-level QFD analysis to get back on track. The use of a Kano analysis to look at needs versus satisfaction would add some guidance to further process-development priorities. The application of Design Failure Modes and Effects Analysis (DFMEA) would ensure that all of the key problematic issues were effectively designed out of the operating systems of the coffeehouse.

If you have never used the tools described here, take some time to learn them before you attempt your first QFD. Additional review of Chapters 1–4 is advised.

21.3 DISCUSSION QUESTIONS

- How do you currently measure customer satisfaction? Are there additional ways to obtain customer feedback on your products and services? Is there follow up?

- What new products and services are you planning to launch this year? Next year? Which customer needs are being addressed? What are the priorities? How do you know for certain yours will be better than competing alternatives?

- After launching new products or services, how do you obtain customer-satisfaction information? What measures do you use to determine performance of new products and services? How well are they aligned with customer measures of success?

- How do you set the targets? How are new products and services aligned with customer needs at present? Is QFD used? Why or why not?

- How much effort is put forth to determine and develop alternative product and service concepts once the customer needs are established? Could more be done to ensure successful concepts?

- How is technical feasibility established for new concepts? Financial feasibility? Is each of these just one measure, or are multiple measures used? How are trade-offs highlighted, and how are they resolved?

- When is customer feedback solicited during the design and development process? How often should it be included?

- How successful were the last three product or service launches? What was learned? Is there objective evidence that lessons learned are improving future launches?

Index

Information in figures is denoted by f.

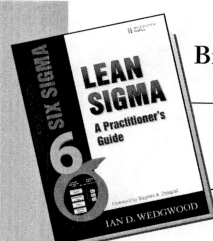
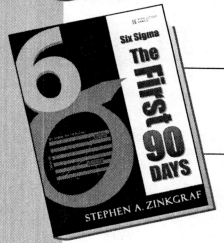